ENDLESS NOVELTY

# ENDLESS NOVELTY

## SPECIALTY PRODUCTION AND AMERICAN INDUSTRIALIZATION, 1865–1925

*PHILIP SCRANTON*

PRINCETON UNIVERSITY PRESS

PRINCETON, NEW JERSEY

Published by Princeton University Press, 41 William Street,
Princeton, New Jersey 08540
In the United Kingdom: Princeton University Press,
Chichester, West Sussex


*Library of Congress Cataloging-in-Publication Data*

Scranton, Philip.
Endless novelty : specialty production and American
industrialization, 1865–1925 / Philip Scranton.
p. cm.
Includes bibliographical references and index.
ISBN 0-691-02973-3 (cl : alk. paper)
1. Manufacturing industries—United States—History.
2. Specialty stores—United States—History. I. Title.
HD9725.S39 1998
338.4'767'0973—dc21 97-11479 CIP

This book has been composed in Palatino

Princeton University Press books are printed
on acid-free paper and meet the guidelines for
permanence and durability of the Committee on
Production Guidelines for Book Longevity
of the Council on Library Resources

http://pup.princeton.edu

Printed in the United States of America.

10 9 8 7 6 5 4 3 2 1

*To Virginia McIntosh,*

WITHOUT WHOM, NOTHING,

AND WITH WHOM, EVERYTHING

# *CONTENTS*

*TABLES*

*PREFACE*

THIS BOOK exists only because of the coolheadedness my spouse, Virginia McIntosh, displayed in mid-January 1995 during the hours following a fire in our Philadelphia home. The blaze destroyed my third-floor study, incinerating thousands of books and records, my computer, and stacks of notes and files. The floods of water that saved the building left the lower floors largely a soggy mess. Just days earlier, I had reached the three-quarters mark in drafting and rewriting the manuscript and had carefully tucked a set of backup disks into my study's closet. Now what remained of the room's and closet's contents was perhaps five tons of debris that firemen had shoveled out the windows, forming two truck-sized piles on the lawn. Late that evening, Ginny remembered that a few days earlier I had printed out the entire manuscript, fastened its five sections with large binder clips, and tucked it on top of a corner cabinet in the study. The next morning, she told the first wave of demolition men that somewhere in the piles might lie the draft of a book, and asked that they look for clumps of it as they loaded their dumpsters. In two days, the construction crew located all five sections, each still held by its heavy black clip, sopping wet and burned around the edges to roughly an oval shape. But for Ginny's quick thinking and the crew's diligence that manuscript would rest in a landfill and this book would never have been completed.

Subsequent skilled labor at the Conservation Center of Philadelphia yielded photocopies of the partially burnt pages. A grant from Rutgers University funded the rekeying of the text (with hundreds of ellipses) into new computer files during summer 1995, while I read a stack of monographs and re-created teaching notes for fall courses, replacing materials that had also been vaporized. At last, turning to the wounded manuscript late in the year, I began replacing its missing phrases and notes, then sent the repaired result to Jack Repcheck at Princeton University Press. Once we agreed on a strategy for completing the final sections, I gathered such research files as had survived and scrounged for material to fill the gaps, writing the closing text in summer 1996.

All scholars know, and many use prefaces to acknowledge, that our final "products" are the result of an elaborate and collective effort.

Here the list of those to whom I am indebted stretches well beyond archivists, librarians, considerate colleagues, and generous funding agencies. This special, extended group merits first consideration and my heartfelt appreciation: the Philadelphia Fire Department's Mt. Airy station and Assistant Fire Marshall Daniel Williams; BroRock Builders' extraordinary demolition and construction teams, BroRock head Gary Rothrock and construction manager John Martin; Movers' Specialty Services and its president, Tim Hughes; Stephen McGill of McGill Realty, who provided nearby temporary housing; Michael Hagan of Hagan/Griffiths and Associates, adjuster for our wondrous insurance carriers, The Philadelphia Contributionship; the Conservation Center of Philadelphia; Provost Walter Gordon of Rutgers University, Camden, and department secretaries Loretta Carlisle and Lena Ng; neighbor Kathy Ivens and her friends at Mantis Computers; and Princeton University Press's incomparable Jack Repcheck. To these, in the same vein, must be added the scores of colleagues who sent books to help begin reassembling my library, along with timely words of sympathy and encouragement. For their collegial support I am deeply grateful.

This study had its beginnings nine years ago, as I was completing *Figured Tapestry*, the second of my monographs on Philadelphia's distinctive textile industry. A commentator on a paper drawn from that work pressed me to think beyond the fabric trades and to consider other industrial sectors in which specialized and flexible manufacturing represented regular business and technological practice. Extended contacts with the Historical Alternatives to Mass Production working group, cochaired by Charles Sabel and Jonathan Zeitlin at the Maison des Sciences de l'Homme in Paris, established that in British metalworking, French plastics and optical goods, and German cutlery, production formats comparable to those in "my" textile districts had long operated successfully. Seeking American analogies was an obvious step toward widening the scope of the textile arguments; it propelled me into the work process these pages finalize.

Along the road, I received fellowship and grant support from the National Endowment for the Humanities, the Mellon Foundation, the Woodrow Wilson International Center for Scholars, the American Historical Association, Rutgers University's Center for Historical Analysis and Center for the Critical Analysis of Contemporary Culture, the university's research grants and sabbatical programs, and the Provost's

Office in Camden. These institutions' generosity afforded me the time for research and travel so necessary to sustained scholarly inquiry. My thanks to each and all of them.

Travel brought me into contact with a marvelous array of libraries and archives whose staff members transformed the research grind into weeks of intellectual excitement. My appreciation goes to Bill Creech at the National Archives, Steve Wright at the Cincinnati Historical Society, Kevin Grau at the Rhode Island Historical Society, Gordon Olson at the Grand Rapids Public Library, Michael Nash, Chris Baer, and Marge McNinch at the Hagley Museum and Library's archives, Susan Hengel of the Hagley Library, and accomplished professionals at the Library of Congress, the University of Pennsylvania's Van Pelt and Engineering Libraries, Rutgers' Alexander and Robeson Libraries, the Free Library of Philadelphia, Brown University's John Hay Library, and the Historical Society of Pennsylvania.

Fellow historians and geographers provided suggestions for and critical appraisals of the work in progress over the years. Whatever its shortcomings, this book would be far less presentable without cordial aid and inspiration from Howell Harris, Richard Walker, Jonathan Zeitlin, Michael Storper, Ava Baron, Merritt Roe Smith, Robert Post, John Staudenmaier, S.J., Carroll Pursell, Thomas Hughes, James Livingston, John Ingham, Häkon Andersen, Bernard Carlson, Steven Tolliday, Steve Lubar, Laurence Gross, John K. Brown, Thomas Heinrich, Gerald Berk, David Jeremy, Jean Saglio, Yda Schreuder, and Rosalind Williams. At Rutgers-Camden, Rodney Carlisle, Andrew Lees, Allen Woll, Janet Golden, and Jeffrey Dorwart stepped in to shoulder workloads during my fellowship leaves. Their support signals the best of collegial relations. Without the History Department's administrator, Loretta Carlisle, my usual level of confusion would descend to total chaos. Departing Rutgers this autumn to join the history faculty at the Georgia Institute of Technology, I convey my thanks to each of these special individuals.

Finally, a special word of appreciation must go to Glenn Porter, Director of the Hagley Museum and Library. In 1992, Glenn asked me to head Hagley's research arm, the Center for the History of Business, Technology, and Society. A year later he provided me the opportunity to succeed him as editor for the Hagley-supported series, Studies in Industry and Society, at the Johns Hopkins University Press. Working with Glenn, with the center's seminar and conference presenters, with authors in the Hopkins series, and with CHBTS's Associate Director Roger Horowitz and Coordinator Carol Lockman has sharpened my thinking and materially enlivened the last five years. Naturally, none of the above are responsible for *Endless Novelty*'s deficiencies and omissions, yet all, in varied ways, have contributed to such value as this study may have.

# ENDLESS NOVELTY

*Chapter 1*

# INTRODUCTION

THIS BOOK represents an effort to recast the history of the Second Industrial Revolution, that epochal transformation of productive capacities and organizational forms which spanned the half-century after 1870 in the United States. As in a foundry, before something can be recast, both the original design and its flaws must be understood. The following discussion will strive to do just this, to characterize, without caricaturing, the received understanding of the transformations that are summarized as "The Second Industrial Revolution," and to suggest the problems inherent in that reigning account. A framework for thinking about varied approaches to industrial production will then be outlined as a basis for the ensuing chapters, which seek to remedy the enduring silence that has enveloped manufacturing activities falling outside the conventional industrial narrative from the Civil War into the 1920s.

The story that follows depicts the deployment of specialty manufacturing as an industrial and institutional dynamic that paralleled, complemented, and at times conflicted with the achievements of this nation's celebrated mass production corporations. Specialty sectors not only crafted the hardware that made mass production feasible and the styled goods that helped define an American consumer society, but also initiated technological and organizational transformations distinct from, but comparably significant to, the creation of routinized assembly, bureaucratic management, and oligopolistic competition. This "other side" of the Second Industrial Revolution is complex and diffuse, neither tidy nor reducible to formulas. In terms familiar to economists, the discussion privileges realism over elegance, for thick description well fits the historian's devotion to difference, detail, and pattern.

This is also a book about the possibilities of capitalism. In making a place for long forgotten enterprisers and workers, it reminds us that the industrial past is rich with examples of variety and versatility, and that how we think about industrial history lays an implicit base for conceiving diverse presents and futures. Being utterly incomplete, yet structured in its partiality, the past can always be reworked in the service of the poet, the policy maker, or indeed the historian.[1] If in this post–Cold War era, there seems to be nothing but capitalism, this hardly means that capitalism is a unity or that there is (or ever was) "one best way" to generate goods, jobs, and supple economies. For those who would imagine importing European or Asian models of industrial flexibility to an American terrain, this work will indicate that such capacities were once a critical element of our own manufacturing capabilities. For those with a long view of industrial change, it should also suggest that their erosion may be related substantively to the fraying of this nation's manufacturing fabric. The route specialty production traced to profit and prominence will be sketched below, along with some of the signals that, by the 1920s, show a number of its sectoral components entering decades of crisis and decay. The policy significance of this account for firms and governments is for others to assess.

To rework the customary understanding of America's industrial transformation, it is necessary first to outline its main elements. What most scholars regard as central to the dynamics that extended across the four decades after 1880 is not just the triumph of speed and scale, but also the durable technical and institutional outcomes. According to studies by leading researchers, these impacts seem to have materialized along six dimensions of economy and polity.[2] First, many American "big businesses" resulted from the harnessing of "first mover" technological advantages to national marketing efforts, as with the Bonsack cigarette machine and American Tobacco. Others represented amalgamations of former rivals during turn-of-the-century and 1920s merger waves that yielded U.S. Steel and General Motors, among others. Though there would be additions and deletions, the set of corporations regarded as the industrial economy's core for most of the twentieth century was in place before the Great Depression.

As well, across the five decades after 1870, entirely new and propulsive manufacturing sectors flourished throughout the land. Electrical power restructured production, lighted streets and homes, and, linked with photographic technologies, gave America the movies. Telephony profoundly altered communications, as did radio, which soon swelled into a medium for entertainment and advertising.[3] Like the railroad before it, the automobile shrank distances and reconstructed time, becoming an individualized cultural referent for mass production *and* mass consumption. Synthetic chemicals brought the nation new fibers and plastics. King Gillette's Blue Blades changed men's shaving practices and images of proper public faces, while cash registers, typewriters, and adding machines reshaped retailing and office work. From the trivial to the systemic, the scope and pace of the new and the modern grew in tandem with the expansion of the nation's center corporations.[4]

Meanwhile, older industries transformed themselves. Steel companies designed new mills to achieve straight-line production, using overhead cranes and internal railways to speed materials handling and enlarge output.[5] "Automatic" machinery daily extruded thousands of bottles and cans that other machines filled and sealed. With the introduction of sophisticated "cracking" towers, oil refining shifted technically from kerosene and lubricants toward gasoline, fuel oil, and a host of petrochemicals. Drab, bulk-marketed consumer staples like soap, flour, and biscuits acquired flashy brand names, flogged relentlessly though ads in mass circulation magazines to achieve the high volume, fixed-price sales necessary to amortize heavy investments in large-scale technology and marketing.[6] Generally invisible, except for the ads, these shifts affixed the seal of modernity to a host of ordinary goods even as they created or sustained big business in America.

As a result, industrial work changed. Center firms installed new technologies that lessened the need for traditionally-skilled workers and brawny laborers, though there might be places for them elsewhere in the economy. Hundreds of novel job titles appeared, demanding limited technical ability and close attention to detail. Eventually dubbed "semiskilled," workers taking such positions tended machines, assembled parts, monitored flows, or inspected the outputs of capital-intensive systems. The extent to which this routinization of labor deskilled and dehumanized workers was widely debated, then and more recently, but it seemed a pervasive corollary to the organizational, technical, and market imperatives of managerial capitalism.[7] In parallel, company leaders and their advisers erected layered managerial hierarchies to process and control flows of information crucial to profitable production and marketing, even as educational institutions labored awkwardly to train engineers and managers to fill the positions thus created.[8]

These shifts also intersected with finance, for the first merger wave surrounding the turn of the century shifted investors' and bankers' attention from railways to manufacturing and made the house of Morgan and its colleagues pivotal to the rationalization of competition and production. Moreover, increased vending and trading of industrial securities gradually ended a sleepy era on Wall Street. Ultimately, the 1920s surge of RCA, GE, and other glamour stocks generated a nationally mesmerizing run-up in common shares and equally broad sales of preferred stocks and bonds. Securities syndicators' requirements for financial data helped reshape accounting practices along the way, and the spread of shareholding installed the separation of ownership and control that in part defined modern managerialism.

Last, the enormous market power such institutions wielded fueled political struggles over government regulation of industrial businesses. One arc of this trajectory ran from the vague Sherman Act of 1890, through its desultory, then aggressive, enforcement, to the ambiguous "rule of reason" decision in the Supreme Court's 1911 Standard Oil case. A second commenced with Wilson's Clayton Act and the creation of the Federal Trade Commission, extended through the War Labor and Industry Boards, then petered out in the FTC's curious course during the 1920s.[9] The Great Crash and its aftermath triggered renewed regulatory fervor, including the widely maligned National Recovery Administration, the Securities and Exchange Commission, the National Labor Relations Act, and much else. Even so, the politics of federalism left to the states the responsibility for chartering corporations, despite occasional appeals for a national incorporation statute, a stance that precipitated rampant "charter-mongering" in a race for minimal oversight won by New Jersey and Delaware. Until the Great Depression, notwithstanding much caterwauling, center corporations and their interests overmatched the capacities of governments at all levels.

Thus runs the conventional tale. It displays the triumph of giant managerialist firms exercising technical ingenuity, organizational refinement, marketing savvy, and the power derived from pursuing efficiency and economies of scale. This restructured industrial economy relegated to a quiet periphery those firms and sectors which did not achieve throughput, sustain mergers, increase minimum effective size, raise public capital, venture internationally, and/or move resolutely to manage markets along with employees and production. It is a compelling story, but it is seriously incomplete and fundamentally flawed.

The difficulties lie in five areas. Most obvious is a fallacy-of-composition problem. The story's framing did not result from an inclusive review of American manufacturing's many facets, styles, and formats, but focused from its inception on the biggest, most readily and heavily studied institutions in the industrial system, whether examined by

Berle and Means in the 1930s or by Chandler's seminal *Strategy and Structure* thirty years later. Yet the nation's several hundred largest firms accounted for first a tiny, then a modest, fraction of all manufacturing employment and value added as the Second Industrial Revolution proceeded. The silence enveloping the other 80 or 90 percent of the nation's industrial capacity suggests that closure was too quickly achieved, that the uninterrogated have been simply excluded from consideration. Firms differing in organization and strategy from the "leaders" were hardly a host of backward sweatshops headed by unimaginative tyrants. Indeed, it was the giants who were peculiar and unrepresentative, and the "others" who constituted the bulk of American production.

"Ah, yes, but these exceptions were the wave of the future" is the ready reply. They were more efficient, more effective, and their excellence prevailed over run-of-the-mill companies in the struggle for market advantage. The best of them grabbed markets by the throat and reshaped them, using the visible hand of managerial expertise to erase the hazards of profit-searing price warring. Yet, as acute observers have noted,[10] such accounts reek of teleology, determinism, and functionalism: what happened at a particular point in time resulted because it represented the best system-outcome. Of course, it matters considerably at what temporal point one takes such a judgment and how one bounds the system. From the vantage point of the 1990s one cannot celebrate corporate giants' optimization of factor efficiencies as uncritically as was possible a generation ago, now that even managerial jobs are hemorrhaging amid a sea of debts and restructurings. Similarly, tributes to national dominance slip out of joint once global competitive advantages shift their valences. Naturalistic efficiency accounts also elide the exercise of state or market power, presuming rule-observant fair play in markets and among institutions, and afford precious little room for the contingent surprises that fill historical studies of war, politics, or science and the biographies of persons and institutions.

This last thought leads into another difficulty: ascribing rationality to the process of core firms' emergence. It is not only that what may be rational for the individual actor can prove irrational when gauged at the level of the firm, sector, or nation, though this matters substantively to specifying decision contexts and the planes at which outcomes are assessed. It is also that the interplay of rational actions produces results at variance with any particularized logic—unexpected consequences that do not yield easily to systematization and that, given radically imperfect information, cannot be anticipated. If we code whatever works as rational, we are substituting redescription for analysis and masking the need to account for historically—and socially—variant understandings of the meanings of "works" and "rational." There is a further ambiguity. As Anthony Giddens has explained, plans, decisions, and actions combine raw materials drawn from both "discursive consciousness" (articulable reasoning) and "practical consciousness" (subarticulate, culturally derived knowledge of how to go on), as well as unconscious desires. The recent vogue for discussions of "corporate culture" implies some recognition of these nonrational dimensions to action and suggests the reductionism of any account that imputes an overarching logic to historical developments.[11]

Fourth, the prevailing narrative minimizes the situational and processual diversities of the leading institutions that fill its field of vision. As one commentator has noticed, this derives from an "underlying assumption of homogeneity among all large-scale enterprises" that results in treating them as "one single block."[12] Such compression flattens empirical variations that themselves have gone unevaluated. For example, in 1900, a third of the fifty largest manufacturing plants in the United States made custom and specialty goods, not throughput commodities; and three decades later nearly half the nation's big businesses remained under family control, rather than having dispersed, anonymous shareholders.[13] These may be incidental "factoids," but that judgment must be established after inquiry, not assumed beforehand.

Last, the mainstream story of the Second Industrial Revolution silently introduces canons of significance that obstruct the reconsideration of period and process. Because of their presumed consequences, great managerial enterprises and mass production formats have been deemed strategic, financial markets crucial, and forward integration fundamental to the achievement of American industrial preeminence. But, to venture an ecological metaphor, this constitutes a business environment in which elephants' lifeways are taken as more important and are more closely examined than those of baboons or antelopes, who display different behaviors owing to their sociability or adroitness. Survival, success, and significance within patterns of ecological (or economic) balance and change cannot be adequately appreciated if elephant criteria are used to evaluate all creatures, nor can productive complementarities be perceived. Huge corporations and their triple investments in management, technology, and marketing cannot be trivialized, but neither ought their practices to become standards for significance until we have carefully investigated other formats for industrial development.

Beginning such an investigation is the aim of this book. In what follows, the reader will tour infrequently visited sections of the American industrial terrain, few of which conformed to the leading-edge imperatives of the received narrative. Sectorally, some of those byways can be considered strategic for the extension of industrial prowess (machinery and machine tools), whereas others operated at the center of expanding consumption (furniture, styled textiles and apparel, jewelry). Sectors well outside the "mainstream" of mass and flow production grew prodigiously, created or seized technological innovations, employed millions, and banked substantial profits across the same decades that featured the deployment of managerial capitalism. Equally important, their industrial story arose from the ideas and actions of "ordinary" proprietors and workers by the thousands, not from the intersection of "factors" (technology, capital, geography, organization, et al.).

Of course, with rising population, tariff protection, and national economic integration, all markets were expanding steadily. But if the story of the Second Industrial Revolution revolves around emerging center corporations, why should the putative periphery have done so well? More interesting, following unprecedented growth, why did some specialty sectors falter badly in and after the 1920s, and, equally intriguing, why did others persist and continue to develop? Further, what were the complementarities and antagonisms between the throughput leaders and the rest? What "ways of going on" underlay the latter's successes and, in changing environments, informed their disasters? Is it not plausible that multiple paths to industrial profit and accumulation long coexisted, each of which could be blindsided by historical contingencies? The paths toward specialty production, which I first explored in studies of textile manufacturing,[14] had distinctive implications for business practice regarding technology, management, marketing, finance, and related dimensions of enterprise strategy. They also conditioned the creation of collective institutions, interactions with the state, and gendered constructions of the individual. This battery of questions and issues structures the following discussion.

We may begin with a conceptual claim: in America's developing industrial economy, ca. 1870–1930, there were four broad approaches to the business of making goods and meeting needs—custom, batch, bulk, and mass production.[15] In custom work, an item was individually crafted for a purchaser, made singly to discrete specifications. These specifications might recur, as when the colleague of a steam engine buyer was sufficiently impressed by its performance to order another just the same for his plant; or they may vary, as when an associate of a custom tailor's client hauled his different body and different taste in fabrics to the shop for a fitting. In either case, the product was manufactured "one-off" and its price reflected the care and precision necessary.

In batch manufacturing, goods were made in lots of varied size, often on the basis of aggregated advance orders. Here producers canvassed a host of possible buyers with a range of samples or, alternatively, entered into contracts for manufacture of a number of machines, components, or final goods whose characteristics derived from purchasers' needs or designs. Firms usually manufactured such specialties only to the extent of the calls for them, for extras logged into inventory might sit a long time before fresh demand for them materialized. While batch production permitted economies in shop-floor work flow not available in custom manufacturing, it differed little in the unpredictability of demand and goods' specifications, as styled fabric makers and foundry proprietors well appreciated.

Bulk manufacturing yielded staple goods in large quantities, using swift but relatively simple technologies and lower-skill workers than in custom or batch efforts, relying on cost-saving efficiencies to realize profits from markets filled with essentially identical goods. Whether in cotton sheeting, nails, or dimensional lumber, the bulk format entailed often-fierce price competition, manufacture for inventory, a fairly stable product array, and a perennial search for means to avoid these conditions' depressive effects on profits. The relative technical simplicity of bulk production and middling costs of entry encouraged sectoral new starts in periods of market expansion, further threatening the earnings of existing firms, as many staple cotton goods companies discovered after 1910.[16]

Mass and flow production, by contrast, involved more elaborate technical and capital-intensive deployments for making standardized goods in similarly huge volumes. Here deep investment in dedicated production systems generated immense capacity that had to be profitably used; hence the salience of struggles over market share, efforts at international extensions, and dedication to refining processes and cutting costs. In such domains the basic steel, auto, and oil-refining sectors formed durable oligopolies that became the most visible elements of America's industrial system.

Custom and batch production demanded a capacity to shift outputs constantly. Together they might well be characterized as flexible or *specialty* formats for manufacturing, a domain where practice had to contend with diverse, fluctuating demands. Bulk and mass or flow formats constituted *routinized* production, where the challenge was to increase output and systematize the processes for making a narrow roster of goods for more stable markets. These terms refer to *approaches*, not firms or industries. They could be combined within firms, reflecting either techniques and markets for different product lines or elements within a complex parts and assembly system. They were often mixed together in manufacturing sectors, and a particular company could shift from one to another over time. Nor do they correlate reliably with scale, for some custom and batch operations were huge and capital-intensive, as in ship and locomotive building, just as many bulk producers were of quite modest size in sawmilling, brewing, and distilling.[17]

Given these caveats, how do specialty and routinized approaches map across industrial lines? Table 1 suggests the pattern evident in the early twentieth century. Custom, batch, bulk, and mass manufactured items appear in each of the broad classes: producer and consumer durable and nondurable goods. How important were the specialty lines in the maturing economy? Tables 2, 3, and 4 examine the 1909 Manufacturing Census reports at a moment between the first wave of new technologies, mergers, and oligopoly formations and the postwar surge of auto, radio, and chemical industries.

Ninety-three sectors, each with production above $25 million, together representing over 90 percent of all output, were selected for review. As aggregated manufacturing statistics blur a number of the distinctions made above, the tables only approximate the distribution of formats at the time. Three groupings seemed plausible: sectors clearly dominated by custom and batch work, those similarly committed to bulk and mass efforts, and a "mixed output" class in which all approaches blended together. At the extremes, there was little slippage across the divide between specialty and routinized production. For example, there was no bulk production of locomotives. Philadelphia's

TABLE 1
Product Groups by User Class and Producer Formats (Selected Examples, ca. 1890–1920)

| | *Custom* | *Batch* | *Bulk* | *Mass* |
|---|---|---|---|---|
| Producer durables | Turbines<br>Special machinery<br>Conveyor systems<br>Cast metal patterns | Most machinery and machine tools<br>Pumps and steam engines<br>Pyrometers and precision instruments | Belting<br>Chain<br>Mill hardware<br>Wire mesh and screen<br>Pipe and tubing | Standard steel rail<br>Typewriters<br>Telephones (leased) |
| Producer nondurables including intermediate goods and fuel | Cutting tools<br>Ship plate<br>Single-use foundry patterns<br>Linotype and offset plates | Ferrous and nonferrous alloys<br>Special yarns<br>Metal parts and components<br>Stationery<br>Specialty chemicals | Screws, rivets<br>Staple yarns<br>Files<br>Packing waste<br>Wrapping paper<br>Lubricating oil and grease | Fuel oil<br>Mild steel<br>Newsprint |
| Consumer durables | Special-order cabinetry, furnishings, jewelry, or carriages | Household furniture<br>Carpets<br>Silverware<br>Musical instruments<br>China and glassware | Door hinges<br>Window glass<br>Kitchen accessories<br>Linoleum and oilcoth | Household wiring and switches<br>"Dollar" watches<br>Sewing machines |
| Consumer nondurables | Tailored clothing<br>Wedding cakes<br>Catered foods | Magazines and journals<br>Costume jewelry<br>Ready-to-wear styled clothes<br>Silk hosiery | Work clothing and shoes<br>Cigars and pipe tobacco<br>Sheeting and towels | Soaps and cleansers<br>Cigarettes<br>Canned foods<br>Lightbulbs<br>Refined sugar and flour |

Baldwin Locomotive Works director described his trade thus: "Sharp fluctuation of demand; impossibility of manufacturing in advance of orders; impracticability of mass production; necessity for employing [skilled] specialists; high cost of each completed item and varying conditions under which it is sold and used call for unusual adjustability, flexibility, ingenuity, and resourcefulness by executives."[18] On the other hand, special orders for turpentine or cottonseed oil were

TABLE 2

Specialty Production Sectors, 1909: Value Added, Employment, and Value-Added Increase since 1899

| | *Value Added (millions)* | *Employment (thousands)* | *V-A Increase (%)* |
|---|---|---|---|
| Foundry and machine shops | 688 | 531 | — |
| Printing and publishing | 536 | 258 | 84 |
| Railroad shops construction | 206 | 282 | 90 |
| Women's clothing | 176 | 154 | 136 |
| Wool and worsted goods | 153 | 169 | 61 |
| Furniture | 131 | 128 | 79 |
| Electrical machinery and products | 113 | 87 | 162 |
| Medicines and drugs | 92 | 23 | 61 |
| Hosiery and knit goods | 90 | 129 | 101 |
| Marble and stone products | 76 | 66 | — |
| Brass and bronze products | 51 | 41 | 85 |
| Dyeing and finishing textiles | 48 | 44 | 79 |
| Musical instruments | 46 | 38 | 95 |
| Leather goods | 45 | 35 | 64 |
| Jewelry | 44 | 30 | 83 |
| Shipbuilding | 42 | 41 | 3 |
| Millinery and lace | 41 | 39 | — |
| Men's furnishing goods | 38 | 38 | — |
| Cutlery and tools | 35 | 33 | 90 |
| Carpets and rugs | 32 | 33 | 51 |
| Hats, fur felt | 26 | 25 | 80 |
| Lamps and fixtures, gas and electric | 25 | 19 | 107 |
| Fur goods | 24 | 12 | 108 |
| Silverware and plated ware | 24 | 17 | 65 |
| Paper goods, not elsewhere classified | 24 | 19 | 135 |
| Street rail cars | 17 | 22 | 223 |
| Locomotives | 17 | 15 | — |
| Total | 2,840 | 2,328 | Avg.[a] 82% |

*Source*: Department of Commerce, Bureau of the Census, *Thirteenth Census of the United States*, vol. 8, *Manufactures—1909* (Washington, DC, 1913), 40–43.

[a]Average V-A increase for all sectors providing figures for both 1899 and 1909.

TABLE 3
Mixed Format Sectors, 1909: Value Added, Employment, and Value-Added Increase since 1899

| | *Value Added (millions)* | *Employment (thousands)* | *V-A Increase (%)* |
|---|---|---|---|
| Iron and steel rolling mills | 328 | 240 | 59 |
| Men's clothing | 271 | 240 | 74 |
| Cotton goods | 257 | 378 | 58 |
| Boots and shoes | 180 | 198 | 82 |
| Autos and parts | 118 | 76 | 3,900 |
| Paper and pulp | 102 | 76 | 80 |
| Silk fabrics and yarn | 89 | 99 | 99 |
| Copper, tin, and sheet iron products | 87 | 74 | — |
| Agricultural implements | 86 | 51 | 50 |
| Leather tanning | 80 | 62 | 62 |
| Carriages and wagons | 78 | 70 | 9 |
| Brick and tile | 69 | 77 | 71 |
| Glass products | 60 | 69 | 50 |
| Chemicals | 54 | 24 | — |
| Pottery and terra-cotta | 54 | 56 | 68 |
| Confectionery | 53 | 45 | 112 |
| Stoves and furnaces | 50 | 37 | — |
| Rubber goods | 46 | 27 | — |
| Railway cars | 45 | 43 | 56 |
| Food preparations, not elsewhere classified | 41 | 15 | — |
| Fancy and paper boxes | 29 | 39 | 85 |
| Cooperage and wooden goods, not elsewhere classified | 23 | 26 | 27 |
| Rubber boots and shoes | 20 | 18 | 9 |
| Corsets | 18 | 18 | 117 |
| Wire rope and cable | 18 | 12 | 94 |
| Firearms and ammunition | 17 | 15 | 75 |
| Mattresses and spring beds | 15 | 11 | 98 |
| Totals | 2,288 | 2,095 | Avg. 75% |

*Source*: See table 2.

TABLE 4
Routinized Production Sectors, 1909: Value Added, Employment, and Value-Added Increase since 1899

| | *Value Added (millions)* | *Employment (thousands)* | *V-A Increase (%)* |
|---|---|---|---|
| Lumber | 648 | 695 | 64 |
| Malt liquors (brewing) | 278 | 55 | 50 |
| Tobacco products | 239 | 167 | 40 |
| Distilled liquors | 169 | 6 | 106 |
| Slaughtering and related | 168 | 90 | 63 |
| Bread and bakery goods | 159 | 100 | 98 |
| Flour milling | 116 | 39 | 58 |
| Illuminating gas | 114 | 37 | 108 |
| Iron and steel blast furnace products | 71 | 38 | –7 |
| Food canning | 55 | 60 | 55 |
| Paints and varnishes | 46 | 14 | 85 |
| Copper refining | 45 | 15 | 5 |
| Soap | 39 | 13 | — |
| Butter and cheese | 38 | 18 | 75 |
| Petroleum refining | 38 | 14 | 79 |
| Fertilizers | 34 | 18 | 119 |
| Cement | 34 | 27 | 92 |
| Coke (from coal) | 32 | 30 | 99 |
| Manufactured ice | 32 | 16 | 202 |
| Cottonseed oil and cake | 28 | 17 | 107 |
| Coffee and spices | 27 | 7 | — |
| Mineral and soda waters | 27 | 13 | 84 |
| Clocks and watches | 24 | 24 | 81 |
| Wire | 24 | 18 | 894 |
| Cane sugar | 22 | 9 | — |
| Beet sugar | 21 | 7 | 727 |
| Turpentine | 20 | 40 | 44 |
| Explosives | 17 | 6 | 155 |
| Sewing machine cases | 17 | 19 | 44 |
| Lead smelting | 15 | 7 | –51 |

TABLE 4 *(cont.)*

| | *Value Added (millions)* | *Employment (thousands)* | *V-A Increase (%)* |
|---|---|---|---|
| Glucose and starch | 12 | 5 | 27 |
| Zinc refining | 9 | 7 | 83 |
| Oils, not elsewhere classified | 9 | 2 | 24 |
| Molasses | 9 | 4 | — |
| Malt | 8 | 2 | 70 |
| Nonpaper bags | 8 | 8 | 165 |
| Linseed oil | 6 | 1 | 105 |
| Tin plate | 6 | 5 | 18 |
| Other smelting | 5 | 2 | 160 |
| Totals | 2,669 | 1,655 | Avg. 65% |

*Source*: See table 2.

improbable. In the middle, census amalgams like "cotton goods" mingled the bulk manufacture of staples with seasonally styled women's dress goods, upholstery, and curtain fabrics, or mixed, as "electrical machinery and parts," customized forty-ton generators with lightbulbs made by the hundreds of thousands. Though hardly ideal, these rosters suggest the state of play in production five years before Ford systematized Model T production at Highland Park.[19]

The measures chosen for display are value added in manufacturing, average total employment, and change in value added since 1899. Value added is more useful here than product value, for it documents the impact of production techniques on materials fed into various labor, machinery, and marketing formats. For example, flour milling in 1909 generated $900 million in outputs, only 13 percent of which represented processing gains, whereas foundry and machine shop operations made $1.2 billion in goods, 56 percent of whose value was created by transforming inputs. Contrasting output values would obscure the differential significance of what was accomplished in production—hence the salience of value added. Average employment indicates the weight of sectors on the national labor scene and permits the rough calculation of value added per worker. Changes in value added show shifts in sectoral performance in two relatively prosperous years a decade apart. When compared with all industry averages, they also highlight leaders and laggards but cannot in themselves account for differences in performance.[20]

The 1909 national totals were as follows: value added, $8.5 billion; industrial employment, 6.6 million; average increase in value added since 1899, 76 percent. Batch and custom sectors from foundry products to locomotives contributed a third of value added and just over a third of employment. Further, of the twenty-two specialty sectors for which 1899 comparisons are possible, sixteen showed value-added gains greater than the national average (table 2). Mixed sectors constituted another third of employment and 27 percent of value added, but thirteen of twenty-three sectors lagged behind mean value-added increases, including the largest three (table 3). Bulk and mass production trades accounted for a fifth of industrial jobs but 31 percent of value added—testimony to the propulsive force of a generation of technical and organizational innovation. Sectoral increases in value added almost balanced, as nineteen were higher than and seventeen below the national average (table 4). In routinized sectors, value added per worker averaged over 30 percent higher than in specialties ($1,604 versus $1,220), again suggesting the effective substitution of capital for labor. If total value-added figures in all specialty and routinized sectors for which figures are available in 1899 and 1909 are compared, the batch and custom group's rise was 82 percent versus 65 percent for bulk and mass formats.[21] Though these are awkward approximations, at a minimum they underscore the range and the significance of specialty production. Perhaps half the value added in American manufacturing derived from this approach, and growth in specialty sectors was substantial through the close of the twentieth century's first decade.[22]

Specialists' competitive strategy involved differentiating products and marketing capacities for novelty or quality. The ability to make many different goods well and/or meet complex specifications often helped make price a secondary consideration in sales. In styled textiles or machinery, the goal was to obstruct perception of specialist goods as substitutable, thereby blocking price-centered competition so that firms could reap high unit profits.[23] This effort had correlates in firms' selection of technology, labor, financial, and marketing tactics, and frequently had locational implications as well.

Productive flexibility was most often achieved through investment in general-purpose technologies that could be readily altered or adjusted with variations in goods ordered. Successful operation of such setups depended on firms' ability to hire and retain skilled workers whose capacities for shop-floor problem solving and mastering changing tasks represented an essential company asset. Management's relations with such workers were often personalistic rather than bureaucratic, as the firms' vitality and reputation depended as much on the mobilization of talent as on the wage-effort bargain.[24] The diversity of work in process pressed managers to devise systems for tracking the progress of orders and particularizing costs, challenges that often led them to exchange ideas on job-ticket or accounting schemes with others in the trade. Selling nonstandard goods routinely involved the creation of samples or preparation of estimates, and makers highly prized close contacts with clients, for they supplied critical information and minimized errors and, if long-term, could buffer pressures for price bargaining in slack times.

Specialists had to contend with unpredictably irregular demand at the firm level and sharp seasonal and business cycle fluctuations at the sectoral level, each of which had financial and locational implications. The lower a company's fixed costs, the simpler it was to scale back production to match slipping orders and the easier to avoid the temptations of either price shaving or manufacturing for inventory. Batch operations long remained private firms or closely-held corporations, indifferent to stock and bond flotations and reliant on retained earnings for expansion or bridging hard times and short-term borrowing for working capital. Many utilized extensive contracting networks, rather than investing in integrated production. They shed labor or shortened hours in slow periods and ran overtime when orders jammed their ledgers. In this context, managerial innovations focused on means to systematize, rather than standardize, production, information processing, labor recruitment, and marketing. Proprietors adopted and shared schemes for improving work flows through the shop, refining accounting to provide cost data for diverse outputs, or, through collaborative efforts, creating hiring bureaus and common terms of sale. "System" became the specialists' buzzword, whereas "standardize" permeated the discourse of staple commodity production.

These tactics in turn had spatial consequences. For a company that sought to place accommodation notes, let contracts for patterns, parts, or services, and partake of an often-replenished pool of skilled labor, there was no better location than an urban industrial district filled with firms practicing comparable specialist strategies. Indeed, contacts among manufacturers in such spaces generated productive complementarities as well as collective institutions to address shared trade or regional interests (industrial banks, clubs, sectoral associations, insurance societies, and schools). Often, such interpersonal and interfirm relations built measures of trust, expectations of reciprocity, and expressions of solidarity among ostensible competitors. These constituted what Alfred Marshall recognized as the special atmosphere of industrial districts, a spatially embedded sociocultural asset renewed through routines of interaction.[25]

These approaches contrast broadly with the routinizing imperatives of bulk and mass formats, but there also was considerable variation *among* specialists' practices, a diversity within diversity that must be appreciated.[26] Take the matter of finished goods inventories. In style-sensitive consumer lines, manufacturers' stocks held at a season's end were "dead" and could be sold only at harsh markdowns to "vultures." Hence mill owners considered it madness to produce hundreds of tapestry styles for stock. Yet book publishers had no alternative but to do just this, setting press runs on the basis of market guesstimates whose unreliability fostered a vast "remainders" trade that both jeopardized the sales of new releases and undermined pricing. Responding to an accumulation of these errors, D. Appleton and Company reportedly destroyed 300,000 volumes of warehoused titles in the mid-1890s "rather than remainder them" and further demoralize a depressed book market.[27] In producers' goods, batch firms might hold inventory of certain parts used in various machines but, given the unpredictable pace of technical change, shied away from stockpiling finished products. Hardware makers in Connecticut's Brass Valley turned out thousands of different items but in general supplied them to jobbers and wholesalers who held stock, issued catalogs, and reordered the sizes and styles most heavily in demand, for manufacturers would carry inventory at their factories only under extreme pressure.[28]

Moreover, in specialty steel and electrical manufacturing, batch and custom work was thoroughly married to mass production techniques, most often as a correlate to corporate integration or system building. Bethlehem Steel occupied an armor plate niche but gradually moved into basic steel, specialty structural forms, and shipbuilding. General Electric and Westinghouse commenced with custom-built electrical generation equipment, and large lamps, then added small motors, specialty switchgear and turbines, household lightbulbs, and finally mass market appliances, made in separate plants. Other "bridge" firms like Allis-Chalmers entered unrelated standardized lines (tractors) while retaining specialty capacities (heavy electrical and mechanical equip-

1. Making the "big stuff" at Westinghouse Electric. Machinist Herman Backhoffer at his 30′ engine lathe in the South Philadelphia Steam Power Generating Division, 1919. Courtesy of Hagley Museum and Library.

ment), or ventured into styled consumer goods from a base in staples (American Woolen, Cannon).[29]

These sectoral contrasts, and others that will follow, strongly indicate that there was no single model for specialty manufacturing, no core around which deviations or innovations can confidently be arranged. Like "mass production" or "managerial capitalism," the term is inductively derived, not analytically prescriptive. Nonetheless, specialty manufacturing connotes a battery of production approaches ubiquitous in postbellum America. To be sure, their very diversity ill accorded with a standardization of economic discourse and business expectations that arose in the first half of the present century. One way of doing business, the routinization approach, came to be regarded as the "one best way," an outcome that had profound consequences for specialty production and perhaps a long-term impact on the sectoral balance and flexibility of American industry.

Specialty manufacturing took place at giant enterprises making the "big stuff" of America's infrastructure (locomotives, heavy machinery) and top-of-the-line consumer goods (Steinway pianos, Gorham silver), firms here termed "integrated anchors." In the same decades, clusters of smaller companies in urban industrial districts offered diverse finished goods to households and enterprises, relying on thick webs of contact and affiliation to organize production and sales. These tool builders, jewelers, and "furniture men" figure here as a second group, the "networked specialists." In their efforts, many of them contracted with a third array, "specialist auxiliaries," for essential intermediate materials and services (metal castings, dyeing), firms the anchors also called upon at times. Together, these three classes constituted the vast majority of industrial specialists in America.

As there were scores of specialty trades, dozens of urban districts, and thousands of firms, I inevitably had to select a subset of these for detailed research and to thin that group's ranks during the writing phase. In the final analysis, Philadelphia represented an essential starting place, given my previous monographs on its flexible textile complex and my coauthored study of its overall industrial structure (*Work Sights*, with Walter Licht). Familiarity with Quaker City metalworking and its great locomotive and shipbuilding enterprises led directly toward makers of machinery and machine tools, and thus to Cincinnati. Thinking outward from textiles and apparel to other specialty consumer goods sectors brought me to jewelry, hence Providence, and furniture, hence Grand Rapids and Chicago. Along the way, discovering that New York City was a specialty production center rivaling Philadelphia yielded a brief consideration of a Manhattan specialty trade, printing and publishing. As I became aware that Pullman echoed Baldwin in railway supply, that Gorham Silver long prospered alongside Providence's more erratic jewelers, and that General Electric and Westinghouse were deeply versed in specialty approaches, these large enterprises joined the story. Other appealing places, trades, and firms ultimately could not be included lest this book double its size and surely halve its readership. Thus I axed an extended discussion of Newark (jewelry, tools, fine leather goods), shelved research on New York jewelers and a half-dozen shipbuilders from Bath, Maine, to Seattle, abandoned work on Milwaukee (heavy machinery), and gave up hope of finding room for Pittsburgh's fancy glass trade. Fat files on Cleveland's foundrymen and metal-forming machinery builders, Paterson's silk industry, and furniture clusters at Jamestown, New York, Gardiner, Massachusetts, or High Point, North Carolina, all dropped off the table. What remains is a not-quite-arbitrary selection from a far larger cohort of specialist candidates, my choices ultimately being made in relation to the depth of available sources and in order to achieve a measure of spatial and sectoral balance. The trades, cities, and companies discussed below do exemplify broader patterns but in no way exhaust specialty manufacturing's many variations. Hence the narrative engages three major cities and five of lesser scale, probes a dozen industries (six consumer, six producer oriented), and examines a larger number of business associations, union drives, and, of course, firms immense and tiny. Doing more seemed excessive; doing less has proven impossible.

The following chapters will detail specialty manufacturing's development roughly from the Civil War into the late 1920s. The flowering of batch and custom capacity in mid-nineteenth-century proprietary firms will first be outlined in Part I through a series of company-level reviews that pay particular attention to machinery building and the social relations of specialty manufacturing. Next follows an evaluation of two early institutions that undertook to share technological knowledge and initiate industrial education, thus introducing issues that recur in later decades. In Parts II and III I conduct an industrial version of the grand tour, beginning at the Centennial Exhibition in Philadelphia, moving to the city's textile and metalworking districts, then north to encounter Providence's jewelry and silverware complex. Traveling south to New York City permits a review of its extensive printing and publishing sectors before our train leaves for the Midwest, with stops at Cincinnati's machine tool district and Grand Rapids' developing furniture companies. Part II treats the elaboration of specialty trades from the Centennial into the early 1890s, working rather more on a sectoral than a firm-level plane and engaging matters of institution building and the structuring of urban industrial districts. It closes with an effort to conceptualize the diversity of specialty sectors and urban industrial districts, elaborating on the categories introduced above.

Part III documents the differential impact of the 1890s depression on batch sectors in the Midwest and East, then assesses the effects of renewed expansion, technological shifts, and changing business practices through 1912. Touring recommences with our arrival at the Columbian Exposition in Chicago, as we glimpse out the window the great Corliss engine that animated the Centennial, now powering the Pullman Works where our parlor car was constructed. A stop at Pullman complements our consideration of Baldwin Locomotive, whereas reviewing Chicago's furniture trades opens the way to retracing our path to Grand Rapids and Cincinnati, for updates on sectoral developments. Moving eastward, we halt near Pittsburgh, then take a side trip to Schenectady, so as to encounter Westinghouse and General Electric, the giant "bridge" corporations that propelled the electrification of America. Returning to Providence, we shall witness jewelry's mounting troubles before rolling back into Philadelphia at the height of its specialty manufacturing prowess. In eight chapters, we will move across territory and time, etching a portrait of the Second Industrial Revolution's "other side."

With Wilson's election and regulatory activism, and with the Great War, a tide of changes in the business environment gathered momentum, culminating in the 1920–21 inventory depression. The charge for Part IV is to explore differentials of decline and growth among custom and batch producers well into the 1920s. Perhaps this time in a rickety, two-seat flying machine, we will drop down on sites that best capture specialists' diverging paths: Providence, Cincinnati, and Grand Rapids. Throughout we shall encounter mutations in custom and practice at the level of the firm, as well as recurring constructions of regional and sectoral institutions. Batch producers' initiatives toward and responses to the labor movement, state policies, and market power relations will also be assessed. The conclusion brings us briefly to our third and final great fair, Philadelphia's awkward Sesquicentennial (1926). It links this chronicle of versatility, skill, prowess, frustration, failure, and endurance to later industrial and technological developments and to the challenges of interpreting industrialization in light of what has been presented here. Our first destination is Providence, Rhode Island. The year is 1865.

# *PART I*

## EARLY YEARS

## *Chapter 2*

## SPECIALTY MANUFACTURING TO 1876

IN THE fall of 1865, John Richards, a "foreman mechanic" for Cincinnati woodworking machinery builder J. A. Fay, traveled east to Providence on company business. There he took special pains to stop in at the works of Messrs. Brown and Sharpe, the renowned makers of precision measuring tools and the Willcox and Gibbs sewing machines, to inquire after an English-made gear-cutting gauge they reportedly used to lay out variously sized "wheels." Though he was a stranger to J. R. Brown and Lucian Sharpe, and his inquiry promised no profit, the elder partner received Richards with "courtesy and interest," providing him both the information requested and a brief account of his "early career in constructing delicate and exact mechanism." Sharpe soon strolled in from the shops. "He had been making his 'rounds,' and had brought away on his shirt sleeves evidence of practical contact with the tools and work." When Sharpe offered Richards a tour through their plant, the latter accepted instantly, finding arrayed in the rooms of "what seemed to be an old grinding mill . . . a revolution in machine manipulation, [for] a great many of the processes carried on were at that time unknown elsewhere in this country." Brown and Sharpe, unlike many of their contemporaries, strove to make "excellence the first characteristic of their work." Sharpe added a geographical contrast: "Our New England tool makers are all the time trying how cheap they can make tools, [but] in Philadelphia the tool makers try how well they can make tools."[1] Richards treasured this experience and recounted it a quarter-century later on the occasion of Brown and Sharpe's relocation into a huge, new Providence plant to house the thousands of workers who then produced its extensive lines of machinery, tools, and precision instruments.

This little story creates a point of departure for the substantive chapters of this book. It raises themes that will be developed and harmonized in later sections and signals the initial postbellum environment encountered by an industrial sector, machinery and tools, that would be crucial to the process of building the American manufacturing juggernaut. It speaks to attitudes and routines of a distinctive factory culture among makers of specialized producers' goods that can be con-

2. The Brown and Sharpe works in Providence, 1870, short years after John Richards' visit. Courtesy of Hagley Museum and Library.

trasted both with those of their colleagues in fields of style-sensitive consumer items and with the expectations and practices of firms devoted to bulk-produced staple goods. Interrogated closely, Richards' visit to Brown and Sharpe opens the first of many windows onto the shifting landscapes of specialty manufacturing during American's Second Industrial Revolution.

At this juncture, however, it is necessary to set out a minimal sketch of the framework within which the ensuing narrative will be located. As the introduction maintained, this book claims that our understanding of American manufacturing's dramatic growth from the 1870s through the 1920s has been limited by the scholarly attention lavished on great managerialist corporations and mass production systems.[2] Parallel to and interacting with the dynamics that generated those institutions, American entrepreneurs also developed an extensive network of batch and specialty production whose variety and significance is just beginning to be appreciated. Even as John D. Rockefeller commenced his "conquest" of Cleveland and the railway wars unfolded into crisis consolidations, hundreds of enterprises initiated and expanded capacities to feed corporations the complex equipment they required and furnish their legions of managers with the stylish consumption goods they demanded. Most, like Rockefeller, started small and many, like him, in partnerships. Some, like Standard Oil, rose to national prominence and became big businesses, whereas many more achieved modest and often durable successes by refining sectorally appropriate variants of industrial versatility.

Brown and Sharpe in the 1860s exemplified the intersection of specialty (custom and batch) production and the "American system" of duplicate, repetition, or mass output. It both manufactured in bulk the Willcox and Gibbs sewing machines (up to 33,000 yearly on contract from the patent holders) and marketed machine tools, precision gauges, and instruments for the metal- and woodworking trades, goods made to custom orders or in batches. Its partners announced to the trade, "We are prepared to manufacture special machines, patented articles in the line of machine work, or fine machinery of any description"; they offered no discounts and demanded cash on delivery yet kept twenty of the company's basic machines "in stock," any of which could be "adapted as required."[3] The firm expanded rhythmically (to 5,500 workers in 1917) and built its international reputation on the enormous diversity and quality of its tools, cutters, and measuring instruments, shunning the developmental path toward narrow-line mass production. Brown and Sharpe became a classic bridge enterprise, blending its gradually less-important standard sewing machine work with thousands of batch production orders for small tools or machinery components, plus the individual assembly and customizing of machine tools.[4] With these outlines penciled in, let us return to John Richards in order to get a better feel for the milieus from which endless novelty flowed into postbellum American manufacturing.[5]

## Reading Richards on Brown and Sharpe

John Richards was an ambitious American mechanic. Born in Philadelphia, reared in southern Ohio, and about thirty years of age when he visited Providence, Richards had set to work learning steam engine operations as a youth. At eighteen, he secured a mechanic's position in a Cincinnati chair-making plant and began fiddling with improving the performance of its machinery. When the shop burned down three years later, Richards determined "to qualify himself as an expert woodworker." To this end, he contracted to pay a veteran cabinetmaker thirty dollars for a year's accelerated apprenticing,[6] then put in twelve months learning to construct wooden waterwheels and gears, plus a third year mastering joinery.

In the late 1850s, a Columbus firm making wood-body planes and screw clamps engaged Richards as a foreman, and again he "immediately commenced redesigning the equipment." By 1860, he rose to general superintendent, holding a small portion of the company's stock. When the outbreak of war disarranged his employers' finances, Richards returned to Cincinnati and joined J. A. Fay's woodworking machinery establishment as "a charter member . . . with the title of engineer," soon making "sweeping changes in the company's designs and methods."[7] Thus an experienced junior manager and shop-schooled machine designer waited on Mr. Brown and Mr. Sharpe four years later with a query about an English gauge. This background helps illuminate several of the key features of Richards' story. One might well wonder why on earth the Providence partners, employers of several hundred men and women and active in national and international markets, would spend hours chatting with this unknown, unannounced Ohioan and escorting him through their works. There was no business to be gained. Surely they had better things to do. But perhaps not.

Richards' gauge inquiry, and whatever business and biographical notes he provided Brown to contextualize it, established the visitor's bona fides as a young colleague in the fraternity of machinery and tool builders. He was both articulate and a practical shopman; their hospitality could foster reciprocity in terms of information on current western conditions or the odd novel thought about gearing or cutting-tool designs. This open-door policy was common practice among machinery firms across the later nineteenth century; violations of the custom were scored in the trade press.[8] Visiting proprietors, partners, or supervisors could call upon a firm's principals and expect to exchange views on business and technical matters informally. Salesmen were often afforded the same courtesy, and workers "off the street" regularly had access to shop floors to ask after openings or request a chance to show their hand at a bit of work that might qualify them for a post then or thereafter.

What is important about this pattern is that it involved face-to-face, noncash information exchanges among individuals of widely differing standings, given a baseline recognition that each party was competent to take up the burden of communication and its reciprocity expectations. A junior outsider's admission of ignorance created an opening for a principal's "kindly counsel and instruction," along with the necessity of its appreciative "acknowledgment," as Richards later wrote of his longer-term contacts with Philadelphia machine tool builder William Bement.[9] In addition, the visitor might offer an unexpected new angle on a shop or market problem, as could a traveling worker or a technically proficient salesman. Fellow proprietors usually wrote ahead of their intentions to stop by and were often treated to a tour and a dinner. If relations were especially cordial, they would be offered a guest room at their host's home for the night. What was despised, however, was fakery and pretense. In technically demanding trades, opening verbal gambits rapidly screened out inept visitors from those like Richards, and a little shop-floor problem solving separated able salesman and all-round workmen from sample peddlers and half-trained posers trying to talk a better game than they could play.[10]

Behind the fact of Richards' chat and tour lay the customs of the machine trades' factory culture, in this instance interpersonal practices that softened the distance between strangers, the gaps of age differences, and, to a degree, the barriers of class and status. Another element in this habitus was the honor accorded to the "practical" manufacturer, the proprietor with dirty hands or in this case greasy shirtsleeves. Practical manufacturers had risen from the shop floor to ownership yet retained their direct engagement with the tangled complexities of production. Their authority derived from experience, which was especially salient in specialty operations where products constantly changed, far more than from their power to hire and fire. An owner or partner's ability to indicate ways to make ratty yarn run in the loom or detect why a cutting tool chattered and jumped signaled his mastery of the trade's mysteries. His willingness to plunge in and invite the grimy consequences narrowed differences of station and laid the base for respect and reciprocity.

Building on that base took more than grubby hands, however. Manners mattered a great deal. Intervening in the shops in a fashion that demeaned workers upon whom a misery had fallen reflected poor judgment; collaboratively lending a hand and an idea or forthrightly acknowledging a shared confusion was preferable by far. The practical man remained enough in touch with the balkiness of materials and the cussedness of machines to cooperate in taming them, rather than venturing to dictate and demand. The effective exercise of authority thus depended on both technical mastery and individual character, each of which structured interactions with workers (and clients). Trade wisdom had it that owners who belittled or drove their "hands" lost their best operatives to others who found a balance point between indifference and arrogance; but of the two extremes, the first was the greater evil.[11]

J. T. Langdon, a woodworking plant owner, put the issue crisply: "A great many think they can manage a business in kid gloves and highly-polished boots, and once in a while going around among their men and machines with a hop, skip, and a jump. . . . But business isn't handled that way successfully. You want to be in and around and through it, and if necessary, draw on a pair of overalls and frock, and take hammer, cold chisel, and wrench . . . and lead the way. Men do and dare with a prompt, wide-awake leader, when they could not be driven into it with a pair of kid gloves . . . [or with] whacking or driving."[12] The practical manufacturer, giving "personal attention to business" with a right attitude, could lead in a fashion that opened a path for his workers to "do and dare" as they faced the challenges that accompanied specialty production. Richards appreciated and remembered this, as did thousands of others. Though personal vagaries and sectoral contexts spread performance across a wide spectrum, successful players understood the value of cooperative relations to productive versatility and their contrast with the rule-bound hierarchies of routinized manufacturing. Whereas the latter's pyramids were shaped by the imperatives of their machine and information systems, in batch production social relations of respect and reciprocity were often a key to exploiting the potential of a firm's technologies and design innovations.

A third dimension of Richards' account worth reviewing is its evocation of international and interregional contexts. The British gauge he pursued was one item in the transatlantic flow of techniques, individuals, and styles that fueled batch production's growth. For decades, shiploads of skilled workers and prospective entrepreneurs arrived familiar with Jacquard looms or Queen Anne ornament, which was as significant for versatility in fabrics or furniture as were immigrant laborers, Carnegie, and the open hearth process for bulk steelmaking. The reverse flow also heavily featured batch specialists' innovations. After the 1867 Paris exposition, Brown and Sharpe instruments, Corliss steam engines with Porter governors, Fay's woodworking machines, and metal-shaping machine tools from Philadelphia's William Sellers were copied widely by appreciative makers in Britain, France, Germany, and Russia, as were features of home and industrial sewing machines.[13]

Moreover, Sharpe's comment about the differing goals of New England and Philadelphia toolmakers ("how cheap" versus "how good") and his firm's commitment to excellence suggest both regional differences and the strategic choices confronting tool builders in the 1860s. It is not clear which Yankee firms Sharpe was criticizing, but as prices were often figured by machine tool weights then and later, cheapening meant "taking metal out," which in turn reduced rigidity and precision and brought users' complaints.[14] Providence's Corliss and Brown and Sharpe rejected this approach. By contrast, Philadelphia builders (Sellers, Bement, I. P. Morris) had focused on heavy machinery for railroad shops, including the Pennsylvania, and on steam engines for regional factories and shipyards, given the Delaware basin's negligible water power. For these clients, durability and quality were the selling points; and, as orders were generally for single units scattered across many firms and drawn to varied specifications, performance over time was critical in building a reputation that could draw new business as well as contracts for replacements or additions. As Richards noted, Philadelphia firms made engines and tools that took "cognizance of practice all over the world, and consist[ed] of original designs, massive, well-fitted . . . [whereas] the New England type was, on the contrary, a 'manufacture' at that time."[15]

In using the term "manufacture," Richards referred to the effort to establish economical "duplicate" work from standard designs, as was done with wooden clocks, staple cotton spinning frames, and plain looms. Eventually successful for sewing machines, this quest failed in most other wood- or metalworking lines. Richards explained that after 1870 New England machine-building methods "changed greatly in the same direction" Philadelphia had taken and "Western or Middle State makers" echoed this shift. Profits came from situationally opportunistic pricing of their specialty goods, not from volume sales, a strategy well expressed in William Bement's pithy motto, "Make good work and ask a good price for it."[16]

Sectoral development followed from these contrasting premises. In those few subdivisions of the machinery trade where sizable and relatively standardized demand could be located, bulk manufacturing of a modest range of models predominated, as at Draper in textile looms, Singer or Wheeler and Wilson for sewing machines, and United Shoe Machinery. However, custom and batch production predominated in the rest of the industry, from paper-making Fourdriniers to printing presses, from metal-cutting and forging machinery to all varieties of woodworking, from ore crushers and cement grinders to power hammers and drop presses. As the century waned, the general machine builder gave way to the versatile specialist, often devoted to a single industry or industrial process, yet technically alert and adept, spinning out new models and variants at clients' instigation or through the continuous sifting of information and experimentation with possibilities for novelty.[17] Richards' visit caught the first upward thrust of this trajectory, and his later career would exemplify the machinery sector's struggles toward dynamic growth.

Two years after his Providence sojourn, John Richards left Fay and Co. to return to his native Philadelphia for a design post at Bement's works. In 1869, he traveled to Britain as an independent woodworking machine designer, married a teacher, then removed to Sweden where he opened a firm for manufacturing his innovations. Late in 1870, Richards returned to Philadelphia to form the Atlantic Works, a machinery partnership, but when depression curtailed its prospects, he set off for England a second time. There he secured contracts to draft machinery plans for the imperial Russian arsenal and other clients, and organized Manchester's Richards and Atkinson as "importers and makers of woodworking machines." Retaining his partnership interest, Richards abandoned Britain for a return to Philadelphia in 1877 to open the American Standard Gauge Works, leaving his European firms in his son George's hands. Reportedly "broken in health through overwork," he repaired to California in 1880 and turned his hand to machine tools and hydraulic devices, founding the San Francisco Tool Company.[18]

Evidently reinvigorated, Richards inaugurated the Industrial Press in his spare moments to print and distribute a half-dozen of the books he had written since 1874 on industrial practice and shop calculation, as well as a German-language primer his wife had first published in Britain. From this press in 1888 issued San Francisco's first "mechanical" journal, *Industry*. Richards' editorial appeals triggered the formation of a regional engineering society shortly thereafter. Meanwhile Richards had returned to Britain in 1885 for a year to assist George in expanding the family business, stopping for a month's work on his return route "in Pittsburg with the Westinghouse company, assisting in the development of . . . steam engines and centrifugal pumps." Into his seventies, he focused on hydraulic machinery, developing a specialty as a consultant to water companies, while contributing trenchant pieces to other journals after shutting down *Industry* in 1896.[19]

Richards' later activities are as suggestive as was his visit to Brown and Sharpe. At a minimum, they highlight the tricky situations of machinery builders in times of economic malaise, the value of novel design at all times, the extension of the nation's industrializing spaces to the Pacific rim, the growing significance of technical literature, trade publications, and proprietors' efforts to form collective institutions, and the opening of profitable spaces for the consulting expert. Moreover, in the context of a transatlantic technical fraternity, Richards' travels underscore the mobility of both workers and entrepreneurs within regional, national, and international industrial frameworks. Yet on other aspects of specialty manufacturing, he was quite silent. To understand more fully marketing, accounting, and financial practices, the calculation and defense of prices, relations with the state, labor organizations, and clients, and linkages with the emerging mass production system, we shall have to look elsewhere.

First, to flesh out this portrait of postbellum custom and batch machinery trades, four other figures will come into view: Providence's renowned George Corliss, Oberlin Smith of southern New Jersey, Cincinnati's William Lodge, and Thomas Savery of Wilmington, Delaware. Then turning to the domain of tools and consumer goods, we will encounter the Philadelphia Disstons and Bromleys and the Grand Rapids Widdicombs, active in making saws, carpets, and furniture, respectively. A glance at the origins and development of the Franklin Institute and the creation of Worcester's Free Institute of Industrial Science will introduce the issue of manufacturers' institutions in chapter 3, leading toward the grand Centennial Exposition, both expression of and stimulus for batch manufacturing prowess. At the end of Part I's Richards-like grand tour, the contours of specialty production in the decades on either side of the Civil War should begin to stand out more clearly from the surrounding shrubbery.

## Individuals and Firms in the Machinery Trades: Four Examples

At his death in 1888, a British journal lauded George H. Corliss as "the best known engineer [America] has ever produced," a man whose name "ranks next in familiarity to that of Watt." However, this eminent mechanic never spent a day as an apprentice.[20] As a youth in rural New York, Corliss first clerked for a cotton mill's retail shop, then after several years of additional schooling opened a general store on his own account, doubtless with financial support from his physician father. Bored with the routine of his shop tasks, he set to devising a machine that would stitch leather and, while "improving and developing" it, undertook to make the tools necessary for its manufacture. This task led him to a local machine shop, the story goes, bringing him at age eighteen into contact with the complex charms of the steam engine.

Seven years later in 1844, Corliss relocated to Providence with his stitcher plans and an idea for regulating a steam engine's operations that could achieve greater "uniformity of motion and economy of fuel." There he secured a connection with Fairbanks, Bancroft, and Co., a general machine works "which had already begun to specialize in building steam engines." Soon Corliss became involved in creating a metal planer for William Sellers, a former Fairbanks employee, but shortly moved to the Hope Iron Works, where he learned the routines for crafting traditional steam engines. In 1846 he struck out on his own as the lead partner in Corliss, Nightingale and Co., supported by a loan of eight thousand dollars from another New England physician, his Connecticut father-in-law.[21]

Like Richards, Corliss became a skillful machine designer, but unlike his woodworking confrere, he framed a singular "great notion" and secured the family financial support with which to press it home. Corliss's breakthrough enabled the quick mechanical transmission of information within a steam engine, at moments when power demands stepped up or down, through installation of what was called the "governor cutoff." In earlier engines a centrifugal governor, which responded to rising or slackening power needs, was linked to the external throttle, but Corliss's controller was connected directly to cutoff valves that instantly raised or lowered the flow of steam, thus conserving energy and saving fuel.[22] His two partners, Edwin Nightingale and John Barstow, each invested ten thousand dollars to develop the invention, a working version of which was completed in 1848 and patented the next year. Barstow's "liberal wealth" enabled the trio to borrow "all they wanted at the banks" of Providence; when he retired from the firm in about 1851, Nightingale's father stepped in to endorse the company's notes for local circulation. Repeatedly, family connections were a critical element in the materialization of Corliss's novel design.[23]

Marketing a new engine was a substantial challenge, for clever and untested refinements appeared frequently. Moreover, a subset of these had been patented, each with claims the government had ratified and their promoters inflated. Breaking into the market and dealing with potential litigation were two genuine problems. On the first count, Corliss took aggressive steps. In bidding for contracts, he offered potential users two prices and a choice. The purchaser could either accept the factory's quotation or pay the makers over five years the cost of the fuel the engine saved, when calculated against the running expense of its predecessor. Corliss pledged that his innovation would reduce fuel consumption one-half or more, an assertion backed by an ingenious pricing tactic, requiring only that the option be selected before the machine was started up.

A Newburyport, Massachusetts, mill took the bait in 1855, declining to pay $10,500 outright for two engines, choosing to remit "the saving in coal for five years." Corliss collected $19,734, and soon buyers readily paid his baseline quotes. Another Corliss stratagem was the introduction of a penalty price. In an 1852 contract to power a rolling mill, Corliss claimed that his patent engine would save three tons of coal out of every five previously expended. If a lesser efficiency were achieved, the price would be reduced one dollar for every pound necessary over the prescribed two tons. Thus if savings were but 40 percent instead of the 60 percent promised, the firm would refund $2,000 to the user. The prime mover was ordered, but its makers forwarded no refunds, for Corliss engines operated as advertised.[24]

In both cases, Corliss's tactics showed not only his confidence in the quality of his engines but also his expectation of honorable dealing on the part of those with whom he was contracting. This entailed the assumption of a "trust environment" between suppliers and clients, and understanding that agents (in this scenario, engine buyers) would not cheat principals (Corliss).[25] Only because such expectations were usually fulfilled in practice was the engine maker able to confirm the operational value of his devices. Therefore he priced them accordingly, at levels that enabled him to buy out his partners and become sole proprietor of the Corliss Steam Engine Company in the Civil War years.

Patent litigation, on the other hand, seethed with distrust and opportunism. Several rivals claimed prior discovery of principles Corliss engines embodied, and others simply copied his engine's features.

Corliss both defended and attacked in the courts, reportedly spending nearly $100,000 on legal expenses to secure his primacy,[26] a sum that both underscores his need for bank facilities and gestures toward the scale of profitability that enabled him to clear all debts by 1863.[27] When he took a series of wartime engine contracts for the Union, Corliss was comfortable enough financially to be little troubled by the seven years it took the federal government to clear the accounts, for he declined to pay lobbyists to hurry the process. Nor would he "grease the palms" of manufactories' purchasing agents to sell engines, declining orders that could be secured only through the graft and kickbacks all too common at the time.[28]

During the prewar decade, however, Corliss's firm could not rely solely on engine sales, as both lawsuits and buyers' technical caution slowed acceptance of the new design. In the early 1850s while commissioning patterns for the fourteen most commonly required engine sizes and seeking orders for specials, Corliss searched for contracts to outfit factories from Georgia to New York state with gears, shafting, and boilers. As engines then accounted for only 30 percent of revenues, securing "quite a business" in these lines helped him retain the two hundred experienced workers in his foundry, forge, machine, and pattern shops. Meanwhile, the proprietor was devising a patented machine tool to cut accurate beveled gears that were critical to shifting the direction of power transmission with minimal loss of force. In addition, like other steam engine builders, Corliss tried his hand at crafting locomotives and marine engines, making roughly a dozen of each through the early 1860s. Here he overreached his capacities. His locomotives were too heavy, wasting power in driving themselves rather than pulling loads, and his steamers, for which the firm supplied "complete machinery" to contractors building the hulls, decks, and trimmings, were several times rejected by clients after trials. As these vessels remained company property, the firm first leased them to coastal shippers before Corliss transferred them to Nightingale as part payment in purchasing his enterprise share. Finally, like others, Corliss experimented with designing pumping engines in the late 1850s, pursuing contracts with waterworks over the ensuing thirty years, eventually rivaling Brooklyn's Henry Worthington and I. P. Morris of Philadelphia, the trade pacesetters.[29]

By the war's end, the Corliss Steam Engine Company had identified its area of greatest competence: power equipment for industrial plants, especially engines, boilers, and gearing. Despite continued fiddling with pumps (and eventual success at nearby Pawtucket) and occasional contracts for ship engines, this would remain its specialty through the end of the century. International publicity and acclaim for the Corliss engine followed its installation at the Paris Exposition in 1867. Its key features were so widely imitated that the judges at the 1873 Vienna International Exhibition awarded Corliss their top honor, even though the firm had declined to exhibit, an acknowledgment that his innovations were evident in virtually every steam engine shown there. Corliss's tools for cutting beveled transmission gears achieved similar recognition. However, when an agent of Milwaukee's Edward Allis Company, an early "western" builder of heavy machinery, wrote in 1868 to request the price of the gear-cutting machine, he was rebuffed. Corliss's son William, the firm's treasurer, replied that the patented machines were built only for their own use and offered to contract with Allis to make "any size bevel gear less than eight foot diameter." The firm would market its capabilities but not its specialty technology.[30]

George Corliss's management style derived from the almost obsessive self-reliance of an independent proprietor and autodidact. When his firm expanded onto a nine-acre site in 1856, Corliss was "his own architect and engineer in designing and constructing the buildings." At the shops, foremen rarely had much success in suggesting revisions to his plans. "In such matters," one recalled, "it was almost a totally useless undertaking to attempt to effect a change in his determination." Corliss was dedicated and stubborn, "not given to looking up in mechanical works the inventions of others," and thus prone to error in fields outside his core experience. He resided in the drafting room, penning accurate "off-hand" sketches for his draftsmen to detail, personally supervised and inspected work nearing completion, and "scarcely gave himself a day away from his business" until the late 1860s (except Sundays, for he was a strict Sabbatarian). In addition, "he was a wonderfully secretive man and seldom or never divulged his inner thoughts." It is thus no surprise that at Corliss, "there was none of the 'Visitors-Always-Welcome' spirit" that suffused Brown and Sharpe.[31]

This approach "told" on Corliss, producing a "nervous strain" that materialized in alternating periods of assertive cheerfulness and grave moodiness. During the latter episodes, "no one . . . cared to interview him," but at least when feeling grim, "he in an unconscious way 'hung out' a sign and then [employees] fought shy" of him. In order to undercut metalworkers' penchant for binges on Saturday paydays, Corliss disbursed wages on Tuesdays and encouraged workers to leave part of their earnings in the firm's coffers to earn 6 percent interest (and lessen his needs for working capital). Even so, he was known to believe "thoroughly . . . in good workmen and good work and good material," and was not an interfering, hovering shop master. Moreover, he maintained the high labor rates established in the late Civil War period for at least the next twelve years, dealing with ebbs and flows of contracts by shortening hours or slimming the workforce, not by ratcheting wages downward. Corliss was a demanding yet honorable employer, neither a driver or a chiseler. In his late years, one scribe characterized him: "No carpet knight is this, but one of your rough solid workers, who stands no nonsense, who would not thank you for a compliment, and who cleaves his way through every obstacle till finally his object is achieved."[32] Still, like Brown and Sharpe, his firm prospered immensely through attention to precision and quality, through the effective practice of specialty machine building.[33]

Whereas these Providence innovators cut large figures in the American machinery trades of the late 1860s, the three individuals we meet next were then beginners in or aspirants to factory proprietorship. None would ever gain comparable fame, but Oberlin Smith, William Lodge, and Thomas Savery, like Richards, represent a new generation. Respectively, they became noted figures in the development of metal-forming machinery, the creation of the Cincinnati machine tool industry, and the refinement of wood pulp–grinding and paper-making machinery. Together their efforts both helped make mass production and mass marketing possible and further extended the foundations of specialty manufacturing.

Born in Cincinnati in 1840 to English immigrant parents, Oberlin Smith's unusual first name reflected their abolitionist sentiments.[34] Young Oberlin early displayed mechanical aptitude and built a working steam engine at the age of fifteen, most likely while learning metal-working at one of the city's riverboat engine yards. At his father's death in 1857, the family relocated to southern New Jersey near his mother's relatives, who possessed sufficient resources that Oberlin could attend a local academy and Philadelphia's Polytechnic College before entering the employ of the Cumberland Nail and Iron Works in his new hometown, Bridgeton, New Jersey. Five years later, after scoring a fifty-dollar bonus for a pipe-reaming invention, Smith started a tiny machine works and repair shop in town. Soon a talented Philadel-

3. Oberlin Smith, in contemplation, behind his first machine shop, Bridgeton, New Jersey, 1863. Courtesy of Hagley Museum and Library.

phia cousin, later an engineering professor at Indiana, Cornell, and Stevens Institute, joined Oberlin as a partner in 1863–64.[35]

Oberlin Smith and J. B. Webb initially did a general jobbing shop trade, making wrought-iron fence and railing, plumbing and gas fixtures, and refitting or reworking broken machines and parts. Some of their repair business came from the numerous food-canning outfits in the district, and Smith determined that he could design a better foot-operated press for shaping can tops and bottoms than those he was called upon to fix. Only four of these sold in the next five years, but in 1869–71 the partners recorded eighteen purchases, then nineteen in 1872 alone, some shipped to clients as far removed as Tennessee, Maine, California, and Washington Territory. The firm also made dies for use in its and other makers' machines, dies being the hardened steel cutting or embossing tools that sheared or impressed their shapes on sheet metal inserted into the press bed. Dies were no trivial source of revenue, for when presses sold for $75 to $115, individual dies cost customers from $21 to $54 each.

By 1873, Webb departed the Ferracute Machine Works[36] to commence engineering courses at Michigan. Smith brought in his younger brother Frederick as a replacement partner, committed his facilities to the manufacture of foot-driven presses for canning enterprises, and began advertising in industrial periodicals. Within three years, he designed a version for belt-power hookups; seventy-two machines of both varieties and in four models sold during 1874–76, three of which went to international customers in Canada, Australia, and Sweden. In the mid-1870s, Smith set about designing a forming press that would draw (stretch) rather than cut metal, shaping instead of punching blanks. Its patent confirmation, approved shortly after the national Centennial, soon moved Ferracute far beyond the requirements of canning plants.[37]

While Smith was using trade cards, broadsides, and journal advertising to sell machines, William Lodge was still constructing them at John Steptoe's Cincinnati shops. Born in Leeds (1848), the son of a British textile mechanic, Lodge emigrated to the United States in the late 1860s after completing a machinist's apprenticeship, then worked for three years at a Philadelphia machinery establishment. In 1872, he relocated to Cincinnati and Steptoe's three-hundred-man tool-building works. Likewise a British transplant, Steptoe mirrored the J. A. Fay company by specializing in woodworking machines, but "got the reputation of undertaking to build any kind of machine for which he could get an order." Lodge started as a journeyman and joined the local mechanics union, quickly rising to a foremanship at the shop and to the union presidency. However, within a decade he shifted from being his fellow shopmen's spokesman to employing them. During the 1880s, at least seven of Steptoe's workers successfully commenced tool building on their own accounts, including Lodge in 1880 at age thirty-two. Investing his $1,000 savings in a partnership with two machinist colleagues, Lodge soon became the pivot around which the district's tool industry developed. Like Richards, Sharpe, Smith, and many others in machinery and related sectors, Lodge was a workshop graduate who translated his mechanical experience into a specialty manufacturing venture.[38]

So too was Thomas Savery, but the shops he graduated from were those of the Pennsylvania Railroad, one of the nation's most technologically sophisticated enterprises. Savery (b. 1837) descended from a Pennsylvania Quaker family centered in rural Chester County, but at an early age he chose mechanical pursuits over agriculture and commerce. By 1860 Thomas had secured a post at the Pennsylvania's renowned Altoona Shops, from whose metalworking "school" would later also come Samuel Vauclain, director of the Baldwin Locomotive Works. Savery's skills (and perhaps forthright Friendly manner) brought him promotion to a foremanship by 1863, but he resigned the next year, moving back near home, family, and his "darling Sallie." Pusey and Jones, a Wilmington, Delaware, engine and shipbuilding company, had offered him a shop superintendency at $1,200 yearly, an advance in both wages and responsibility over his railroad position. Savery's acceptance was quick; but soon after his relocation, Pennsylvania manager J. P. Laird pursued him with a bid of $1,500 annually were he to return to the railroad.[39]

Thus, only six weeks into his new job, Savery gave two months' notice to firm partners John Jones and William Gibbons in mid-March 1864 and ten days later conducted his "successor *in prospectus*" through their works. Whether this Mr. Hinchell looked to be a disaster or Savery had mightily impressed his employers (indeed—remembering Sallie—whether Savery's resignation was merely a tactical maneuver) cannot be known. In the event, Pusey and Jones commissioned Gibbons to chat with him early in April, match the Pennsylvania's offer, and sweeten it with both a pledge of "partnership in three or five years, maybe less time," and a promise to "see [Savery] safely through the [buy-in] finances." Reconsideration was in order. When Gibbons pushed another pawn, offering "a percentage [of annual profits] which would be worth 6 or 700" yearly until his partnership could be arranged, Savery allowed that he felt "greatly inclined to stay." He soon wrote Laird to ask a "release" from his agreement to rejoin the railroad and strategically restored goodwill with his employers by affirming that he would "remain at $1300 a year" plus the profit percentage (2.5) and assurance of partnership. While Gibbons and Jones offered their "hospitalities," Savery moved promptly to secure Sallie Pim's assent to marriage and purchased a proper Wilmington home.[40]

As in Corliss's market efforts or Richards' afternoon exchanges with Brown and Sharpe, trust was necessary to this outcome. Both Savery and his employers had to have confidence that neither would break faith with the other, for a commitment to a partnership was as serious a pledge as a proprietary firm could make to an employee. Similarly, Savery respected his requirements in giving long notice of impending

4. Thomas Savery, 1862, just prior to joining Pusey and Jones. Courtesy of Hagley Museum and Library.

departure and similarly depended on Laird's good offices to free him from his written acceptance of renewed employment at Altoona. Though negotiating tactics threaded through this sequence, expectations of honorable dealing were its foundation. Savery's alacrity in adding marriage and homeownership to the new arrangement at Pusey and Jones indicated his commitment to its particulars.

In addition, this sequence provides a relatively early illustration of the rival appeals of a promotional ladder within a managerial hierarchy and the challenges of proprietorship. The Pennsylvania's pursuit of Savery (and similar courting by the Baltimore and Ohio) showed that the railroads highly regarded his prospects for advancement. Yet in neither institution would he ever be an owner, and the route to top management would be long, competitive, and chancy. For this ambitious twenty-seven-year-old, the lure of a near-term partnership was far stronger, even if substantial financial rewards would be deferred for five or more years. Savery was not a "pure" entrepreneur like Corliss or Smith, dedicated to starting and controlling his own firm, but instead worked best as a collaborator or team leader. At Pusey and Jones, he expected to share in developing a going concern and to have far more latitude to meld his personal interests with those of the firm than would ever be possible at a railroad. He would not be disappointed.

In 1866, an aging John Jones sold Savery half his one-third share in the firm for $50,000. The new junior partner mobilized his savings, liquidated most of his modest stock holdings, drew down his profit percentage, and borrowed from kin in order to make the initial payment of $15,000. He reduced the balance over the next four years largely from his partnership earnings, while he and his growing family lived comfortably on his monthly $150 salary. By the close of 1866, Savery's partners recognized the interpersonal skills that had led Altoona shop workers to offer him a gold watch as a farewell testimonial (he declined it, preferring "a statement of views, signed by the men"). As a respite from supervising the manufacture of steam engines and general machinery, they sent him on the road to Philadelphia, New York, and regional small towns to gather information on shipyard tools or I. P. Morris's pipe-cutting machines, to collect overdue accounts, and to recruit business for the company. Soon he ranged more widely, submitting bids and gathering orders, including in 1868 one for a papermaking machine, marking his entry into the specialty that would fascinate him for the rest of his days.[41]

From the close notes Savery kept on many of these marketing efforts, a specialty machinery firm's approach to securing orders can be outlined. For ten regular sizes of steam engines, Pusey and Jones issued a price list and offered a New York agency a 5 percent commission for any sales forwarded. In drafting bids for custom engines, Savery estimated their total weight, multiplied by ten cents a pound, then added flat charges for any special features required. This "by the pound" pricing was customary for machine tools and continued to be standard practice among foundries well into the twentieth century.[42] However, in bidding on contracts for bleachers, rag cutters, press rolls (all for paper mills), and other equipment, Savery based estimates on cost records for analogous prior orders. Tracking direct production expenses followed the confirmation of contracts. Calculations included labor, materials, and outlays for subcontracting (at Pusey and Jones, for patterns and some castings) but ignored overhead, power, and depreciation. Again this reflected common practice at works making diverse goods in 1870, for Captain Metcalfe's seminal treatise on shop accounting lay fifteen years ahead.[43] By 1870, sophisticated costing systems *had* been developed by corporations offering staple goods and railway services, and quietly at the Baldwin Locomotive Works, but these were either of little relevance to companies making widely varied products or not widely disseminated.[44]

Savery's jottings on costs show his concern to emphasize the healthy condition of the firm's machinery section and the comparably weak performance of its shipbuilding segment. Of 48 machine contracts he logged for 1871, 42 yielded profits and only 6 brought losses. Together on sales of $81,000, these efforts generated $15,000 above shop costs toward covering expenses (for power, repairs, et al.) and contributing to net profits. Machinery contract prices stood a solid 23 percent above labor and materials expenditures. By contrast, though 26 postwar shipbuilding contracts brought in over $1 million, collectively they generated a loss of $65,000 to the partnership. Similarly, in 1872, 40 machinery contracts netted over $12,000 and the only ship built again lost money. In later years, following Savery's critiques, Pusey and Jones shifted its emphasis ever more to machinery, ended its practice of lowball bids on ships, and chalked up profits on the vessel contracts it did snare.[45]

In 1871–72 Thomas Savery secured his first two design patents for the paper machinery that became his specialty and which accounted for over a third of his firm's machine-building contracts (32 of 88). However, papermaking machines' markups over factory costs were not as robust as those on steam engines and other work (18 percent versus 26 percent); so Savery set out on a series of paper mill and machinery firm visits to deepen his understanding of users' needs and metalworkers' practices. On one of these in spring 1873, he stopped in at an Ansonia, Connecticut, brass foundry, the Holyoke (Massachusetts) Machine Company and nearby papermakers, another machinery firm in Gardiner, Massachusetts, Colt's Machine Shop at Hartford, and, of course, Brown and Sharpe. Soon, unfortunately, issues other than technical self-education demanded his attention.[46]

The 1873 economic contraction reached Pusey and Jones early the next year, threatening to cut off a growth spurt that had doubled sales since 1870. The firm had always done job and repair work in addition to its main lines, but in the slow spring of 1874 it scouted eagerly for business, making several tons of rivets for a Philadelphia client and miscellaneous tools for the Boston Navy Yard. Several sizable contracts were confirmed in the summer, but Pusey and Jones had to borrow heavily through short-term notes to get down to work on them, then struggled with slow-paying customers for the rest of the year. In November, several of these notes were taken to "PROTEST," as Savery recorded emphatically; creditors inventoried and appraised his personal property as the partnership neared bankruptcy. Though they owed over a quarter-million and had little more than a tenth of this sum on hand, the owners evaded liquidation and their situation stabilized gradually during 1875. As his tenth anniversary with Pusey and Jones passed and the nation's Centennial neared, Savery had had a near brush with the humiliation that could befall "embarrassed" partners whose homes and goods went under the auctioneer's hammer to satisfy creditors. Steps were shortly taken to prevent a recurrence. The company incorporated; then Savery sold a portion of his shares, invested the proceeds in diverse securities, labored to capitalize on and extend his patents, sought means to rig contract bidding to ensure profits, and became active in the civic life of Wilmington, thereby building a web of associations that served as a species of career insurance. Never again would he rely solely on his partnership for income or security. Thus did the 1870s depression instigate changes and signal an ambiguous maturity that the Centennial Exposition symbolized more generally.[47]

American machinery building in the decade after the Civil War was more often practiced as a flexible form of proprietary capitalism than through the corporate format of the Lowell Machine Shops or in the nascent "American system" approach of federal armories. Owners and partners might have quite varied relations with clients, rival firms, and their own workers, but a personalized tone permeated their activities. Technical virtuosity and novelty was central to success, but cost figuring was by any measure rudimentary. The object in pricing was through reputation or patents to set figures that reflected and rewarded quality and innovation. Firms assured performance through retention of skilled workers and direct attention by principals to design and inspection as well as to problems that arose in use.[48] Gathering enough trade to keep their works active could lead partners into unknown territory and yield gnawing losses on contracts, but the same uncertainties could create either new avenues for applications of familiar techniques or fresh problems that could rechannel a company's energies.

As the Centennial approached, firms like these were the bone and sinew of the American industrial infrastructure. They provided the tools and machines that underwrote bulk papermaking and canning, the techniques that would inform mass production, the prime movers for factories and railways, and, as well, the hardware for batch production of styled consumer goods (fabrics, jewelry, or furniture). They operated in an environment of historical openness, a context of creative disorder, promise, and threat; and their proprietors devised tactics to achieve a measure of security without loss of momentum.

In addition, such firms had begun to cluster in industrial cities and districts, linked by propinquity and the flow of contracts and information—places like Providence, Worcester, Paterson, Newark, Philadelphia, Wilmington, and gradually, in the west, Cincinnati, Cleveland, Grand Rapids, and Milwaukee. Machinery builders were not quite members of a community; for though they shared unevenly a variety of beliefs and practices, there was as yet a tissue-thin institutional fabric. Nor were they atomized competitors, for neither their products nor their patterns of contact reflected the substitutability or the generalized indifference/hostility that might be required for price-based rivalry. Instead, they represented a loose industrial *network*, an occasionally fractious trade collectivity with informal rules of entry, reciprocity, and propriety, that engaged a broadly shared set of technological challenges and common finance, marketing, and labor process dilemmas.[49]

For many of them, the boundaries of the firm were porous, as their contracting and subcontracting linkages and spawning of factory graduates suggests. In sum, they represented a species of businessmen distinct from the elite Boston merchants who fashioned the Lowell corporations or the proper Philadelphians who underwrote the Pennsylvania Railroad. From these personalized, practical/technical, and cultural conventions stemmed their distinctive role in American industrialization. From similar bases flowed the prowess of their colleagues in fashionable consumer goods, printing and publishing, and specialty intermediates and supplies (fine yarns and leathers, saws, cast type, or alloy metals). In stepping beyond the machinery sector, we will encounter several of these company and consumer suppliers.

## Consumer and Producer Specialties: Three Family Firms

When historians assess American consumer goods manufacturing sectors, or those providing supplies to producers, they usually address the drive toward mass production. This invokes antebellum New England's prodigious issue of cheap cotton fabrics, the shoe firms of Lynn and environs, the trade in simple chairs and tables, or the innovations that brought inexpensive clocks to market. Later decades featured striking advances in meatpacking and canning, continuous flow production of lubricants and chemicals, mass marketing of brand-name goods, and the advance of southern rivals to challenge New England's dominance of staple textiles.[50]

Accurate in their particulars, together these accounts conjure a highly selective universe. Much more than a lust for quantity and cheapness energized sectors oriented to consumer demand and the needs of factory producers in the mid-nineteenth-century decades. Indeed, as early as 1820, a few American textile workshops tried to reorient buyers' preferences for European fancy and styled fabrics, even as urban furniture craftsmen began to wrest top-end clients from importers.[51] Through the 1870s, the task for batch specialists in consumer trades centered on matching or bettering the styles and qualities of imported goods at prices that would command buyers' attention. At the same time, suppliers of small tools, factory belting, or batch-made intermediate goods (fine yarns, alloy metals, builder's hardware and millwork)[52] faced technical difficulties and quality demands comparable to those of machinery builders and also had to meet hundreds more clients' diverse expectations.

Three immigrant British families, the Disstons, Bromleys, and Widdicombs, will here introduce the sawmaking, carpet, and furniture trades.[53] In 1833, Henry Disston and his sister arrived in Philadelphia from Nottingham with their father, a mechanic who hoped to introduce machine lace making to the United States. The elder Disston's sudden death wrecked that plan completely. Henry, then fourteen, was soon apprenticed to a firm of three resident English sawmakers for a seven-year term (his sister was sent into domestic service). At his indenture's end in 1840, Disston invested his $350 savings in sawmaking equipment and took on an apprentice of his own, David Bickley. Fifty years later, Bickley, still with the "Disston House," commented:

> The business was carried on in the back room of a building [near] Arch Street, below Second. There were just the two of us, and in that little room all the work was done. . . . In those days the grinding was not done by the maker, and I have often taken a wheelbarrow, loaded it with saws, and wheeled it out Second Street to Kensington, to the grinders. . . . At that time a good workman did well in the smithy if he turned out one dozen saws in a day; now, with the improved methods and machinery, a man turns out 15 dozen a day and does not work as hard.[54]

Buyers preferred Sheffield saws made from tough crucible steel, so Disston imported his raw material from Britain and concentrated on quality, reportedly withdrawing an entire lot from a dealer when one blade showed signs of softness upon delivery. Disston sold locally and personally, building his reputation for reliable work through the 1840s. By decade's end, his prosperity justified erecting a factory, to which he drew a corps of Sheffield workers who initiated crucible steel production in 1855. That year, Disston issued his first catalog, illustrating the diversity of his output: twenty-one varieties of saws, four lines of knives, and a welter of small tools, springs, and cutters. He soon secured American rights to use a patented British process for recycling crucible scrap, thus economizing on materials, and by 1859 employed 150 men making high-quality steel for on-site fabrication.[55]

In the late 1840s, Henry Disston brought two of his younger brothers and their sister across the Atlantic. Another brother remained in England to assist their widowed mother and serve as a contact for information on British saw and steel developments. (After her death, this brother journeyed to Philadelphia in 1868, but *his* son returned to Britain as an agent for the firm, 1874–93.) Each brother became a "subpartner" in the works, enjoying a life-interest profit share, while the firm also supported the sisters and mother. The founder, however, reserved the right to buy in these interests at his siblings' deaths, assuring ownership succession to his sons, five of whom lived to adulthood.[56]

Comparable measured provision for workers and neighbors exemplified Disston's careful paternalism. He created a free medical "dispensary" for his employees, then opened it to the poor of the Kensington district generally, and funded food distribution in slack times to the area's unemployed and their families. Starting in 1855, Disston also bought land along the Delaware for the eventual stepwise relocation of his works, which began in 1872. As a part of this gradual process, he had over five hundred homes erected for workers' occupancy and created a building and loan association to finance house sales to those inclined toward ownership. That this was a calculated strategy to retain crucial skilled workers while resiting the plant is obvious. More important, it worked and was widely acclaimed as evidence of Disston's regard for his employees.[57]

Funds for such family and workforce provision derived from Keystone Saw Works' hearty expansion during the war decade. After the initial panic, both government contracting and the Morrill tariff enhanced Disston's fortunes. While meeting an increased demand for saws, the company exhibited its versatility by filling orders for bayonets and sabers and by supplying steel plate for local shipbuilder William Cramp's warship contracts.[58] Meanwhile, the tariff slapped heavy duties on imported steel, giving Disston an advantage over other sawmakers who confronted sharp cost increases, even if they shifted sourcing to Pittsburgh's early crucible companies.[59] In saws alone, the firms' sales increased fourfold, 1860–66, to near $1 million annually, far outpacing inflation. Disston invested in expansion, including a new rolling mill, and urgently sought machinery and ideas "that would increase production without diminishing the quality of the product." Profits also underwrote construction of his first residence on North Broad Street in 1864 and a European trip in 1865, during which he discovered and purchased French band saws that revolutionized the firm's techniques for saw handle making. Brother William Disston's 1868 arrival occasioned the addition of a "jobbing shop" to fill accumulating special orders for saws and tools and to experiment and develop new products and processes. Disston analyst Harry Silcox likened it to an industrial "laboratory," noting that jobbing shop subpartners secured over half the company's forty-four late-nineteenth-century patents. By 1872, when the transition began to the Tacony riverside site, Henry Disston and Sons employed nine hundred workers and operated the nation's largest saw works.[60]

Like some early steam engine builders, Disston had started with a single focus and survived and expanded by widening his product types and lines. However, after the Civil War, he gravitated back to his original sawmaking specialty—multiplying variants within this arena, mechanizing processes where feasible without loss of quality, and investing in capacity to fashion industrial circular and band saws along with hand tools for builders, carpenters, cabinetmakers, orchardmen, and householders. In the 1870s, Disston also reached for national markets through targeted advertising, placing splashy eight-page displays in *Iron Age* that featured over a hundred woodcut illustrations.[61] Clearly the company was "scaling up" and specializing, yet it expanded product diversity into every cranny of American woodcutting. Disston courted both special orders and company workers, for each client and each employee was critical to the open-ended learning that kept the firm alive to new market or technical possibilities and able to engage them profitably. It was such practices that underlay the quality and reputation of Disston Saw and in time brought it a flow of orders from abroad, even from Sheffield.[62]

If Disston was a specialty firm that both integrated backward into steelmaking and drew a host of brothers and sons into the expansion of the founder's enterprise, John Bromley's carpet manufactory exemplifies both a batch operation locating within a network of complementary specialists and a family firm creating spin-off companies for second-generation entrepreneurship. Unlike New England's corporate integration, Philadelphia's textile industry by 1850 defined *dis*-integrated manufacturing—the interlocking efforts of separate, partial-process firms devoted to spinning, knitting, weaving, dyeing, and finishing, plus recycling mill waste and rags for spinners' use and sales to papermakers. This industrial structure created opportunities for chiefly immigrant artisans and factory veterans with small capitals to commence manufacturing on their own accounts in rented quarters. One such enterpriser was Yorkshireman John Bromley.[63]

Born near Leeds in 1800 to a wool handweaving family and "educated to the business" as a lad, Bromley toiled in his father's shop until his marriage in 1827, when he began independent work. By 1835, he had secured a ready market for his woolens and joined his father and twenty other weavers in initiating a steam-powered carding mill that provided the collective's members with clean "rovings" for hand spinning and weaving in each of their workshops. This cooperative failed in the depression of the late 1830s; Bromley, his wife Susannah, and their five children departed for Philadelphia in 1840. Within a year, his carding experience landed him a partnership in a small New Jersey spinning mill, but his wife died and the firm foundered soon after. Bromley remarried and returned to Philadelphia. At age forty-five, he commenced handweaving flat carpeting (ingrains) on a single loom, having failed at everything except fathering children. (By 1861, his second family would also number five siblings.) Entering middle age, Bromley appeared to be a classic ruined handloom weaver facing thin prospects.[64]

Thirty years later, Philadelphians regarded John Bromley as the dean of the carpet trades, a man discussed at length in a compendium of the commonwealth's leading manufacturers. Three tactics keyed this remarkable reversal of fortunes. First, Bromley chose wisely in selecting carpeting as his new product line and spending two years retooling as an adult apprentice to a local weaver. By midcentury, the urban middling classes had begun rejecting bare floors, oiled cloths, and rag rugs in favor of patterned woolen carpets that echoed the fine floor coverings of the wealthy. Demand for carpets far outstripped the capacities of the few New York and New England firms possessing variants on Erastus Bigelow's ingrain power loom, a shortfall that provided employment for thousands of Philadelphia handweavers across the next generation.[65] Many of them, like Bromley, were refugees from the collapsing British handwork trades, but two other factors helped him do far better than most.

Bromley, like many area businessmen,[66] was a birthright member of the Society of Friends. Settling in Philadelphia and becoming active in one of its many "weekly meetings" proved a critical asset, for he established enduring and valuable contacts with leading commercial figures in the Quaker community. Soon "Bromley's carpets [were] in demand, especially among the Friends, who did what they could to help these beginners along." For example, the downtown Bailey Brothers served as wholesale distributors for almost all of Bromley's ingrains, offered him advances that funded the purchase of additional looms, and "recommend[ed] him to the favorable notice of other carpet dealers." One of these, Orne Brothers, invited him to "match their foreign colors in carpets"; he subsequently experimented with new styles and stepped up to weaving fashionable Venetians in the 1850s. The Friends' deep commitment to honorable dealing ensured receipt of fair prices for his finished rolls, and his own witness to the same values built his reputation as a worthy supplier and an honest employer. Thus did community and culture undergird Bromley's calculus for success.[67]

Yet these connections would have availed little in the absence of a dedication to quality. As John Maule recalled, "The elder Bromley was extremely conscientious about the quality of everything that was used in making his carpets, being especially particular about his dyestuffs and their proper application, and this enabled him to attain a reputation for the Bromley carpets which made the name 'Bromley' a synonym for reliability."[68] As coloring effects were integral to the marketability of styled goods and Bromley was no dyer, being conscientious entailed developing reliable linkages with nearby dyeworks that could regularly meet trade standards for tinting yarns commissioned from neighborhood contract spinners. Joined at the loom by his three English-born sons in the early 1850s, Bromley could directly supervise fabric quality of an output that rose from about five hundred to over three thousand yards weekly by decade's end. The right line of goods, trustworthy dealing, connections that provided markets and financing facilities, and attention to style and quality—all intersected to boost John Bromley's enterprise.

His ascent was gradual. Only in 1860 did Bromley buy his own plant, moving his two dozen handlooms into a disused dyeworks purchased for ten thousand dollars on a ten-year mortgage. The Civil War did accelerate Bromley's advance, for he paid off the note in 1865, aided both by inflation and likely by a share of the profitable Union blanket subcontracts Kensington carpet works hauled in. In 1868, as the patriarch neared the end of his seventh decade, his three eldest sons removed to form their own enterprise, Bromley Brothers, leaving the partnerships to which each had been admitted. This maneuver neatly averted a potential succession crisis, for John had fathered three younger sons in his second marriage. The latter trio worked their way through factory apprenticeships and became replacement junior partners at John Bromley and Sons in the 1870s. The older group's parting was cordial; Bromley Brothers bought property opposite the family mill and erected their own factory with the stakes from their partnership shares.[69]

John Bromley promptly commenced a technical shift, beginning the purchase of ingrain power looms, 62 of which stood alongside 100 hand models by 1876; at this point the firm employed over 200 workers, one-quarter of them women trained to operate the new belt-driven weaving machines. Bromley Brothers followed right along, adding 36 power looms to its 108 hand looms. Together in the Centennial year, the two adjacent companies turned out 1.4 million yards of ingrain, Venetian, and "Patent Imperial Damask" floor coverings worth over $1 million. The war years and the ensuing decade's swelling market for carpeting built the Bromleys' fortune, but a crisis lurked even as plant and profits grew.[70]

North of Philadelphia, a few large companies, fully endowed with power looms, dominated carpet manufacturing. Philadelphia mills added these speedier devices, whereas small, thinly capitalized firms easily entered into handloom production. The result was a marked enlargement in total capacity between 1865 and 1873. When the economy went sour, the northern giants continued to run full bore, overfilling available demand, wrecking prices, and demoralizing the market. Most Philadelphia firms sold through wholesalers and jobbers who soon demanded substantial price reductions that fragile handloom workshops could not resist. The Bromleys' midsize enterprises and others similarly positioned found themselves squeezed from both sides. Rapidly, style and quality became secondary to price as dealers' orders and cordiality collapsed.

In time, this pattern would recur in batch production consumer goods' sectors confronted by serious economic downturns. Wherever there were numerous makers of fashionable goods whose distinctiveness evaporated in a crisis search for cheapness, a commitment to quality could prove a handicap rather than a virtue. Additionally, in those sectors where entry costs were low, displaced workers' desperate new starts could deepen the trade's miseries. Handloom ingrain carpeting in 1870s Philadelphia was such a case.

The two Bromley companies took different routes to addressing these difficulties. John Bromley and Sons added an upscale line to its range of carpetings, the Damask variety of flat goods, seeking to evade the glut in regular ingrains by moving toward a better class of fabrics that might resist pricing pressures. The patriarch still held to his connection with his old wholesalers, expecting rightly that they would not undermine him with predatory demands for killing reductions. Bromley Brothers, burnt out of their new mill in 1871, tried a different tack when rebuilding. They copied the integrated manufacturing system of their northern competitors by adding spinning and dyeing capacities and sold ingrain goods direct to retailers. In this marketing effort, they relied on the Bromleys' reputation for quality to secure them larger sales and higher prices than inferior producers.

Long-run developments favored the "Old Man"'s tactic—using technical change to restore product and style competition and evade price wars in lines with overbuilt capacity. Yet in the mid-1870s, both firms' maneuvers proved adequate; neither mill failed in the trade depression. With their own spinning frames and dyehouse, Bromley Brothers managed costs, eliminated subcontractors' profits, and gained more direct control over their ingrains' quality. Meanwhile, John Bromley and Sons' upmarket step set an assertive precedent for strategic product shifts that would lead the founder's second set of heirs into successive, massively profitable ventures in lace and hosiery across the next half-century.[71]

These two alternatives represented classic crisis maneuvers by manufacturers: target cost control, including distribution expenses, and/or move toward product novelties that might restore profits. There were other options that could interact with or substitute for this pair, such as activating collective institutions, deepening investment in technology or marketing, or cutting operations until "normal" demand recovered. However, bulk and specialty firms generally made different selections from this array. Lowell-like mills, for example, long found seeking product novelty more daunting than running full, amassing inventory, and paring costs through wage cuts, workload increases, or investment in technologies to speed production. High fixed costs and dividend expectations made short hours and lowered output implausible, for such steps would quickly generate losses. Batch specialists could shift their general-purpose machinery and skilled workers to new product lines or styles, seek out jobbing work in related fields, and operate at half-speed to retain their most valued employees until an upturn materialized. This was the basic strategy at John Bromley, Pusey and Jones, and even Corliss's works, and its adoption helped sustain Michigan's fine furniture industry as well.

Like fashion textiles, the furniture trade was significant in nineteenth-century manufacturing development along the Atlantic corridor, with early production centered in Boston, the Gardiner, Massachusetts, district, New York, Newark, and Philadelphia.[72] At the same time, what would become the nation's most renowned furniture complex, Grand Rapids, began transforming Michigan's hardwood forests into finished tables, chairs, dressers, and bedsteads. Once rail connections southwest to Chicago and to the East were in place, Grand Rapids eclipsed Cincinnati as the "western" furniture hub, becoming by 1900 the industry's reference point for quality and fashionable case goods.[73] Thus central Michigan will provide the last example for this collection of company profiles, introducing both a new sector and another region.

As at other midwestern sites, raw materials and railroads conditioned the first stages of Grand Rapids' woodworking development. Stands of maple, ash, oak, basswood, and walnut covered great tracts of the region. Landowners and speculators promoted railway building after 1840 to exploit these timber resources and enhance the value of their holdings. By the 1850s a half-dozen modest furniture workshops commenced activity along the Grand River, down which floated the earliest harvests of lumber-cutting teams. George Widdicomb's first shop (1857) lay at the east end of a wooden bridge that spanned the Grand.[74]

Born in Devonshire, George Widdicomb worked at the furniture trade in Exeter until his emigration to the United States in 1840. He, his wife, and their four sons settled in Syracuse, New York. George found steady employment in cabinetmaking until the mid-1850s and schooled each of his sons in the woodworking craft. In 1856, William, the second eldest, struck out for Battle Creek, Michigan, where he had been promised a post in a furniture factory. Arriving to discover that the firm had failed, William redirected his trek to Grand Rapids and promptly obtained employment at the Winchester Brothers' shops. The Winchesters needed able craftsmen, so William sent for his family, which proved willing to relocate, for the upstate New York trade had soured. Reunited, the five Widdicomb craftsmen within a year committed their savings to opening their own "cheap" goods works. When it burnt along with the bridge in 1858, the loss was but four hundred dollars. The quintet quickly rebuilt and hired six new employees once William Widdicomb's selling trips began to reap increasing orders. By 1860, the family enterprise added a local retail store, but the war disrupted everything.[75]

At the first call for troops, elder sons George Jr. and William enlisted, Harry and John following a year later. With his salesman and most reliable workers gone, George Sr. continued the business in a reduced way until it "expired" in 1864. Later that year William mustered out and returned to cabinetmaking, first for a few months at another of the town's small shops, then on his own in a single rented room with a "capital of $23.00." In 1865–66, the Georges, father and son, died; but the younger brothers returned to join William as partners, pooling their military earnings to buy a lot and erect a two-and-a-half-story workshop. By the time they expanded and added an outside partner to raise additional funds in 1869, Grand Rapids was fully linked to midwestern markets by four rail lines and the town's upward course as a furniture center was discernible. Employment at Widdicomb Brothers and Richards rose from twenty-five toward a hundred men. By 1873, when the partners incorporated with a nominal capital of $150,000, the number of local furniture firms had doubled to eighteen in three years.[76]

The three Widdicomb brothers divided their managerial responsibilities neatly, William "having the office management, Harry the power plant problems, and John being factory superintendent."[77] The young company started out with a bulk-produced staple, "the old fashioned spindle bed [for which] there was an enormous demand . . . especially in the west and pioneer districts of the frontier. Shipments were made in carload lots to St. Louis and Kansas City 'in the white' and upon delivery, were finished by the retail merchants by dipping the pieces in large vats of varnish."[78] Several local firms made similar low-end goods, with the Grand Rapids Chair Company becoming so committed to plain cheap seats that it was nearly bankrupted by competition from prison manufactories.

The 1873 contraction taught these and other staple-oriented firms the wisdom of the Phoenix Manufacturing company's high-end, style-conscious strategy. From its 1868 beginnings, William Berkey's firm sought specialized city markets for "parlor, office, and library furniture," used black walnut heavily, and supported its British foreman's efforts to bring in "first class mechanics from London." When the depression flattened demand and intensified price competition in cheap lines for rural and working-class homes, Grand Rapids' top-end furniture companies' sales to the urban prosperous and their businesses expanded after only one shaky year. By 1876, Phoenix, Berkey and Gay, and the Nelson-Matter Company had begun to turn Grand Rapids' gaze away from the frontier and toward eastern and midwestern cities. Appropriately, all three leased space at the Centennial Exhibition.

The Widdicombs were slow to respond, however. In 1876, William hit the tracks to contest a change in rate classifications for his cheap bedsteads as a result of through traffic compacts among major railroad lines. His arguments convinced a number of general freight agents to restore the older, lower charges; but when he was rebuffed by a Wabash official, Widdicomb dropped every account in towns serviced by that line, a stance he maintained for over thirty years.[79] Eventually younger brother John overcame William's similar stubbornness on style matters, and the family company joined the rest of Grand Rapids in satisfying the late-nineteenth-century rage for ornamented hardwood furniture.

Grand Rapids' course across the quarter-century after 1850 exemplifies what geographers Brian Page and Richard Walker term the midwestern "agro-industrial revolution."[80] The town's first reach toward external markets derived from building local capacities for turning nearby forest resources into staple goods, a pattern mirrored by Minneapolis flour milling, Kentucky distilling, and various midwestern meatpacking centers. However, by 1876, leading Grand Rapids furniture firms had shifted away from bulk outputs toward flexible batch production. In subsequent decades, stylistic novelty would be the city's watchword and its ticket to eminence among furniture production centers nationally. As smaller shipments of high-value goods to widely scattered retailers replaced carload lots to single clients, William Widdicomb's contentions with railroads on rate questions would become a matter for collective action by Grand Rapids furniture makers. Hence institution building developed into a strategic correlate of the reorientation toward quality case goods production.

This opening battery of company profiles has sketched parts of the landscape of specialty and flexible manufacturing in the nineteenth century's middle decades. Adjacent to the great New England textile corporations, the systematizing railroads, and "American system" precursors of mass production stood hosts of skill-intensive, product-diverse enterprises. Their abilities to craft machinery and prime movers supported the extension of both bulk and specialty production and, through Corliss's, Worthington's, and Morris's pumps, underwrote the hidden infrastructure of urbanization. Batch operations also provided the tools, gauges, instruments, and saws for both specialty and bulk manufacturing of fabrics and furniture, sewing machines and food containers. In finished consumer goods, their capabilities by 1876 were increasingly focused on capturing middle-class markets for styled household items, an effort to harvest secularly rising demand for diversity through a creative amalgam of tariff-sustained pricing and fashionable ingenuity. Spindle beds and cotton sheeting might sell in carload lots, but carpets and parlor furniture promised profits through a version of competition focused on product features rather than on price. Even so, although shared dilemmas made it sensible, collective action by specialist firms was unusual, and institutions to sustain it remained rare. A personal crusade like Widdicomb's to secure better shipping rates was at the time more common than creation of organizations on behalf of local or sectoral interests. There were exceptions, however.

*Chapter 3*

# INSTITUTIONS AND THE CONTEXT FOR SPECIALTY PRODUCTION

EVERY INDUSTRIAL enterprise is of course an institution, but here the term refers to organizations created by manufacturers who define common goals and undertake collective efforts to achieve them. In the United States, few such institutions existed in the middle quarters of the nineteenth century, but hundreds of them appeared by 1900 in an unprecedented burst of associational activity. Specialty producers created associations and other interfirm projects most often on a city or metropolitan scale, whereas in bulk trades multistate or national organizations proved more common. For example, the Boston-based New England Cotton Manufacturers Association (NECMA, 1865) created a forum for discussing the shared interests of staple fabrics corporations. Company officials (mill agents, superintendents, and, by 1867, treasurers) met twice yearly for dinners, exchanging information on "the best methods of manufacturing cotton." NECMA published these reports and occasional sectoral statistics in an annual volume but had neither "permanent quarters" nor a staff until the 1890s.[1] By contrast, with considerable regularity and variable success, postbellum industrial specialists focused locally on technical education and labor relations, and more broadly on marketing dilemmas. Such interfirm efforts aimed both to regularize key "externalities" that figured in specialty districts' competitive advantage and to foster a spirit of cooperation within the larger competitive context. To prefigure these later collaborations, this chapter introduces two institutions created at early centers of specialty manufacturing: Philadelphia's Franklin Institute and the Worcester County (Massachusetts) Free Institute of Industrial Science. The chapter closes with an overview of Part I and a discussion of several questions evoked by this introduction of approaches to and settings for custom and batch production.

## THE FRANKLIN INSTITUTE, 1820s–1870s

The artisan-manufacturers who first gathered in Philadelphia in the mid-1820s were a diverse lot of proprietors and craftsmen who sought to promote and diffuse technical knowledge of economic significance. At the initial meetings for "forming an Institute for the benefit of Mechanics," Samuel Merrick, a merchant's son tasked to revive a distressed fire-engine manufactory, joined toolmakers Matthias Baldwin and David Mason, together with a silversmith, a shuttlemaker, a steam-engine builder, a type founder, a tanner, and a gun maker. Aware that commerce had ill prepared him for the challenge ahead, Merrick had applied for entry to a mechanics association that sponsored discussions of technical questions. Blackballed as an outsider, he "determined to establish his own society, where he and others like him could learn the science upon which their arts and crafts were based and, not incidentally, elevate themselves socially."[2]

The Franklin Institute's founders specified two goals when appealing for a legislative charter: diffusing mechanical knowledge to workingmen (through providing public lectures and amassing a library) and offering prizes for useful inventions. In its first two years, an evening school for draftsmen supplemented the topically scattered lecture series and the institute inaugurated its exhibitions of American manufactures and innovations. Its award evaluations validated authentic novelty and exposed fraud but were feasible only after favorable resolution of a struggle with the Patent Office over the right to purchase copies of patents. By 1826, the Franklin Institute had a thousand members, stable quarters, and a healthy treasury, and it had begun publication of its *Journal*, which reported on patents, canals, and the regional coal and iron industries. Artisans and manufacturers bulked large in the membership; indeed, Baldwin the locomotive builder remained on the board of managers into the 1860s.[3] But the institute's leadership soon was peopled by gentlemen, especially attorneys and academics committed to promoting "professional science."[4]

The exhibitions expanded dramatically, drawing crowds in the tens of thousands. Yet scientific investigation, at times related to manufacturing issues (measurement of water power, analyses of boiler explosions), overshadowed service to mechanics in the 1830s and 1840s. Several science-minded proprietors (Merrick, Baldwin, Isaiah Lukens) remained active, but in becoming one of the elite institutions of antebellum science, the institute edged away from its origins as a "democratic learned society." Its purchase of Masonic Hall expanded exhibition facilities and incurred massive debt just as the 1837 depression struck, forcing the institute to scramble after money for the next two decades. The leadership also had to reconsider their mission shift, for the founding of the Smithsonian and the American Association for the Advancement of Science introduced both a better-funded research center and a truly national scientific society. As its historian noted concisely, "Although it was not yet clear in the 1840s, the Franklin Institute's future vitality rested on an identification with the demands of industry, not science. . . ."[5]

In the 1850s, the *Journal* redirected its attention to concrete industrial issues in a sensible effort to deepen the institute's contacts with Philadelphia's vibrant machinery and railroad equipment builders. By this time, Samuel Merrick had served as the Pennsylvania Railroad's first president and was operating Philadelphia's Southwark Foundry in partnership with his two sons. Belying its name, the "foundry" constructed marine engines, lighthouses, and apparatus for gas distribution. Heavy machinery also proceeded from I. P. Morris's Pascal and Port Richmond Iron Works, the Baldwin, Norris, and Eastwick and Harrison locomotive plants, plus the Bement and Dougherty and William Sellers machine tool shops. Courting these and other proprietary capitalists was key to revitalizing the institute. In the same vein, the annual industrial exhibitions held added significance once institute managers increased "the number and value of awards." In the 1850s, audiences for the two-week shows reached 100,000, but repeated efforts to develop technical education, the third pillar of industrial service, failed completely. Only its drafting classes were durably successful.[6]

This gradual reorientation saved the institute from disaster. The 1857 panic again shattered its finances, which still carried $50,000 in residual debt from the 1830s. By 1860, urgent fund-raising was in order, but the $11,000 pledged by concerned members proved difficult to collect once war intervened. The *Journal* faced a $5,000 deficit (it had never made money), and building repairs had been too long deferred. Late in 1863, local industrialists assumed power in a friendly takeover, when through organizational elections "an almost totally new slate of officers and managers was installed." Sellers became president, John Towne—a manufacturer who would endow the engineering school at the University of Pennsylvania—vice president, and the Pascal Iron Works' head manager, secretary. On the board now sat Charles Cramp, shipbuilder, along with one of Merrick's sons, William Sellers' cousin and (later) partner Coleman, tool builder James Dougherty, and the Southwark Foundry's superintendent. The next year the elder Sellers and younger Merrick erased the institute's debt by squeezing funds from twenty-seven of their industrial colleagues. Internal changes followed, as "Sellers refashioned the Franklin Institute into something like a professional engineering society to serve the technical interests of manufacturers," twenty years before the birth of the American Society of Mechanical Engineers (ASME).[7]

Sellers and his colleagues had little interest in questions about speedy looms or the cotton supply that occupied NECMA. In his most notable *Journal* article, Sellers instead considered a minute but critical issue for machinery makers: screw threads. Machine builders either made or contracted for thousands of screws that held their devices together. Yet firms had differing specifications for the number of threads per inch and their pitch (the angle of the threads' take-up). Screwheads for wrenching them into place were no less variable. Every time a worn or broken machine screw needed replacing, equipment users had either to call upon the machine's makers for a substitute or undertake on their own the tricky business of duplicating this apparently simple, but actually complex, component.[8] Surely, Sellers argued, a national standard for screw characteristics was warranted. Of course, he offered his firm's specifications as a model, but his alertness to nagging technical problems associated with the diversity of machine manufacture warrants comment.[9]

Standardization of elementary components could cheapen the cost and ease the construction and maintenance of specialty products. In textiles, this point of intersection between routinized and batch production had earlier given birth to the "count" system for yarn sizes and had informed the devising of "change gears" to alter spinning frames' outputs.[10] Sellers had convinced leading Philadelphia metalworking firms to adopt his approach to thread specifications and head shapes, and institute experiments confirmed its superiority over rival schemes. Yet the Sellers standard failed to generalize, for its adoption would force costly adjustments by firms elsewhere, changes that could loosen the flows of information and the ties of dependency between machinery builders and their clients without obvious benefit to the former.[11]

These relational issues would recur as debate continued on materials, parts, and measurement standardization across the next half-century. Insofar as establishing standards would reduce cost and increase competition among their suppliers, both batch and bulk intermediate and final goods firms could benefit materially. But batch-producing suppliers would resist arguments that implied intensified price pressures, no matter how elegantly draped with rhetoric of public interest and technical efficiency. Final goods specialists who made their own components and felt that their specifications represented best practice would also reject such initiatives. To them standardization, as in the Sellers case, smacked of particular interests masquerading behind ideologies of efficiency and scientific procedure. Institutions like the Franklin and the ASME could better argue the case than might Sellers or Frederick Taylor in their individual capacities, but opposition would not evaporate. Resisters, caricatured as backward and foolish, understood that standardization was about power and money, not just progress and rationality.

The institute had begun as a forum for mechanics and manufacturers to pool practical knowledge and learn principles relevant to their work. Its turn to science overshadowed these services to the city's craftsmen, but they did not cease. Evening lectures, at times five nights a week, drew sizable audiences from a membership that oscillated between 1,200 and 2,500, including hundreds of skilled workers and small shop masters. Embracing an industrial service mission refocused the institute on issues entwined with its founding. In the *Journal*'s first 1867 number, its editor formalized this recovery, soliciting papers on "important engineering works" that would be of "direct and evident value to practical engineers," preferring those giving "accurate and full descriptions" over "theoretical discussions and mathematical investigations of general principles."[12] Though a few of the latter were still printed, the *Journal* soon featured articles on iron bridges, steam engine indicators, power belt practice, and the Pennsylvania Railroad's repair shops, supplemented by monthly reports on novel industrial devices. In 1870, it began serializing John Richards' "Wood Working Machinery," continuing well into 1872, the year Richards joined the institute's Committee on Arts and Manufactures. Unlike the New England dinner associations, the Franklin Institute encompassed a broad community working on problems in critical areas of technical innovation. Moreover, "[t]he Institute and its community provided an ideal incubator for [future] industrialists and engineers."[13]

Reinvigorated through its capture by representatives of Philadelphia's specialty metalworking firms, the Franklin Institute again functioned as a forum for technical debate and the diffusion of practical knowledge. Its library soon subscribed to hundreds of trade journals, including dozens of international titles, and maintained complete patent records.[14] Yet in one respect, the institute failed. Technical education, its perennial stumbling block, increasingly seemed essential to the extension of flexible manufacturing's capabilities. This task was better handled at a less-heralded site for batch specialization—Worcester, Massachusetts.

## Worcester and Its Institute, 1840s–1880s

Worcester, the self-advertised "heart of the commonwealth," might seem a peculiar industrial center to those who take textile cities as representative of Massachusetts' industrial dynamism. The absence of waterpower barred Worcester from joining the staple fabrics surge, but the city soon prospered through adopting steam engines and attending to leading regional sectors' metalworking needs, as its extensive entries in the 1832 McLane Report show.[15] By 1840, Worcester, in the upper Blackstone Valley, possessed a canal link to Providence and rail connections to Boston and New London. Moreover, it boasted multiple "facilities . . . erected for rent with power to a number of tenants," making it possible for "mechanics to begin business in a small way without incurring the expense incident upon the erection and equipment of a shop." Into these spaces went men like Thomas Daniels, builder of mechanical wood planers, and lathe maker Samuel Flagg—artisan-proprietors who met in November 1841 to form the Worcester Mechanics Association (WMA).[16]

Committed to "the moral, intellectual and social improvement of its members" (three-quarters of whom were skilled workers, the rest industrial employers and small businessmen) and to "the perfection of the mechanic arts," the association also took responsibility for "the pecuniary assistance of the needy." That task moved it beyond the Franklin Institute's boundaries, but in other respects the association followed its lead precisely. The WMA soon offered public lectures and drafting classes, leased quarters for meetings and a library, and began by 1848 a series of mechanical fairs. Worcester's industrial structure echoed Philadelphia's, for networks of founders, forge shops, pattern makers, machine shops, and assemblers generated Furbush and Crompton looms, Goddard and Rice's paper machinery, and Daniels' planers (plus "improved" versions of them proffered by at least three local rivals).

By 1854, an expansive prosperity brought this comment: "Without the aid of a single act of incorporation [a sly slap at the textile giants], mechanical business has increased in this city by individual enterprise alone more than tenfold." In a generation's time, mechanics had become "nearly a majority of the population . . . of twenty-two thousand," "owners of nearly or quite half of the taxable real estate," and producers of $6 million in goods annually. Local firms' "reputation for variety, excellence and finish on all labor-saving machines and implements extends far and wide." The association celebrated these achievements by raising funds for a great hall, begun with Ichabod Washburn's $10,000 donation and finished in 1857—a $148,000 edifice whose auditorium seated two thousand.[17]

Plans for a WMA educational branch that could reach beyond mechanical drawing to chemistry and engineering practice were also made in 1857, but abandoned once secession and war followed that year's financial panic. Initiation of the Worcester County Free Institute of Industrial Science was delayed until 1865, when a nearby tin-goods manufacturer offered the city's industrial leaders $100,000 toward the expense of starting a tuition-free practical academy open to the county's aspiring young men. At first the school was planned as a traditional books-and-lectures enterprise, but Ichabod Washburn had a different notion in mind, along with the funds and influence to see it realized. Washburn had commenced business as a metalworker, building first textile machinery, then wire carding-engine "clothing," and ultimately drawing wire in hundreds of varieties. Throughout, like Trenton wire maker and bridge erector John Roebling, he had designed and built his own machinery by trial and error, sustaining his deep faith in the value of hands-on experience as a base for innovation. To his mind, practical education in a metalworking district entailed a mix of regular classes and shop labor, the latter in a full-scale machine works, not a toy laboratory. Moreover, the institute's Washburn Shops, built and endowed at his expense, would be operated as a business.[18]

Though at times overlapping at individual firms, routinized and specialty production implied needs for different sorts of technical expertise. In tonnage steel, flour milling, or petroleum refining, engineers knowledgeable in the principles of physics, chemistry, and mechanics were critical to materials testing and the incremental refinement of flow and throughput. Here was the domain for engineering as a species of laboratory science: professionals carefully testing multiple factors bearing on a relatively stable problem set. But in machinery and tools, as in fashion textiles and furniture, the need was for novelty, even genius, in design, in organizing the work process, in dealing with uncertainty, and in engaging others' knowledge to address collectively far more disparate clusters of problems.

Science seemed to link almost effortlessly with routinization and systematic instruction. But advances in machinery and tools and in their profitable manufacture derived from rude technologies, scattered insights, and "practical" awareness of human and mechanical relations. How might it be possible to institutionalize the reproduction of this art and craft, given that apprenticeship was waning and, in its old form, was inadequate for the preparation of imaginative workmen and prospective entrepreneurs? The extension of bulk production called for legions of quickly trained and carefully supervised workers, but the multiplication of specialty manufacturing, not least in machinery, demanded the efforts of more skilled workers than had ever been available. Ignoring the problems of bulk enterprises, Washburn's answer to the specialists' dilemma was experimental realism, a school operated as a business.

Worcester Institute students would for three years learn principles in half-day classes (mechanics, drafting, etc.) and learn practice in the shops of a real, if captive, business. They would work without pay at lathes and planers under the guidance of skilled veterans and a superintendent who solicited orders to pay the bills. Their output would have to stand up in price and quality to clients' expectations; and profits, if earned, would be recycled to support at least eight of the most needy and able student workers. Thus would they come to appreciate the twin disciplines of collective effort and the market, the limits of machines, the character of materials, and the means by which formal knowledge might be translated into factory practice, technically and socially. The best of them would lead or start enterprises, and the rest would form a new class of skilled operatives more deeply educated than any in the nation's history.[19]

The Washburn Shops were a puzzling success. There, lathes were designed and erected that won gold medals at mechanical fairs in Baltimore (1869) and New York (1871). By 1873, sales of lathes, drawing

5. The Worcester Institute's Boynton Hall (left) and Washburn Shops, ca. 1875. Courtesy of Hagley Museum and Library.

stands, models, tables, and "iron and wooden apparatus for chemistry laboratories" neared ten thousand dollars annually. By 1876, the business had begun to manufacture hydraulic elevators, the production of which "would bring it great credit." Yet its achievements little impressed some students, who bemoaned the management's "incompetency," and others who complained that the shops' Centennial exhibit included machine tools the most difficult parts of which were not entrusted to them but finished by already-skilled workers. Truly, the shops *were* run as a business, which peeved some learners no end.[20]

Eight years later, there was similar consternation when institute head George Alden reviewed the Worcester program at the national meetings of the ASME. His paper occupied but fifteen pages in the published *Transactions*, but a transcription of the extended, occasionally splenetic, "Discussion" filled thirty-five more. Professors from the "school culture" faction in engineering education blasted the 2,400 hours of shop work Worcester students performed over three years as leaving too little class time for preparing a theoretically proficient professional. Their model engineer was an expert and manager, one who gathered scientific knowledge and gave orders (or cleared the way for others to do so). Such men wasted their time grubbing among machines, subordinated to skilled workers, as were Worcester's students. Gray-whiskered shop veterans and clean-shaven Oberlin Smith demurred, Smith asserting: "It will not do for a mechanical engineer who may have numbers of men under him, and who may have to contrive new processes which must be commercially cheap, not to know [both] the theory of his work and a good deal of the practice also. . . . He should simply know enough of the theory of these operations to guide the other men." Smith's engineer was a direct actor in and around the shop floor who had to deal with novelty, costing, and workingmen, and whose ability to persuade, innovate, and "guide" demanded (and defined) *practical* mechanics. Such men could master the obstacles confronting specialty manufacturers and, like Smith, join them as entrepreneurs.[21]

The Worcester model for technical education informed later attempts to weld shop and book work. Perhaps its most noted successor was the cooperative engineering program at the University of Cincinnati, which alternated paid employment at regional manufacturing firms with class sessions, extending the learning sequence to five years while keeping the door open to poor but capable young mechanics and draftsmen.[22] Yet already by the 1880s, academic engineering spokesmen were measuring their students against professionals in architecture and law. The Worcester notion of creating a shop-experienced proto-entrepreneur, a designer-manager intimately connected with the realities of metal-bashing, spoke to a differently contoured world of expertise and ambition, the world of specialty production.

The distance between the office and the shop would widen as the authority of science and school credentials devalued grit and grime, as well as their shop-floor innovation and human relations corollaries. In metalworking, one product would be Taylor's managerial scientism, an effort to achieve workplace control authored by a batch and specialty shops veteran obsessed with eradicating variation and uncertainty. Yet as John Richards reminded *Industry*'s readers, the arrival of "book" engineers at specialty works either led them into emergency apprenticeships in the folkways of shop practice or generated sharp conflicts between contrasting embedded claims to knowledge, triggering four-way struggles among shop-floor workers, foremen, engineers, and proprietors or top managers.[23]

There proved to be no "one best way" to address these tangles. By the turn of the century, in some sectors, huge firms would create their own schools and retrain college engineers. In others, apprenticeship programs would be redesigned or revitalized. Elsewhere, sectorally based educational institutions would attempt to graft shop experience onto technical and design capacities, even as many specialty trades, like printing, furniture, or jewelry, ignored engineers almost universally. However, the key points for the 1870s are the diversity of circumstances facing actors across the spectrum of specialty manufacturing possibilities, the contemporary uncertainty as to what technical or engineering education entailed, and the batch veterans' appreciation, epitomized by the Washburn Shops, that understanding production in its greasy details was fundamental.

## Summary and Context

This opening narrative has attempted to portray the environment for and practice in specialty manufacturing on either side of the Civil War. Giving particular attention to metalworking, machinery, and textiles, sectors in which the system of repetition production was most advanced, it argues that batch and custom firms were articulating a more complex dynamic of skill-demanding, diversified industrialization than their generic-product colleagues. These practitioners were as technically innovative as any in America, many of them active in patenting and participants in formal and informal technological networks. They were proprietary capitalists, operating as individuals and in partnerships; and their businesses directly manifested their individualities, displaying interpersonal and work relations as diverse as were their products. They clustered and prospered at nodes of specialty prowess, which their presence reinforced, making certain cities magnets for custom and batch capacity. At such sites, clear-eyed property developers built incubator spaces for new firms, networks of productive interdependence formed among sectoral subdivisions, and hosts of skilled workers found employment.

By the 1870s, some firms with batch origins grew huge and mixed bulk production of heavily demanded items while continuing to explore novelty, as with Washburn's staple versus specialty wires, Disston's handsaws versus experimental band and circular saw innovations, and Brown and Sharpe's standard sewing machines versus machine tools and instruments. Other companies remained devoted to specialties, as at Bromley, Sellers, Corliss, and Smith, or learned their virtues from colleagues, as did the Widdicombs. In marketing they relied on reputation and quality, not simply price, and they ventured to construct collective institutions for "democratic" technical exchange and education at an early date.

However, four questions this narrative provokes merit further attention. How different were midcentury specialists from bulk output proprietary firms? As this industrial format was not a paradise, what about the workers? Given that all the firms profiled were successful, what of failure? Last, how were state policies linked to their operations in the war era and the succeeding decade?

The New England cotton corporations, expanding railroads, and rising producers of oil products, standard shoes, looms, or sewing machines were unusual figures on the industrial scene in the 1870s, for most staple goods flowed into the American economy from thousands of less visible proprietary and partnership enterprises that scurried to shore up weak capitalizations and keep in touch with technical advances. Their owners' biographies often crowd the compendious city and county histories the Centennial sparked. Proprietary bulk producers inhabited much the same cultural space as batch specialists—a personalized, relational milieu; but many of their production processes were vulnerable to firms seeking scale economies. Thus distant rivals could penetrate their regional markets as the rail system advanced. The most adroit among them, like Grand Rapids' Widdicombs, could prosper by shifting into specialties, but others faced extinction if capital-intensive consolidated works hammered down prices and captured market share. In iron, meatpacking, or soap, their numbers thinned steadily as the century wound down. In other bulk sectors then lacking throughput possibilities or governed by locational particularities (cigars, lumber, bread baking, ice), thousands of firms persevered quietly.[24] They provided wages to millions but occupied an entrepreneurial sphere different from that of batch specialists, generating few patents, industrial districts, or institutions.[25] Reputation surely mattered among them, but versatility was not salient to their quests for profit. Most of them and their trades still await interested researchers.

For workers, specialty manufacturing was a maze, for as always they were by turns essential and expendable. Bulk production jobs might be routine and tedious, but employment was relatively stable. In specialty lines, work depended on orders secured; and seasonal or business cycle slumps were not commonly bridged by production for inventory. As studies of Massachusetts have shown,[26] unemployment was pervasive for all classes of workers in the last half of the century, but in batch operations its incidence was surely greater. Why then seek work in such companies?

It would be hard to establish that specialty manufacturers were better bosses than were proprietors of staple goods establishments. After all, many a nail works master respected "his men" and their abilities and evinced no reluctance to get filthy or lend a hand when problems burst forth, whereas some specialists like Corliss were distant, harsh, and secretive.[27] Perhaps the more challenging and varied work mattered a good deal, for such plants were schools for skill enhancement, the best of them offering repeated chances for experiment, learning, and recognition. Many workers toured these factory schools, gathering experience in a dozen or more companies, whether of their own volition or after layoffs. Such wanderings created the all-round workman and sustained his willingness to walk out the door when he felt himself to be ill used. Others grew in place, becoming part of the core workforce that represented a key asset in the firm's bundle of capabilities, and were kept on in the worst of times. Of course there was contention and conflict, but batch specialists offered positions that stretched and strengthened workers' abilities, rather than tasks that merely tested their capacity to endure routine. Arbitrary dismissals and stubborn foremen abounded, but these were universal industrial hazards. For skilled men and those in the process of building skills, batch producers of machinery, tools, and styled consumer goods arguably offered many of the nation's best jobs.

Such firms were training grounds for entrepreneurship as well, but this should not be overstressed; for only a small proportion of workers ever started their own firms, and most of these, then as now, went broke. Though the evidence is scattered, pay rates at specialist firms may have been somewhat higher than at bulk operations where there were comparable occupational slots (machinists, woodworkers). This too should not be pressed, given the probability of layoffs and their effects on overall earnings. Still, in specialty firms the most proficient workers did have chances for elevation to their trade's most demanding tasks, jobs that paid appreciably more than regular journeymen's rates.[28] This said, gender issues were surely equally important, once what might be called the "manhood question" developed in the postbellum era.[29]

The preceding factors (on-the-job learning, chances for entrepreneurship, higher wages) all pertain to an individual's rational calculation of options, but gender concerns delve into affective and cultural matters that are no less constitutive of personal choices. In the early nineteenth century, David Leverenz has argued, among the multiple American constructions of masculinity, several generations of skilled workers exemplified and reproduced an artisanal form centered on craft mastery and autonomy. By the 1850s, the economic spaces for sustaining this version of gender propriety in its original guise narrowed, as men increasingly worked at other men's pleasure for life, if indeed their craft skills were not simply devalued (as in shoemaking, handweaving, or hand-printing presswork). Though the Civil War may not have been a watershed for the recoding of working-class masculinity, it certainly afforded men in the industrial North opportunities for a hazardous invisibility within huge organizations, the discipline, grit, and boredom in camp and aboard ship alternating at wide intervals with chaotic demands for heroic action. For some, these experiences may have made the safer tedium of wage labor in bulk sectors more palatable, even attractive by comparison. But for other men, and for a separate stream of skilled immigrants, the social relations of batch and custom production could engender a reworked version of artisanal manhood.

In shops and mills observing canons of mutual respect and reciprocity, a web of rituals and performances sustained skilled workers' self-esteem. There, the proprietor was the "Old Man," and workers related to him in a watchfully fraternal way. This was not the stereotypical father-to-child paternalism (more common in masters' and skilled workers' relations with women, youths, and the unskilled) but instead represented the institutionalization of customary bonds between adult sons and their graying, often wise, and obviously experienced fathers. For such men, the rival masculinity of force and bombast belonged to the stockyards, to railway navvy camps, among teamsters, or on the docks. When an overseer or foreman adopted the ways of the brute, resistance if not revolt ensued. When manifested by a master, or, worse, an heir-apparent stripling, such behavior called workers to display a "manly bearing" to reaffirm boundaries. If this failed, unless they were barred by kin obligations, debt, homeownership or hard times, men packed their tool kits to seek better situations or gathered together in strikes for dignity. Yet when "heroic work" had been achieved, overcoming the obstacles of complexity, tight deadlines, or uncertainty in novel or huge-scale contracts, ceremonies of celebration followed and routinely were captured for the camera and the historical record.[30]

This masculinity of "competence respected" had lateral referents as well, conditioning relations *among* skilled workers that idealized the man who could help another without demeaning him yet also would condemn shoddy work, the ill-treatment of tools, or hasty shortcuts that risked injury. It underlay the collegiality and coercion of the "stint," which Taylor so despised, that collective and adult recognition that workers were not machines, that employees' interests were *linked* with those of owners rather than being identical or subservient to them. This reframed artisanal manhood was hardly egalitarian. Pecking orders abounded, and although most highly skilled workers knew better than to show up their less competent colleagues, disciplinary rituals kept them from getting "big heads" and drove the show-off, the faker, and the utter individualist from the shop. Manhood was, as ever, an interactively shaped practice, but this version's expression and reproduction was far more feasible in specialty manufacturing than amid the tedium of bulk processing or the rule-bound routinizations of "American system" producers, and workers knew it.[31]

Of enterprise failure, perhaps less need be said. Most industrial history is written from the records of firms that had a measure of success over some years. Batch specialists, however, did have a particular vulnerability. Such establishments were consistently founded by independence-minded skilled workers, most of whom in postbellum years had had some experience as foremen or shop superintendents. Their mastery of a division of the trade's labor process (in textiles—dyeing, carpet yarn spinning, or Jacquard weaving) gave promise that they could meet demands for quality but guaranteed nothing in regard to design characteristics, ability to price goods profitably, or capacity to reach markets effectively. Moreover, most new specialists were as capital poor as they were rich with practical experience.

Such nascent firms needed help. Before 1876, the lucky ones hit an expanding market and had clients besieging them with orders. However, few new starts had supportive distributors or contractors, as the Bromleys did; and thus many an enterpriser was forced to return to the shops in search of a place, burdened by debts difficult to clear.[32] A partnership match between a shop man and a salesman had better chances, for the latter at least knew well the tortuous byways of pitching goods and negotiating prices. An obsessive bookkeeper was a fine early acquisition, provided that one of the partners understood how to evaluate accounts. Still, were a novice specialist firm lodged in a vital industrial district and its proprietor known as a good workman, openings for custom from sectorally related companies might rapidly appear, even from the works he had left.[33] The new operator might well take advantage of the area network's facilities—commissioning independent designers, if in consumer fashion lines, or contracting with batch suppliers (foundries, pattern makers, etc.) for goods and services integrated into mechanical novelties.

There was, however, another sort of enterprise that the Bromley case highlights, one that would recur in other trades as market and technical conditions shifted: the entrepreneurship of desperation. When a sectoral format faltered badly, expelling sizable numbers of workers, as in the carpet transition from hand to power looms and concurrent shift to female weavers, some of those pushed out commenced manufacturing on their own, singly or in tandem. Using cheap and old machines, these "garret" or "mushroom" firms came instantly under the sway of distributors who could supply yarns and patterns, thereby reconstituting the putting-out system of cottage industry days. Reviled or pitied by established manufacturers and pressed relentlessly by wholesalers on the price of their work, they rarely lasted long, swelling the number of failures. Several hundred such enterprises surfaced briefly in Philadelphia carpets during the 1870s shift; all but a few were gone at decade's end.[34]

Ample evidence of failure hardly deterred ambitious workers from initiating businesses in skill- or labor-intensive sectors of the economy. "Rags to riches," as in the Bromley case, had an enduring appeal; and cultural and pecuniary aspirations routinely overcame rational expectations about the slender possibilities for success. Even so, as the above discussion suggests, there were real differences between the chances of the displaced desperate and those of others lodged in some version of a support network. Particularly in styled consumer goods and in specialty subdivisions of machinery and intermediate products, a subset of the latter group would continually reinvigorate sectoral capacities, despite the odds. The rest were lost and forgotten.

Regarding the national state's connection to the deployment of batch manufacturing through the 1870s, only three matters seem directly relevant. Most historians view the century's middle decades as a period during which federal (and state) engagement with manufacturing drew back from the domestic development promotionalism of the early republic. This is true enough, but insofar as patents, tariffs, and government contracts are concerned, both structural and occasional supports for specialty firms materialized. If they could be demonstrated to the Patent Office, then defended at law, innovations that keyed profitability could be produced under a time-bound monopoly and priced accordingly. Corliss, Disston, and Savery, among others, found patent protection immensely valuable and fought resolutely against interlopers and imitators in order to reap their proper returns. A comparably aggressive strategy rarely availed in fashionable consumer goods, for novelties turned over seasonally and were "dead" long before any litigation over copying could be concluded. (Establishing Corliss's primacy in steam engine innovations, for example, took fifteen years of court skirmishing.) In books, textiles, clothing, and furniture, neither copyright nor patents for products appealed, especially in that American firms then were duplicating European products shamelessly.[35] The relevance of patents within batch specializations was sectorally specific.

The tariff was different, for the Civil War's Morrill tide raised all boats, those of tonnage iron and baled sheeting makers and those of machinery builders and carpet weavers alike. Tariff provisions restricting imports of preferred European finery, whether mechanical or stylish, delighted specialists and enhanced their prices and profits, notwithstanding the protests of Democrats, dedicated free traders, and export-oriented agriculturists. Republican protectionism and denunciations of European "pauper labor" became hymns of faith among American industrialists for more than a generation. Eventually, the throughput masters and others would see tariff reciprocity as a gambit for securing export openings, once their cost advantages and capacity excesses matured.[36] This would break a Gilded Age solidarity, but the state's power to obstruct importations and thereby ease the path to profitability for all manufacturers should not be undervalued for the later nineteenth century.

Federal contracts were a tiny element of domestic demand both before and after the Civil War. However, during the conflict they provided a rich harvest for specialists able to shift readily to meet the needs of the Union for cannon, ships, blankets, stockings, and tentings. One of the war's lessons may have been that it was simpler for versatile batch firms to rework their operations to produce long runs of "government goods" than for bulk enterprises to move toward outputs differing significantly from their regular lines.[37] War production did solidify shipbuilders' technological shift to iron vessels and, in textiles, financed the expansion of small and midsize mills.[38] This stipulated, it is hard to view such contracts as other than temporarily sustaining for specialists. Over the longer haul, patents and tariffs were more salient to innovative firms and European-menaced sectors, respectively, whereas government contracting was an expedient during the war and nearly vanished thereafter.

Part I has provided a structured if unsystematic evocation of custom and batch manufacturing as Americans approached the Republic's hundredth year. This essay in difference underscored distinctions between specialists' approaches to enterprise and those of routinized industry, as well as the diversities among firms deriving profits from novelty. In making machinery, specialists' proficiency underwrote the refinement of productive capacity in multiple sectors. In intermediate goods, it added tools to the kits of a comparable array. By 1875, aided by tariffs, it had begun to wean middle-class Americans from assumptions that only importers could provide stylish personal and household goods. The Centennial Exhibition would bring these streams of innovation and flexible capacity together with one another and with exemplars of European excellence. It would indicate both the achievements and the deficiencies of American specialty manufacturing. There's no need to buy a ticket to the show; just turn the page.

# *PART II*

## CENTENNIAL TO COLUMBIAN:

## SPECIALTY PRODUCERS, 1876–1893

*Chapter 4*

# THE 1876 EXPOSITION AND PHILADELPHIA MANUFACTURING

PHILADELPHIA'S Centennial Exhibition will serve as our point of departure for viewing specialty manufacturing's expansion through the early 1890s. Though we shall not lose sight of individuals and firms, greater attention will be given to sectors, industrial districts, and the institutions they fostered. In a series of chapters, we will move through much of the American industrial belt, from Philadelphia to Providence and New York, then west to Cincinnati and Grand Rapids, arriving in Chicago just prior to the Great White Way's opening and the crushing depression that haunted its fabulous sights. Along the route, we will encounter trades both familiar (machinery, textiles, furniture) and fresh (jewelry, silverware, publishing). Toward the close, an assessment of emerging institutions will be paired with an evaluation of specialty manufacturers' status within the larger industrial economy in 1890 and a tentative typology of firms and sectoral formats within this approach to production. Specialty manufacturing and its correlate flexibility took a variety of forms, a point suggested in Part I that will be more fully documented and discussed below. At the outset, one additional issue, the relationship between space and specialty manufacturing, must be introduced.

Just as our customary categories of "industries" imperfectly capture the contours of batch and custom production (given the many mixed output sectors), so too can "cities" be faulty containers for its spatial deployment. Although propinquity and agglomeration had real value for textile and jewelry companies in Philadelphia and Providence, their nodes of concentration were neighborhood-sized and linked through contract and contact across spaces shaped by other sectors and other interests; in the case of jewelry, they leapfrogged state lines within a regional complex. Moreover, distinctive characteristics of sectoral markets and technologies, as well as differing development trajectories, generated positive effects from spatial concentration in some trades and decades and sharply negative effects on others. Though Alfred Marshall highlighted and perhaps romanticized the advantages such "industrial districts" held for diversification and flexibility, the record is more complex and thick with unintended consequences. Indeed the concept, "industrial district," is itself awkwardly limited, as when once-neighboring metalworking specialists relocated to disperse across a city-region, as increased needs for space (the results of "district" success) pulled them out of older clusters of location and association.[1] Indeed, truly scattered specialists like northeastern silverware firms could achieve a conditional, strategic solidarity that eluded those in other trades operating cheek by jowl in well-developed manufacturing districts. An old caveat is worth recycling: political boundaries poorly enclose economic activity.[2] Writing here of neighborhoods, networks, districts, cities, and regions simply acknowledges the spatial diversities that accompanied specialty production's sectoral variations.

The later 1870s and 1880s were critical years for American custom and batch manufacturers. Despite many difficulties, it was in these decades that they solidified the quest for novelty as a vocation. The twists of an economy that repeatedly fell into recessions, the inconstancy of an electorate that twice favored a Democratic president who menaced the tariff, the surge of labor organizing that challenged proprietors' customary authority, and the growing complexity of technical change and national marketing together pressed specialists into their own version of the "search for order." Sectorally and spatially uneven, it had two prongs. First, at the level of the firm, principals sought means to manage the uncertainties of diversity by adopting varied schemes for shop management, cost figuring, sales, and worker payment—a drive toward systematization distinct from the promotion of standardization (e.g., Sellers' efforts toward universal screw thread specifications). Second, in trades and regions, manufacturers formed organizations to address issues of politics, labor relations, technical education, information flows, and marketing, a process capped uneasily by the launching of the National Association of Manufacturers in 1895. Amalgamating these themes with accounts of place and practice will shape this overview of seventeen busy years bracketed by America's greatest nineteenth-century expositions.

At the close of Part II, the detail and dynamics of these accounts will be distilled into two linked typologies, characterizing on one hand the firms and sectors that constituted specialty manufacturing, and on the other the kinds of urban places they inhabited. The first will suggest that we think of specialty manufacturers as constituting three operational categories: (1) integrated anchors—makers of high-end producer (and a few consumer) durables facing a relatively narrow client base, demands for quality and performance, sizable capital investments, and small numbers of nationally competitive firms; (2) networked specialists—wider-market makers of industrial equipment and consumer goods (durable and nondurable), present in larger numbers, facing modest entry costs, selling through distributors and/or to retailers, and developing notable spatial concentrations, often at multiple sites; and (3) specialist auxiliaries—providers of intermediate goods and services who developed contracting networks sustaining the first two groups' growth.[3] In addition, a small proportion of batch operations may be regarded as "outliers"—individual (or small clusters of) niche market specialists located in scattered cities and small towns.[4] A second typology addresses the variations among cities that hosted significant levels of specialty manufacturing. It suggests that such urban areas can be differentiated as interactive, parallel, derivative, or narrow focus sites. The first two terms refer to cities in which specialist trades either intersected in generative ways or operated in adjacent spaces, but on unrelated developmental lines. Derivative clusters of specialists took shape in cities whose dominant format for manufacturing focused on bulk or mass production approaches, whereas "narrow focus" signals those secondary cities in which one or two specialty sectors led industrial growth. Each typology stands as a tentative effort to draw pattern and process out of the multiple streams of batch and custom production that flowed through late-nineteenth-century America.

## Celebrate and Reflect: Specialists at the Centennial

"At the fairs and expositions popular after the Civil War manufacturers of all types entered their products in competition for prizes, hoping to widen their markets through the education of the public."[5] Though cities from Newark to St. Louis mounted a half-dozen national industrial fairs before 1876, the Centennial Exhibition was the first to secure extensive international participation.[6] European specialists in "art manufacturing" reportedly chose to send chiefly "their less ornate products, because they thought Americans would prefer the plainer things," but their displays excited admiration and imitation nonetheless. Indeed, as a French visitor noted, Americans had already gone some distance in "borrowing the methods and skilled processes of continental workmen" to make jewelry, bronzes, silverware, and furniture that exhibited both "solidity and good taste."[7] Writing in the *Revue des Deux Mondes,* Louis Simonin did regret that Centennial organizers had taken their love of "democratic equality" too far in giving only one category of awards and had slighted skilled labor by making "no mention of the foreman or master mechanic, without whose inspiration, without whose aid so many beautiful things would never have been produced." But the goods captivated him. Of course, the huge Corliss steam engine was "the most remarkable thing in Machinery Hall." All visitors agreed on that. But more striking to Simonin were other specialties: "In carriage making, cabinetwork, glasswork, and pottery the United States is almost the peer of France and the other great nations. In other things they have got ahead of us; and all this in spite of the high prices of labor. It may be said that we are their instructors and masters, as Italy was for us at the Renaissance, and that they are destined to surpass us someday, as we did the Italians."[8]

Perhaps intended as much to prod Gallic complacency and score France's thin showing as to honor American capacities, Simonin's remarks were warmly received on this side of the Atlantic but not widely echoed by other foreigners.[9] To be sure, observers lauded American machinery as world-class and recognized that accurate, inexpensive watches had supplanted sewing machines as exemplars of mass production. Yet, pace Simonin, in style and execution most lines of American consumer specialties seemed awkward and unrefined when contrasted with their European counterparts.[10] The Grand Rapids furniture exhibit provides a typical instance.

The Michigan city's Phoenix, Nelson-Matter, and Berkey and Gay firms shipped elaborately carved and ornamented bedroom suites to Philadelphia. Nelson-Matter showcased its patriotism with an extraordinary bedstead. "Of massive oak, the head and foot were provided with numerous niches for statuettes; figures of George Washington, Columbia, Christopher Columbus, and Gutenberg were mounted in them and there were at least five others.... Surmounting the whole affair was a huge wooden eagle with wings widespread."[11] Visitors found it charming, but appalled American critics penned sharp reviews published in New York, Chicago, and St. Louis newspapers. As trade journalist Arthur White later noted, early Grand Rapids proprietors doubled as designers, but in styling their "incompetency was equaled only by their audacity.... While the workmanship of these suites was fairly good, the designs were incongruous and expressed nothing of value in art. In comparison with the foreign exhibits of furniture . . . the Grand Rapids collection was sorry to contemplate."[12] Chastened Michigan furniture makers began both studying style guides (Spofford, Eastlake, et al.) and recruiting experienced designers from the East or Great Britain. The Centennial had provided a significant, if unpleasant, stimulus toward tasteful, rather than grandiose, novelty.

Other consumer goods specialists, particularly in textiles, learned similar lessons from the international displays and strove to elevate the quality and variety of their lines. Asked in the Industrial Commission's 1899 hearings whether American consumers' old preference for imported finery still prevailed, department store magnate John Wanamaker estimated that it had diminished to a quarter of its former scale. "I think the great turn came in the Centennial Exhibition at Philadelphia, where we had such large foreign representation. The whole country went in and said, We can make those goods; and they did it, and [are] doing it now still better."[13] Philadelphia mill proprietor Frank Leake agreed. Queried about ventures into "higher and finer grades" of fabrics since 1890, he responded: "I think your date a misleading one to judge from. There has been a marked advance since 1876. The Centennial Exposition [*sic*] gave a great impetus to all this sort of thing. . . . Prior to that time we had been living in [an era] of black broadcloth and haircloth furniture."[14] Though Leake exaggerated the primitivism of earlier specialty weaving, he forthrightly credited the Centennial with facilitating the trade's shift to a higher plane.

Ten million people strolled through the exhibition's gates in 1876 to view tens of thousands of artworks, curiosities, machines, and products of industry. Their numbers included Thomas Savery, Oberlin Smith, and, of course, George Corliss, but none of them could have enjoyed the Philadelphia festivities had the new leaders of the Franklin Institute not geared up seven years earlier. Though the nation's hundredth anniversary clearly demanded a celebration, it was far from obvious that Philadelphia would be the site. Advocates for locating the Centennial in New York or Washington made strong arguments, but tool builders William and Coleman Sellers headed an institute group that proved to be "better organized." They memorialized Congress in 1869, secured supporting resolutions from city and state governments thereafter, and helped push the bill that created the Centennial Commission through the House and Senate in 1871. Though the institute's

*Journal* would later rue the clumsiness of preparations, deplore the fair's "trivial shows and productions," and echo Simonin's complaints about the absence of competition for graded prizes, the exhibition came off with few hitches and no scandals of note.[15]

Savery's Pusey and Jones company was doubly involved in the grand affair. Still struggling with the economic slump, the firm bid on the structural ironwork for both Machinery and Memorial Halls in 1875. Its initial price of $82,000 for Machinery Hall's building members was negotiated down to $78,000, "under unjust compulsion" in Savery's view, before the commission finalized contracts in February 1875. Savery's pique at being squeezed soon evaporated, however, if the diary entry "Space! Centennial" two weeks later is any indication. Pusey and Jones would lease square footage in the hall its ironwork supported, sufficient to exhibit a horizontal and vertical steam engine, a paper machine, a rag and stock cutter, roll-grinding and paper-glazing machines, finished paper calender rolls, several ship models, and sundry paintings, photographs, and drawings of finished goods. With occupancy confirmed the following February, just as the works finished Machinery Hall's eight-ton gates, Pusey and Jones set its exhibit in place during April.[16]

Savery regularly trolled among the stalls being assembled by rival and related firms in the weeks before the public opening, gathering information on others who made "chilled rolls" (E. P. Allis and a New Jersey company), visiting Swedish and Swiss machinery emplacements, and scouting for new business with fair success. The day after attending the May 10 inaugural ceremonies, at which President Grant and Brazilian emperor Dom Pedro set the Corliss prime mover to work, Savery escorted the emperor through Pusey and Jones' shops, underlining the fact that Brazil was its best Latin American iron steamship client. Paper machine sales to Germany followed in the summer, and an agent marketing capital goods to Russia agreed to promote the company's specialties on commission. The dollar value of Pusey and Jones' Centennial information searches and product displays cannot be determined precisely, but 1877 sales reached a half-million, double those of the previous year.[17]

Oberlin Smith likewise hired floor space to show and sell his metal presses. Smith had first ventured into the exposition game in 1869, when he sent a display of gas fitters' equipment to New York's American Institute fair and secured a prize medal and certificate. For the Centennial Ferracute built nine presses, including new belt driven

6. Oberlin Smith's Ferracute Machine Works display in Machinery Hall at the Centennial, 1876. Courtesy of Hagley Museum and Library.

models, and crafted a variety of dies to indicate their versatility. His machinery garnered a medal, but none of it sold, though he credited several international orders in the later 1870s to this exposure. Smith salved his disappointment by picking up a bargain at the exhibition's close. Buying the octagonal building that served as Horace Greeley's *New York Tribune* headquarters, he hauled it back to Bridgeton for recycling as his factory office.[18]

If Oberlin Smith was peripheral to the exhibition, George Corliss and his 1,400-horsepower engine represented its centerpieces. Thomas Savery may have provided Machinery Hall's skeleton, but Corliss installed its heart and circulatory system, the prime mover and power transmission network that drove eight lines of shafting and vitalized hundreds of machines. Exhibition planners well knew that operating machinery fascinated ordinary visitors and could unclasp potential users' pocketbooks. Early debates about means to supply power and cover the cost ended when Corliss volunteered to erect at his expense a single central driver, providing that the commission would fund the auxiliary gears and shafts. Rival steam engine builders briefly demanded chances to build the driver and boilers but would not countenance donating them, leaving the field and the publicity harvest to Corliss. Completed in his Providence shops late in December 1875, the 600-ton engine's components occupied six freight cars when shipped to Philadelphia and included a 56-ton, 30-foot-diameter flywheel and a five-ton forged gunmetal crankshaft. Under the proprietor's supervision, Corliss workers assembled the four-story "leviathan," its boilers, pumps, gears, and so forth, and set it in motion a month before the Centennial's scheduled opening date, precisely according to contract. *Scientific American* described the Corliss engine as "a masterpiece of mechanical engineering" and sagely opined that it "will add greatly to [Corliss's] professional reputation."[19]

Many reporters wrongly claimed that Machinery Hall's Corliss was the most powerful steam engine yet built, but all rightly celebrated the precision that made its operations "noiseless." While the machines it drove were "pounding, screeching, rumbling, and crashing, the 'Corliss' turns its vast fly-wheel as quietly as a lady wields a feather fan," wrote one. A machinist offered a more penetrating comment, however. Silent running was a function of accuracy in the cutting of the beveled gears that distributed power from the flywheel to the shafting, long "one of the most difficult operations of heavy machine tool work." Corliss's gear cutter, the main feature of the firm's own dis-

7. Erecting the Corliss engine in Machinery Hall, Philadelphia, spring 1876. Courtesy of Hagley Museum and Library.

play, was to the practical metalworker far more significant than the engine.[20] Small wonder that years earlier he had declined to duplicate it for E. P. Allis. The towering engine, however, did not lack critics. Charles Porter, a rival builder, claimed that identical horsepower could be generated from a horizontal engine (of *his* design, naturally) far less hulking and overweight.[21] On target mechanically as usual, Porter lacked any understanding of the emotional and symbolic force the "monster" embodied for the crowds that stood silently, dwarfed and enraptured, as its massive flywheel spun round thirty-six times a minute. A smaller, swifter, more efficient engine might have ably supplied mechanical power but could never have both celebrated its maker's machine-building prowess and encapsulated the nation's burgeoning industrial capacities.[22]

## Networks of Specialization: Philadelphia Textiles and Metalworking

More than any other American city, Philadelphia made visible the achievements and potentials of batch and custom manufacturing from the Centennial through the 1890s. Its set of fashion textile districts, a vast metalworking and machine-building complex, and significant printing and publishing enterprises all expanded with demand for specialty consumer and producer goods. In Kensington, Frankford, Germantown, and Manayunk, baled cotton, wool, and silk were cleaned, spun, dyed, and knitted or woven into thousands of varieties of carpets, rugs, dress goods, suitings, braids, stockings, jerseys, towels, draperies, and upholstery fabrics. At Southwark, Spring Garden, Nicetown, and in the city's core, firms fabricated thousands of tons of brass, iron, copper, tin, and steel into millions of castings, machine components, tools, or saws, as well as lamps, ornamental ironwork, valves, pumps, hardware, scales, machinery, ships, and locomotives. Nearly eight thousand printers worked in Philadelphia, turning out lithographs and posters, broadsides, books (scientific, medical, and popular), pamphlets, engravings, music, trade journals, neighborhood weeklies, and a host of daily newspapers. Sellers, Baldwin, and a score of prominent textile and printing establishments exhibited at the Centennial, but they were only the apex of the pyramid.

As we step beyond the exhibition grounds into the city's industrial districts, textiles and metalworking in the 1880s exemplify developing practice in batch and custom production. Extension of styled-fabric capabilities met a challenge from the Knights of Labor, along with attempts to reorient marketing, shifts to new product lines, and collective efforts to organize regional manufacturing interests. Metalworking leaders concerned themselves with systematizing their work processes, a matter that energized Frederick Taylor, and tried to manage shifts in labor relations, markets, and technology to their advantage.

In 1870, the Philadelphia textile industry supported nearly 600 firms and over 26,000 workers chiefly manufacturing tariff-protected styled goods for clothiers, dressmakers, house furnishers, and decorators. Immigrant proprietors from Britain, Ireland, and German states operated most of the trade's mills and dyeworks, having translated their factory and shop skills into enterprises that specialized in one step of the fabric production sequence. Few plants followed the British Dobson brothers (and, shortly, Bromley Brothers) into integrated manufacturing, for both the capital costs and market risks were daunting. Thus the dominant pattern was for knitters and weavers to contract with independent spinners for yarns and dyers for services that together yielded final goods for sale. Product variability demanded skill at all levels. Hence adult men constituted over 40 percent of Philadelphia's textile workforce in 1870, a far greater proportion than in staple mills.[23] Marketing proceeded largely through jobbers and wholesale agents like those the Bromleys depended on; and firms secured working capital by issuing promissory notes[24] or securing advances against yardage delivered to dealers.

The mid-decade depression put serious stresses on this network. As orders flagged, companies laid off workers and shortened hours or, in extremis, cut weekly wages and "prices" for work tied to output volume. Hard-pressed jobbers demanded reductions in makers' wholesale rates for fabrics, and note holders became less tolerant of "first names" begging extensions. In Philadelphia, the economic slump that surrounded the Centennial fostered the Knights of Labor's rise, manufacturers' interest in direct marketing, and the creation of industrialists' banks and other institutions.

Despite the downturns of the 1870s, Philadelphia textile sectors expanded mightily during the decade, accounting for a third of the city's industrial jobs in 1880 (ca. 60,000), a third of its total output, and 37 percent of value added in manufacturing. Excluding consideration of smaller custom and batch sectors, the four chief specialty trades (textiles, metalworking, printing, and furniture) constituted half of all factory positions, 46 percent of product value, and 57 percent of value added. By contrast, five city bulk production industries (sugar, cigars, slaughtering, brewing, and iron and steel) supplied only 4 percent of employment, one-seventh of product values, and 7 percent of value added. By 1880, Philadelphia had solidly established its reputation as a center for industrial versatility.[25] Its textile operatives were among the best-paid workers in the trade nationally, but they suffered from seasonal cycles of layoffs or short hours that worsened in depressions. In the later 1870s, persisting hard times encouraged workers to seek a form of collective organization beyond the individual mill, where shop committees had long spoken plainly to owners. In textiles, the Knights of Labor first gained favor in an ingrain carpet sector doubly squeezed by excess capacity owing to technical shifts and by the depression's impact. During the early 1880s, the labor organization's initiative spread beyond floor coverings to woolens and some segments of spinning and dyeing.

Confident that the doubled scale of their operations since 1870 confirmed their wisdom, textile manufacturers rejected the Knights' appeals to resolve labor disputes through District Assembly officials who neither possessed textile experience nor were their employees. They also distrusted binding arbitration proceedings that would depend on the judgment of ostensibly neutral parties. Workers called for sectorally standard payment schedules that would stabilize wages and center competition on product qualities. Unsure that they could control rate chiselers and offended at this interference with their prerogatives, a cluster of textile manufacturers created counterorganizations of carpet and cotton firms between 1878 and 1880. In addition, another proprietors' group, less troubled by labor conflicts and more concerned with tariff politics and technical education, formed a citywide Textile Association in 1880, having first gathered the previous fall to host the National Association of Woolen Manufacturers' annual banquet.[26]

I have elsewhere explored area mill owners' successful struggles with the Knights and their crisis associations' formations and dissolutions,[27] but the Textile Association merits further attention here. It rapidly enrolled about a hundred proprietors employing a quarter of the city's textile workers, joined a spirited 1880 attack on proposals to revise tariffs downward, then participated informally in a drive to reform local government, securing election of one of its leaders as city treasurer in 1882. As assaults on the tariff persisted, association delegations routinely entrained for Washington, while at home members sponsored the inauguration of technical education in textiles (1884) and reported local and national matters of sectoral interest in their monthly *Bulletin*. Reaching out to proprietors in other trades as labor and political controversies boiled in mid-decade, the association reconstituted itself as the Manufacturers' Club of Philadelphia, renamed its journal, and secured comfortable quarters for meetings, dining, and deal making.

The Textile School project derived from manufacturer Theodore Search's visits to the Centennial Exhibition and his judgment that only achieving European standards of excellence and originality in design would sustain American styled-fabric development. As a substitute for the common practice of duplicating popular foreign fashions, Search proposed emulating the English and Continental design academies that fostered stylistic creativity. This effort drew only mild support among local factory-schooled proprietors, who discounted the value of classroom training. Search, equally stubborn, commenced the school's first courses at his own expense, then rallied a score of association members to underwrite this institution for educating practical designers and manufacturers. In short order, hands-on experience with varied cotton, wool, and silk machinery supplemented design courses in night classes for workers and day sessions for owners' sons and others. The school soon secured regular funding from the Commonwealth by providing the governor with patronage scholarships, one for each of Pennsylvania's sixty-seven counties. An alliance with the Centennial-descended Museum School of Industrial Art[28] strengthened the Textile School's design capacities and paved the way for an 1893 relocation into spacious downtown quarters near the Manufacturers' Club.[29]

The creation of a durable regional trade association, the club, and the Textile School, along with parallel ventures in factory insurance, banking (the Textile Manufacturers' National Bank), and marketing (the Philadelphia Bourse) together illustrate the organizational impulse that permeated a maturing specialty manufacturing center. Bulk sectors in these decades commonly created national associations to defend the tariff or contest railroad rates but rarely started schools or banks to provide collective services to regional enterprise groups. Such lateral initiatives, in a sense attempts to mobilize or capitalize on positive externalities, will be found in each of the specialty districts here reviewed. Their extensiveness and relative success helped define the degree to which specialists perceived shared interests and proved capable of acting on them, whereas failures indicated the boundaries to collective action framed by technical and market conditions, free riders, shifting sectoral fortunes, and collisions with law and public policy. The few local specialists' associations that reoriented to a national level and created powerful networks among scattered firms and districts will hold particular interest.

For Philadelphia textile proprietors, marketing in the 1880s was as serious a problem as were labor organizers. Styled goods entered two spatially dispersed markets. As intermediate products, fabrics for clothing and upholstery and all varieties of braids and trimmings had to be merchandised to thousands of apparel makers and furniture firms for further use. As final goods, carpets, rugs, hosiery, and knitted outerwear (jerseys, mittens) had to reach enormous numbers of retailers by some means. During the decade after the Centennial, most mills sold their styles through middlemen who either ordered directly from samples or gathered and forwarded orders based on samples they circulated among potential users or retail resellers. Classic principal-agent problems haunted this mode of marketing.

If jobbers or wholesalers confirmed orders at quoted prices, took title to delivered "pieces," and cleared their accounts in sixty days, all went well. However, during repeated strained spells in the 1870s and 1880s, middlemen resorted to what makers termed "trade abuses" to shift risks to producers. They canceled confirmed orders when estimates of what styles could readily be sold seemed wrong, or returned goods, complaining that they were inferior in quality to selling samples. Middlemen also delayed remittances yet deducted discounts for prompt payment long after they had expired.[30] Manufacturers could sue for specific performance of sales agreements, which would terminate relations with that wholesaler and incur the risks and costs of finding a more reliable replacement, or absorb losses in the hope of better results the next season. In hard times, jobbers also tried to shift away from direct purchasing and toward being pure agents who toured sample sets among possible clients and sent their orders along. This both expanded the likelihood of cancellations, returns, or "slow pays" and fostered unwelcome price concessions by agents who forwarded contingent orders to makers on "take it or leave it" terms.

When scores, then hundreds, of jobbers adopted such practices, principals had to choose between certain threats to profits and uncertain expenses of searches for alternative channels of distribution. The smallest specialists were stuck "in the hands" of wholesalers, but larger operations could afford to try direct selling. In fabrics for clothing or home furnishings, the best strategy was to establish or share a New York office, buy advertising space in sectoral journals,[31] and flog style samples in person to the bigger clothiers or upholsterers in Baltimore, Cincinnati, Rochester, or Chicago. An "outside" partner versed in sales thus became an increasingly valued firm asset. In ready-for-retail lines, hitting the road was more challenging, but equally important. Seasonal samples of carpetings easily weighed over five hundred pounds, filling huge trunks that had to be carted from store to store, train to train, city to city. Direct sales of styled, finished textile goods also demanded attention to both metropolitan centers and county seats. Not surprisingly, Philadelphia firms in the 1880s often contracted with commission travelers who handled several noncompeting lines (carpets, draperies, hosiery) to show their samples in specific regions. Given the time constraints of seasonal stylings, without the telegraph the effort would have been impossible.

Contemporary developments in production and retailing eased manufacturers' night terrors as direct selling increased. First, in menswear (much less so in women's clothing), a number of sizable apparel firms ventured into modest variations on the conventional fabric set. Woolens for solid color black or blue suits were far from dead, but conservatively patterned worsteds with novelty touches might be sold in fair quantities to cutters in Rochester, St. Louis, or Cleveland at prices above those producers quoted to wholesalers. An added benefit to such efforts was that fabric makers could directly gauge users' predilections and undertake to style yardage that matched them. Second, in both piece goods and finished textile lines, increasingly potent department stores offered millmen another source of block purchasing, consistently for cash rather than for sixty- to ninety-day invoices paid after six months or longer. Philadelphia carpet and hosiery specialists could score winning seasons by selling their lines to Marshall Field, Wanamaker, or Macy, and labored to do so.[32] Firm members attended to such accounts directly; and if price concessions were at times necessary to secure substantial sales, at least principals were positioned to negotiate them and eliminate unpleasant surprises from agents.[33]

Getting buyers to come to sellers was even better than canvassing the nation—hence the Bourse. Distant clients did visit Philadelphia to contract for fabrics at "mill prices" and bypass middlemen; but they had to travel through the city's scattered textile districts to do so. Many instead took through trains to New York and avoided tiresome, dusty treks across Kensington and Frankford and west to Germantown and Manayunk. By the late 1880s, Manufacturers' Club leaders in textiles and the newly associated machinery trades conceived the notion of a central market, a Bourse—again echoing European precedents—that would be lodged amid the splendors of downtown Philadelphia. It would house selling spaces for several hundred firms and, with free stopovers negotiated from the Pennsylvania Railroad, could skim demand from New York jobbers and agents. Opened in 1895 near Independence Hall, the nine-story Bourse represented the fullest manifestation of nineteenth-century Philadelphians' shared efforts at specialty marketing.[34]

Notwithstanding these stratagems, the city's textile sectors reached a plateau in the 1880s. As table 5 shows, their 1890 levels of employment, output, and value added were little different from those a dec-

TABLE 5
Leading Specialty Sectors in Philadelphia, by Employment, Output, and Value Added, 1880 and 1890, with Citywide Totals

| *1880* | *Employment* | *Output (millions)* | *Value Added (millions)* | |
|---|---|---|---|---|
| Textiles | 59,818 | 101.1 | 45.2 | |
| Metalworking | 19,516 | 28.7 | 14.9 | |
| Furniture | 2,989 | 4.8 | 2.4 | |
| Printing | 8,069 | 9.9 | 5.9 | |
| Totals | 90,392 | 144.5 | 68.4 | |
| City totals[a] | 179,278 | 314.9 | 120.6 | |
| Four-sector share of city totals (%) | 50 | 46 | 57 | |
| *1890* | *Employment*[b] | *Output (millions)* | *Value Added (millions)* | *V-A Change (%, '80–'90)* |
| Textiles | 58,058 | 102.4 | 43.1 | –5 |
| Metalworking | 30,029 | 55.5 | 32.6 | +119 |
| Furniture | 2,395 | 4.8 | 3.2 | +36 |
| Printing | 11,941 | 33.0 | 23.7 | +300 |
| Totals | 102,423 | 195.7 | 102.6 | |
| City totals[a] | 212,000 | 522.0 | 234.1 | +93 |
| Four-sector share of city totals | 48 | 38 | 44 | |

*Sources*: Department of the Interior, Census Office, *Report on the Manufactures of the United States at the Tenth Census* (Washington, DC, 1883), 421–24; idem, *Report on Manufacturing Industries in the United States at the Eleventh Census*, pt. 2, *Statistics of Cities* (Washington, DC, 1895), 434–53.

[a]Totals have been revised to exclude five construction sectors: carpentry, masonry and brickwork, painting, plastering, and plumbing.

[b]1890 employment excludes 3,214 office workers and firm principals whose 1880 equivalents were included in workforce figures.

ade earlier, while other local specialty trades advanced dramatically. The immense expansion of the 1870s would not be repeated, but collective ruin hardly menaced local textile mills. Indeed, their average profitability was higher than that of all other textile centers in 1890, even though workers' mean earnings had risen 12 percent since 1870 (36 percent in real terms).[35] Among the four specialty sectors, both metalworking and printing had become leading growth poles, bettering citywide averages for increased employment and value added, but furniture was relatively static, if not contracting, suggesting the rising significance of midwestern woodworking centers. This diversity in sectoral trajectories is exactly what we should expect to find, for there was (and is) both temporal and spatial unevenness to patterns of industrial growth and decay.[36] Moreover, the four sectors' shrinking share of all value added did *not* reflect a dramatic upsurge of bulk or flow production in Philadelphia. The five bulk-oriented trades mentioned above increased their contribution to local value added only from 7 to 9 percent. With chemicals and petroleum refining tacked on, the seven bulk sectors' share reaches 10 percent. Instead the 1880s saw modest growth in scores of highly specialized local trade divisions, such as billiard tables, Jacquard card cutting, and dentists' materials. Philadelphia census reports for 1890 recorded 298 distinct lines of industrial business, perhaps a benchmark for urban productive diversity.[37]

Philadelphia "metalworking" refers to more than Sellers and Disston's machinery and tool sectors, for the table 5 category summarizes the activity of over twenty trade divisions, including brassware, copper and tinwork, gas fixtures, and architectural iron, plus their supporting foundries and general machine shops. (It does not, however, reprise other, more staple, metallurgical trades: making pipe, smelting, or basic iron and steel.)[38] Still, Sellers was local metalworking's most visible and celebrated exhibitor at the Centennial. His firm's twenty-one awards filled four pages in the machine tools report, which commended his enterprise for its steam hammers, grinders, lathes for iron and brass work, gear cutters, borers, slotters, and shearers. Sellers plainly continued the tradition of general machine building, in contrast to the deepening specializations that Cincinnati enterprises would lead.[39] Only a year later, another Philadelphian who would become both famous and controversial assumed a foremanship at Midvale Steel. The two, William Sellers and Frederick Taylor, epitomized the complex challenges facing specialty metalworkers in the 1880s and their contrasts with issues that styled-textile companies confronted.[40]

Like their textile colleagues, batch and custom metal tradesmen activated networks of interlaced specialist enterprises and generally eschewed integration. They also encountered diverse demands on their capacities and contended with cyclical and seasonal surges and dips that created comparable problems with finance and labor relations. Yet there remained a core difference in materials and work processes that led "metal bashers" toward an intense focus on shop-floor dynamics, systematization, payment schemes, and cost finding rather than marketing, banks, and schools.

Put simply, textile products had a transparency and their work processes a determinacy absent in metals. Certainly, many things could go wrong in weaving or knitting, but they were visible and subject to repair. Once a loom was set for its pattern and shuttles readied with yarns, a weaver's close attention to its operation and knowledge of the design usually sufficed to minimize errors. Skilled women inspected fabrics to detect minor flaws and remedy them ("burling"). Moreover, the steps in producing carpets or dress goods were standard: spinning, dyeing, weaving, and finishing. Payment was either by the hour or coded in piece-rate schedules framed to balance earnings from slow, exacting work with those from quicker, easier styles. Contention and conflict abounded, to be sure, but constructing fabrics was a straightforward business.[41] Metalworking had little of this clarity.

To make a part for a boiler, machine, or locomotive, firms had to confront and overcome issues of three-dimensionality, strength, design, composition, and precision. Metals of often uncertain composition, melted and cast in molds, on cooling might shrink to unusable dimensions, exhibit obvious defects ("blow holes"), or contain external hard spots and interior fractures or cavities.[42] Reheating for repairs could damage the material's strength or make it impossible to machine it to specifications. Fashioning the part to finished form could easily demand its circulation among lathes, planers, borers, and drills. Any error in setup or cutting work would render it useless scrap. A single screw made to the wrong pitch could stall assembly, as could a hole drilled a thirty-second of an inch off its center. Workers might do everything apparently well, only to have the item fail in use because of poor design or invisible faults in the metal.[43] Such complications informed the Franklin Institute's early obsession with the causes of boiler explosions and encouraged experiments and information sharing about tests for metal structure or strength (while textile testing remained in its infancy).

Specialty metalworking in the 1870s and 1880s thus represented a more contingent and mysterious set of industrial processes than manufacturing styled fabrics. Local capital goods firms and their network contacts already sold direct, either to heavy tool users, builders, and railways or to purchasers of castings and other components. They remained ambivalent about the value of formal education and founded no schools.[44] Their key problems concerned the organization of production, and to these they turned resolutely.

Philadelphia metalworks strove to establish "system" in industrial practice well before it became a modern management buzzword. Indeed, the antebellum Baldwin Locomotive Works led the way. No specialty product was then more complicated than a railway engine, and few were as heavily taxed in use. Yet locomotive buyers forwarded wildly varied specifications for the engines they desired, which Baldwin built singly or in small batches. Compounding the challenges, each engine contained thousands of parts. To manage the assembly of reliable locomotives, Baldwin focused its systematizing efforts on the drafting room, which became the "brains" of the works. Draftsmen detailed plans for every part of every engine with precision and lacquered their drawings for permanence; then shop orders for initial and replacement parts were based upon them. Managers also devised rudimentary schedules for components production so that final erecting would be smoothed. Elements used in one locomotive could often be applied to another of similar gauge or motive power, making stocking a variety of axles, boiler tubes, or driving wheels sensible as a means to save expense through producing larger batches of parts than immediately necessary. By 1880, the Baldwin approach had diffused throughout railway engine manufacturing.[45]

Locomotive construction was thus systematized, but not standardized, in the later nineteenth century. This distinction is important, for it highlights a practical difference between specialty and routinized manufacturing. System meant organizing the chaos of diversity without diminishing flexible capacity. It might entail formal procedures and specifications, but not a broader standardization of products and their components. No single engine, lathe, or press, no model fixture or structural member, and no definitive set of these would satisfy all needs, present or prospective. System could increase reliability, reduce errors, speed batch throughput, and hence build reputation and enhance profits; but if it went so far as to freeze flexibility and innovation, it would become a burden and a constraint.[46]

Specialists could well argue for standardization among their suppliers, calling for steps that would reduce uncertainty and perhaps materials costs as well. Yet over twenty years after Sellers' 1868 screw thread campaign, James See observed that there had been "no attempt at uniformity of threads, or even of sizes," and for bolt heads and nuts, only "a lame standard of sizes" existed. He reported similar disarray in nearly forty other classes of goods from tool couplings to builders' hardware and millwork. In response, Oberlin Smith called for government oversight of standardization efforts, explaining, "One advantage . . . gained by such supervision would be an occasional early death among some of the numerous wire gauges which are constantly springing into life." See's claim that "[t]he idea that it was good business policy to make things that nobody else could fix" still reflected specialty metalworks' strategies in the late 1880s.[47]

Such disorder nettled Frederick Taylor. Unlike Smith, who devoted considerable energy to achieving system in nomenclature, parts identification, and depreciation,[48] Taylor struggled with taming the flow of diverse work in progress, refining the effort-wage bargain, and experimenting with new tool steels that would speed metal cutting. In over a decade at Midvale, Taylor riveted his attention to the shop-floor dynamics of an alloy steel specialist firm committed to heavy metalworking for government munitions, shipbuilders, and railways.[49] There he devised his differential piece-rate system, commenced time studies of workers' performance, and undertook "laboratory" analyses of metal-cutting practices. In the 1880s Taylor also designed machines and attachments, including a durable steam hammer, a reliable tool grinder, and two boring and turning mills, while completing his engineering degree by examination (rather than attendance) at New Jersey's Stevens Institute.[50]

In 1886, both Taylor and Smith responded to a presentation on a "Shop Order System of Accounts" by Captain Henry Metcalfe, one of the founders of product-specific costing and work-flow tracking. In managing metalworking at Philadelphia's Frankford Arsenal, Metcalfe often had at any time "about a hundred orders under way, of different kinds," and devised a shop ticket system to monitor expenses and output for each item. Costs aggregated and averaged across many orders frequently deceived managers, Metcalfe argued, all the more "as the product of an establishment is diversified, so that the more miscellaneous is the product, and hence the more necessary the knowledge of its difference in cost, the more difficult is this knowledge to obtain." Building on locomotive shop practice, Metcalfe developed three classes of cards: the shop order, a "service" or labor record, and a materials receipt. The order card authorized the work and listed the steps necessary. Foremen handed it to workers, who returned it upon task completion. Forwarded from one foreman to another through departments until fully checked off, the order card recycled to the office.

Metcalfe also issued each worker bound pads of service cards, printed with his name, hourly rate of pay, and payroll number. For each task, foremen inscribed the order number along with a brief work description, then later certified the time expended. These tickets were "as good as money," for their collation in the office authorized payroll disbursements. Redistributed according to order numbers, they documented each job's direct labor cost. Material tickets recorded raw or part stocks used at each stage of the order's fulfillment. Linked with the service cards, they summed up the prime costs of production and identified materials stocks needing replenishment.[51]

Metcalfe noted that no such elaborate records were needed for costing in a "blast furnace" or comparable bulk production enterprises. Yet where product diversity ruled, manufacturers usually relied on aggregates or practiced "thumb-sailing," the cost/price guessing based on production history that Thomas Savery used. Metcalfe's system offered versatile firms greater precision in tying expenses to particular products than had previously been feasible.[52] Taylor and Smith had opposed reactions to Metcalfe's efforts. Consistent with his fastidiousness and distrust of workingmen, Taylor objected both to giving operatives service cards to fill out (they would become grease-covered and illegible) and to having foremen's certification of work completed handled by use of a hole-punch. Foremen should initial their approval, for "[a]nyone who gets hold of a punch can punch the authority of doing work of any extent or variety that he chooses, but hand-writing is much more difficult to counterfeit." Taylor boasted that at Midvale *his* system mandated posting shop orders on glass-covered bulletin boards so workers could neither soil nor fiddle with them, and that his work-process monitoring scheme used "two hundred varieties of printed cards."[53]

Oberlin Smith mocked Taylor's concerns over grime: "In regard to the dirtying of the shop cards, I do not think it amounts to anything." His workers had handled them for years without trouble from "their being lost or dirtied or too much torn for practical use." Smith did have difficulties, however, with setting out one card for each matching task, for at his shop a hundred men might average five jobs a day, which would yield three thousand tickets weekly for office processing. Might not the expense of gathering admittedly important information offset the value of the knowledge secured? Metcalfe responded that a daily job-time form with spaces for up to ten different tasks could be workable, but of course this would complicate the

8. Oberlin Smith (far right) and his engineers sharing a celebratory luncheon on the bed of a new model press, ca. 1900. A companion photograph in the Ferracute Machine Company records shows this press draped with two dozen of the workers who built it, a more common expression of shop culture in machine construction. Courtesy of Hagley Museum and Library.

reshuffling to calculate labor expense by item. Smith also raised a broader point. Metcalfe's approach was important and instructive, but Smith urged his colleagues to think of devising "systems" in the plural—"not *one* system only, because we cannot apply one to all kinds of shops. The shops of this country want classifying into so many classes, and the best possible kind of organization for each will be ascertained only by careful study and by the collation of the experience of a great many persons."[54]

At that moment Smith offered an insight that could have laid the foundations for engaging diversity systematically—an effort to conceptualize difference, establish domains of common circumstance, and pursue discrete strategies for bringing multiple forms of organization to bear on clusters of sectors and subsectors sharing technical and market characteristics. Instead, rival universalizers, among whom Taylor is merely the best known, pressed their cases for the "one best way" to implant "scientific management." The impetus to system infused custom and batch manufacturing, but Smith realized that among specialists there were genuine situational variations, in addition to the gap between them and routinized producers.

Moreover, as debates on accounting practice, bonus pay schemes, and cost finding continued through the 1890s into the new century, two additional points became evident. First, the appeal of system spread very gradually and irregularly, even within metalworking, and scores of idiosyncratic variations multiplied in defiance of trade and professional associations' efforts to achieve sectorally appropriate uniformities in practice. Second, this diversity itself was related to proprietary personalism and company resources. Installing paperwork systems entailed added expenses for "nonproductive" office staff and supplies and inserted both a layer of personnel and a mass of statistical information between the "Old Man" and his shop workers or clients. Hence many manufacturers retained their rules of thumb, ignored incentive pay plans, and used customary semiannual inventory and trial account balances to discover whether they had made or lost money. Such men followed the prescriptions of management and accounting experts no more readily than those of labor organizers, at least until the long depression of the 1890s changed many minds on the first count, if not the second. Ultimately, these dimensions of difference were lost to sight by the 1920s, when an ideology of management as scientific practice and of accounting as a financial tool overwhelmed the more open industrial discourse Oberlin Smith had urged.

Many such discussions in the 1880s took place under the aegis of another institution whose creation involved Philadelphia metalworking leaders: the American Society of Mechanical Engineers. Oriented nationally, the ASME counted Sellers, Bement, Metcalfe, Taylor, and Smith as founders and early members (several later as presidents) yet reached out across the industrial crescent from Worcester to Cincinnati and Chicago for participation. Indeed, John Richards kept in touch with eastern matters from his Pacific-edge outpost and lobbied successfully to host the ASME's 1892 meetings in San Francisco. Organized at an 1880 gathering in the New York offices of the trade journal *American Machinist*, the society aimed to provide a forum for the exchange of technical information by tool, machine, and engine builders, foundrymen, and basic metal producers that would parallel existing organizations among civil and mining engineers. New York specialists like pump manufacturer Henry Worthington joined Bessemer-process innovator Alexander Holley, a cluster of prominent engineering educators, and scores of metalworks' proprietors and superintendents to plan for twice-yearly conferences and publication of the technical papers presented. An effort was made to induce George Corliss to assume the presidency, but Corliss, who "did not cooperate easily with colleagues by temperament," refused curtly, and the post went to Robert Thurston, another steam engine specialist and professor of engineering at New Jersey's Stevens Institute.[55] Oberlin Smith and others also urged political activism, calling for efforts to reform the Patent Office, establish a national university for science, and create a standards bureau. Three of the society's first nine presidents (Smith, Coleman Sellers, and Cramp shipbuilding's Horace See) represented Philadelphia interests.[56]

The ASME unified individuals, not firms, but many members' early concerns derived from specialty production challenges. Papers in the first decade focused on the variable characteristics of materials, analysis of production processes, systems for tracking information and work in the shops, ways to exchange findings that might be generalized among enterprises, and means to organize everything from pattern-drawing files to catalogs of machine components. Members' activities also generated both social networks and distance between engineers (as professionals establishing and controlling applied scientific knowledge) and the machinists whom they sought to manage. Between their semiannual meetings, these engineering careerists also corresponded with sectoral and thematic committees, but as membership topped a thousand (1889) an ever-smaller proportion attended conventions. Instead, ASME loyalists formed or joined independent local societies and clubs that reaffirmed the connection between place and practice, provided continuity and contact, and published regional journals of proceedings in Philadelphia, Pittsburgh, Chicago, and San Francisco.[57]

The ASME thus was a nationalizing element in the organizational surge among specialty manufacturers in the 1880s, one with strong ties to Philadelphia and connections to multiple centers of batch and bulk metalworking from coast to coast. Still, one of its founding members, steam engine designer Charles Porter, had a cautionary Philadelphia experience that underscored the enormous tensions between engineers' claims to expertise and the social relations of management and shop-floor practice.

Porter, who had devised a reliable engine governor and compact high-speed horizontal engines, spent years in Britain developing their capabilities. He returned to the United States in 1868 to reap the fortune his devices promised, only to become enmeshed in a series of undercapitalized New York City firms, none of which realized its possibilities. Tapped to serve as a judge of mechanical contrivances at the Centennial fair, Porter made an extensive set of Philadelphia contacts. These yielded an 1879 invitation to join in reviving the city's idled Southwark Foundry, originated by Franklin Institute stalwart F. V. Merrick as an engine-building concern. Though possessing no capital to invest, Porter anticipated that the new firm would both devote itself to producing his patent engines and install him as president. Neither expectation materialized. Rather, the board of directors selected the elder Merrick's son William to head an incorporated, closely-held firm devoted to general machine work, in which Porter's engines would form but one segment. The directors swiftly tabled Porter's demand of $100,000 for new machine tools, and he soon fell to battling with the firm's patent attorney. Both issues were resolved in 1880, but Porter's determination to be master, based on his engineering expertise, soon embroiled him in conflicts with his employers and the company's machinists.[58]

Initially a lawyer, Porter had gained a reputation as a paper engineer ("never did a day's work in a shop in my life"),[59] indifferent to the work culture of proprietary manufacturing. He gave peremptory orders to veteran "Philadelphia mechanics," alienating them and slowing work in process. Merrick soon barred him from the factory floors and replaced Porter's shop manager with a superintendent not responsible to the irascible vice president. Porter then descended into shouting matches with machinists, battled the "amateur president" who also lacked shop experience but trusted his workers' skills, and found his designs sabotaged by enemies in the drafting room and the shops. In 1883 he resigned under pressure, revenging himself a quarter-century later by pouring a memoirist's calumnies on the "Philadelphia phalanx" that had undone him.[60]

Porter's hubris and rage captured something terribly important about the social relations of specialty metalworking in the 1880s. He was not of, and had no respect for, the factory culture familiar to Merrick and his associates. As an inventor, engineer, and expert, he expected his word to be law in the factory and became apoplectic when resisted. Like Taylor, he had thin appreciation for the gap between technical expertise and its realization through a collective labor process. Porter's presumption that scientific reason ruled was ill-founded, for he never understood the need to build collaborative relationships with those who would execute his ideas in metal. Merrick lacked the expertise to critique Porter's plans but recognized the disastrous effects of the latter's autocratic methods. Accepting responsibility for the shop-floor chaos, Merrick resigned alongside Porter; and after some hard years, Southwark Foundry and Machine secured a new chief officer from William Sellers and Company. James Brooks had both "brains" and an understanding of the factory culture Porter had dismissed. The firm soared profitably through the miserable 1890s. Porter, returning to the plant in 1905, was "filled with amazement" at its expanded facilities and obvious prosperity.[61]

The Porter episode captures the conflict between practical knowledge and science-based engineering that permeated specialty metalworking after 1880; it also reflects the tension between rival forms of expertise. Porter assumed workers' stupidity and fractiousness, but Merrick rebuffed his appeals in part because he had failed to craft a modus vivendi with workers whose skills critically affected the firm's prospects. As Merrick well knew, his own minimal metalworking experience meant that he could not dictate practices to his employees; but Porter had no similar qualms—the designs of reason should prevail. Porter's plight also suggests the dilemma of the ambitious innovator lacking capital. Though he could imagine better engines, Porter had neither the skills to build them nor the funds to underwrite their construction. This made him an unwilling dependent of capitalists like

Merrick and of workers he deemed his inferiors. Corliss evaded this impasse, as did scores of innovators in machine tools and textiles; but they worked with, not against, the grain of specialty manufacturing's factory culture. Technical ingenuity was just not enough.

In the 1880s and early 1890s, the Philadelphia "phalanx" moved steadily forward. Baldwin built a thousand locomotives a year and welcomed Samuel Vauclain as superintendent, having induced him to leave the Pennsylvania Railroad's Altoona Shops. Disston Saw relocated from Kensington to larger quarters north along the Delaware and came to employ a thousand workers as sales exploded. The two Bromley companies became exemplary carpet producers, and one of them poised itself for a novel sidestepping venture into lace manufacturing. Stetson Hat spiraled toward national prominence in styled felt and straw headwear, and area steel fabricators fed Cramp's shipyards custom-forged plates and crankshafts that meshed with Morris engines in building scores of federal and private vessels. Meanwhile, hundreds of printers, gas fixture makers, lamp globe blowers, and iron or brass foundries quietly reinforced the base of Philadelphia's specialty capacities.[62] While boosters proclaimed the Quaker City as the world's workshop, comparable activity also materialized at smaller centers like Providence, our next stop.

*Chapter 5*

# PROVIDENCE AND NEW YORK: JEWELRY, SILVERWARE, AND PRINTING

BEFORE THE Civil War, American jewelry making diverged from watch and household silver production, activities with which it had mingled in the early republic's craft shops. In and after the 1860s, watch companies moved strongly toward standardization and mass production of inexpensive, reliable timepieces. After a brief fling at a trust to control pricing in the 1880s, they engaged in intense price competition that demoralized markets through the next decade.[1] Silverware, made in many styles and qualities from sterling to low grades of plate, emerged from a small cluster of sizable, chiefly New England firms led by Providence's Gorham and Connecticut's Meriden Britannia, while Tiffany in New York City drew accolades for imaginative design and workmanship.[2] In jewelry, neither standardization nor a stable roster of competitors appeared. Instead, three durable spatial concentrations of small enterprises developed around antebellum beginnings in lower Manhattan, Newark, and the Providence-Attleboro district. Like Tiffany, which bridged silverwork and personal adornments, New York and Newark jewelers ruled the market's peak, whereas "eastern" shops worked on cheaper lines.

In 1860, at least 75 jewelry manufacturers operated in Providence, employing 1,750 workers to create products worth $2.2 million (of $10 million nationally). The Civil War wrecked business for two years; a third of the shops vanished by 1864. Those with gold and silver stocks realized large profits without manufacturing by selling metals in a rising market, then "retired" when gold stayed high and demand low. Area employment fell to 750 before reviving once vogues developed for patriotic, martial, and funerary styles, fabricated from brass and other base metals. Thereafter, Providence firms crafted jewelry and ornament from silver, low gold (ten-karat or under), and nonprecious alloys. Borne up by a strong postwar recovery, a group of specialist auxiliary firms emerged by 1870: refiners, platers, engravers, gemstone cutters, and tool producers. Whereas Brown and Sharpe had been early makers of jewelers' tools and specialized machinery, renewed expansion brought others into the field.[3]

Despite the mid-seventies depression, the Providence jewelry industry included 142 firms with nearly 3,300 workers by 1880 (three-quarters of them adult men), generating an output worth $5.4 million, of which $2.9 million was value added in manufacturing (54 percent). Specialty services were the province of 32 other companies employing another 300.[4] As in fashion textiles and metalworking, most proprietors were craftsmen ("bench workers") who had served five- to seven-year apprenticeships, many becoming deft designers. In jewelry, the path to proprietorship was relatively easy for those with a flair for style, a full set of skills and tools, a few hundred dollars in savings, and a sound reputation, which yielded access to rented workspace in the jewelry district and modest start-up credit accommodations. Production techniques were then gradually shifting away from slower casting processes toward die presswork that shaped soft brasses and German silver into brooches, cuff links, pins, and the like, which were subsequently ornamented with stones, wirework, or enamel and fitted with clasps or chains. In 1885, local jewelers ran nearly 700 presses, an equal number of polishing "heads," over 500 jewelers' lathes, and 200 small drop hammers for forging. Most presses were foot operated and locally constructed, but the Providence market drew on Oberlin Smith's capacities as well.[5] Firms like Foster and Bailey, which stood ready to make any of several thousand patterns, adopted metalworkers' systems of job tickets and detailed specifications on pattern drawings, duplicates of which were kept in safes that also held stocks of silver and gold plate. Far less celebrated than Brown and Sharpe or Baldwin Locomotive, the larger Providence jewelry enterprises were just as fully up-to-date technologically.[6]

Marketing, from midcentury through the 1873 crash, meant twice annually showing makers' style samples to New York jobbers, "men of capital [who] bought manufacturers' goods for cash and dealt on long terms with the retailer." The money squeeze of the mid-seventies altered this relationship. Many of the old wholesale houses folded, and survivors now invited producers to provide *them* credits, to sell on a consignment basis, and/or to accept long delays for settling accounts. Salesmen discharged from failed distributors formed new jobbing firms with minimal resources, then asked manufacturers for comparable "concessions." Desperate for business in difficult times, manufacturing jewelers complied; but by the 1880s those emergency terms of trade had settled in as standard practice: "small orders by post card," returns of unsold goods, cancellations of confirmed orders, expectations of free repairs for damaged items, and demands that makers produce inventories of all styles for immediate shipment at wholesalers' calls. Each imposed costs on manufacturers and added uncertainty to marketing, as did jobbers' predilection for paying bills late yet deducting the discounts allowed for timely remittance.

Before eastern jewelers devised counterstrategies, these tactics produced three troubling effects. First, buyers' market power forced substantial inventory risk back onto those manufacturers who built up stocks of seasonal styles. Second, jobbers developed a bent for "shopping" one firm's samples to another maker, particularly a new and eager one, to have them duplicated at a lower quote, perhaps with slight design changes. As die-sinkers and firms making components (chains, clasps) enlarged the auxiliary network, this end run grew simpler. A novice company could often closely match a veteran's styles by calling into service the district's dis-integrated productive capacities. Third, this rage for copying contributed to the intense secrecy that style originators maintained and the hostility older firms manifested toward "garret" upstarts.

Knockoffs could often be produced in two weeks, killing reorders for hot novelties unless their creators had anticipated the market's vogues and built substantial inventories. Even then, the network's flexibility and swift response time facilitated rapid copying of seasonal hits, thus flooding the trade with cheap imitations, devaluing originators' stocks, and leaving imaginative firms bemoaning their lost returns for novelty. As one Providence jeweler fumed in 1886, "one of the greatest evils in the trade [is] the everlasting copying of good styles in inferior materials and workmanship, and cutting of prices." Finally, at seasons' ends, wholesalers circulated among the shops seeking bargain lots of dead stock, goods made up for orders that had been canceled or in anticipation of calls that never materialized. By the late 1880s, manufacturing jewelers were launching bitter complaints against distributors' manipulations.[7]

This situation, distressing to established companies but advantageous to wholesalers and fresh entrepreneurs, offers several insights into specialty production. Under certain conditions, it was entirely possible for a sectoral network to be *too* flexible and spatially compact for its own good. Distributors learned in the 1880s that they could reap the network's economic advantages better than could veteran manufacturers by working up a variation on the putting-out system of early industrialization.[8] Providence's rapid responses to custom orders, high skill levels, and able auxiliaries facilitated the knockoff game, later long a feature of Manhattan's garment industry.[9] Critically important were the low entry costs for new establishments and the trade's sharp seasonalities, which routinely pressed workers into months of idleness between two annual rush periods. Together these pull and push factors refilled the pool of fledgling shops with skilled workers commencing on their own account. Further, adroit second movers chasing seasonal successes held substantial cost advantages over style initiators, who routinely crafted several hundred new samples, only a fraction of which would draw sufficient orders to repay outlays for designing, tooling, and dies. Despite lower selling prices, imitators could score sizable opportunity profits.

Second, in this environment of extreme flexibility, price rivalry could readily displace specialists' beloved product competition. Established firms wearied of jobbers who presented them with close facsimiles of their samples and offered them a choice between matching the imitator's price or seeing the order go to the copyist's shop. The reduced price might bring failure to the garret entrepreneurs jobbers used as foils, but others would take their places, whereas refusing the cut simply slashed the originator's total sales and transferred business to the scrambling newcomers. Such exchanges heightened the traditional tensions between buyers and makers, threatened the latter's profits, and fueled the antagonism between clusters of veterans and climbers in the eastern jewelry trade.

Though both styled textiles and furniture emerged from comparable industrial districts, replete with new starts, auxiliaries, seasonal swings, and short-lived fashions, their contemporary marketing problems paled in comparison with those of Providence jewelers. Why this difference? First, the turnaround time for copying fabrics and furniture was far longer than for brooches and bracelets. Duplicating complex fabrics involved deconstructing their weave-formation, creating patterns and often commissioning the cutting of Jacquard card sets, securing yarns, and matching dye colors, before looms could be set up and weaving begun. This sequence could easily take four to six weeks, more than double the jewelry cycle, and any delays jeopardized apparel cutters' own delivery deadlines. Followers more commonly remodeled others' seasonal best-sellers for the next opening, a maneuver that echoed American textile specialists' reworking of the previous year's European style leaders. In furniture, the lag was far greater, owing to the need to draft plans from sketches of showroom samples before one could devise methods to match constructions and reproduce details.[10] When the long rest periods for drying glued joints and veneers and varnishing are added (and more time allowed if upholstery had to be commissioned as well), the minimum delay reached toward three months. Here again, design theft was attempted, but its effects trailed a season if not a year behind a successful line.

Second, by the 1880s specialists increasingly sold their worsteds or walnut bedsteads direct to cutters-up or retailers, rather than through jobbers.[11] This obstructed the spread of information about seasonal vogues, occasioning further delays for copyists. Though leaks and gossip about trends were constant, fuller and more reliable information surfaced late in seasons and informed planning for the next round. Jewelry jobbers, however, controlled all but a tiny fraction of the popular trade in the mid-1880s, circulating their selections from makers' samples to hundreds of regional retailers and thereby directly appropriating timely news on what was taking hold in the market. Along with the quick reproduction cycle, jewelry making and marketing exhibited an ironic efficiency absent in the other specialty consumer goods trades. Manufacturers would have to struggle against distributors' market power or else become their pawns.[12]

Like styled-textile firms, jewelers gradually articulated two responses: direct selling and trade organization. Reaching retailers individually promised considerable advantages, but the task's difficulties explained wholesalers' existence. Direct sales could protect style secrecy and delay piracy. Hiring roadmen would also put larger firms on quite a different footing from tiny competitors who could not afford to support travelers.[13] Department stores' rising prominence sparked Providence jewelers' first steps, for these retailers bought in fair quantities, paid promptly, and reordered their best-sellers from the original suppliers. In addition, producers' salesmen targeted the best-known independent jewelry retailers, skimming the cream from the top of the trade, and sought out local and national fraternal, business, and sporting associations. Some companies supplemented such campaigns by printing catalogs, mailed to small-town shops, and others hastened their seasonal designing so as to get samples into the market before the New York scrambles opened. These countermeasures gathered enough force by 1887 to put jobbers on the defensive through the 1893 crash.[14]

Both hoped-for and unintended consequences followed. Middlemen began making preseason trips to Providence, seeking fresh styles, and behaved rather more equitably on trade terms. Jobbers' threats to boycott direct-selling jewelers abated and profitability reportedly strengthened, but manufacturers' selling expenses rose steadily as well. By 1890, reports filtered in that retailers were tiring of roadmen's repeated visits and often declined to examine samples. Collections also proved a headache to manufacturers who sold direct to some fraction of the ten thousand independent stores. In a sense, such retailers reunited the three segments of the antebellum trade, vending jewelry, watches, and silverware, but they tended to settle their accounts with the latter sectors' large enterprises before sending remittances to jewelry producers. Even so, the counterattack gave the leading Providence companies more leverage in defending prices and more control over production and inventory than had been possible since the early 1870s. The complexities of direct sales and collections convinced many New England producers that trade organization was essential.[15]

Providence-area firms created several institutions in these years, as did their New York counterparts. Potentially most important was the New England Manufacturing Jewelers Association (NEMJA), modeled on the Silver Plate Association that from the early 1880s had worked successfully "to regulate prices, time of selling, and the rating of concerns" purchasing silverware. Neither NEMJA nor any other jewelry group ever mastered these capabilities, which helped stabilize the marketing of diverse silverware products. The trade's organizations did address other matters collectively: life and theft insurance, pursuit of robbers and burglars who plagued roadmen and shops, litigation against bankrupt jobbers and retailers, and, to a degree, schooling for designers and craftsmen. Of the key Silver Plate capacities, NEMJA and the linked Jewelers Board of Trade managed only to sustain a credit-rating service, failing in attempts to set common seasonal opening dates, curb design piracy, establish standard trade terms, defend prices, or secure adoption of uniform cost-accounting procedures. The market struggles of the 1880s divided the regional industry into groups of firms either selling direct or dependent on jobbers, groups whose interests were opposed on each of the five above-mentioned trade practices. Realizing that NEMJA could neither attract nor destroy their "mushroom" competitors, members abandoned it. At the end of its first decade, NEMJA's roster of firms had fallen 60 percent below its mid-1880s peak, and the association gave up its club and meeting rooms. Only an urgent campaign prevented collapse and retooled the association into an occasional banqueting society and source of pro-tariff petitions.[16]

Notwithstanding these tangles, the Providence jewelry sector achieved two sorts of growth in the 1880s. Its cohort of firms expanded to 170 and their workforces reached above 3,900, roughly a 20 percent increase. Meanwhile, product value rose 43 percent, value added jumped 55 percent, and the ratio of value added per dollar of product swelled from 54 to 59 cents. Workers' earnings also grew faster than their numbers. These gains reflected both technological changes noted above and declines in materials costs. Expenses for brass and other inputs fell from 46 cents per product dollar to 41 cents, one result of nonferrous metal producers' increased capacity and intensified competition, which ultimately led to consolidations after 1893.[17] Thus Providence jewelers quietly profited from upstream rivalries and technical advances.

Because the 1890 census tables detailed a host of costs beyond labor and materials, rough estimates of profitability can be ventured. Once all recorded expenditures are deducted from product values, Providence jewelers enjoyed a margin of $1.8 million on output of $7.8 million, or 23 cents per dollar of goods. This sum would be reduced by defaulted accounts, returns, or improper discounts taken, but it represented a 30 percent return on reported capital, only one-fifth of which was fixed in plant and machinery. Rhode Island jewelers overwhelmingly rented quarters and put their capital into materials, designs, and tooling, an effective investment-minimizing strategy.[18] Yet, as citywide data and figures from other sectors suggest, area jewelers had established a niche that they could not manage or control. Manufacturers could not define and promote style trends; rather, they had to guess about what novelties would capture the uncertain sensibilities of jobbers, retailers, and final purchasers. This helps account for the profusion of their product lines, for makers lacking means to anticipate or shape fashion trends had to rely on their nimble response capacities. Despite gains in value added per dollar of output, their overall performance lagged behind that of Providence's other sectors.

Table 6 contrasts the jewelry sector in the 1880s with local machinery and foundry operations and the staple cotton and specialty wor-

TABLE 6
Changes in Output, Workforce, and Sectoral Shares, Four Sectors and All Manufacturing, Providence, 1880–1890, in Percent

| | *Jewelry* | *Foundry and Machine Shops* | *Cotton Goods* | *Worsted Goods* | *All Mfg.* |
|---|---|---|---|---|---|
| Product value | +43 | +85 | +67 | +147 | +81 |
| Value added | +55 | +101 | +94 | +122 | +89 |
| Workers | +20 | +69 | +55 | +136 | +71 |
| Wages | +27 | +104 | +106 | +142 | +84 |
| Sectoral shares of: | | | | | |
| Product value | –2 | NC[a] | NC | +6 | |
| Value added | –3 | +1 | NC | +3 | |
| Workforce | –4 | NC | –1 | +12 | |
| Wages | –5 | +2 | NC | +5 | |

*Sources*: Department of the Interior, Census Office, *Report on the Manufactures of the United States at the Tenth Census* (Washington, DC, 1883), 428; idem, *Report on Manufacturing Industries in the United States at the Eleventh Census*, pt. 2, *Statistics of Cities* (Washington, DC, 1895), 470–77.

[a]NC = No change, i.e., share in 1890 was less than 0.5% above or below share in 1880.

sted fabrics industries. At both ends of the decade, these four trades accounted for roughly half of Providence's output and value added, but by 1890 jewelry's expansion trailed the mean rates for the other three and the city as a whole. Its shares in all four categories of urban production data declined, whereas metalworking and cottons kept pace with general growth and styled worsteds took off. Led by Brown and Sharpe, the Builders Iron Works, and Corliss, Providence foundries and machine shops showed above-average growth in all but workforce totals. Worsteds for men's suitings and women's apparel had more than doubled since 1880 to become the city's leader in employment (8,789), product ($17.6 million), and value added ($6.9 million) and brought the local textile complex alongside Paterson's silk industry as an important secondary center for fashion fabric manufacturing.[19] Yet jewelry was still the most profitable of the four sectors. Its gross margin (23 cents/dollar of output) rested above metalworking returns (21 cents) and far surpassed worsteds (16.5 cents) and staple cottons (11.5 cents). Put another way, every jewelry worker contributed $457 yearly to a firm's gross margin versus, respectively, $304, $330, and $139 in the other three sectors.[20] Occupying a classic specialty sector, jewelers who could evade the market's many traps and place their styles could reap better returns on investment and sales than their colleagues could, despite being servants of demand and fashion.

Gorham Silver offers another contrast, for, structurally, silverware manufacturing more closely resembled heavy machine building than the jewelry crafts. In 1880 and 1890 there were fewer than a hundred American firms making sterling or plated wares, among which about ten were sizable enterprises with several hundred or more workers.[21] Gorham was the largest, employing in 1890 over 900 workers and realizing $540,000 in profits on sales of $2.5 million, a 20 percent net return on sales, 18 percent on net assets invested, or $600 per employee.[22] Moreover, in silversmithing Gorham had become the style leader in the 1870s; at Tiffany, silver was "only a pretext for jewels and enameled work."[23]

Jabez Gorham began work as a jeweler but shifted in the 1830s to handcrafting silver flatware from melted-down Spanish, French, and American coins, a curious source of material that persisted into the 1860s. In 1841, Jabez took his son John into partnership, though the younger man had earlier failed as an apprentice. Five years later a cousin arrived to handle the books. Late in the forties, John Gorham borrowed $17,000 to build expanded quarters adjacent to the overcrowded shops, installed steam power, and rented to other small firms the extra space not yet required. This pattern of building beyond current needs and leasing would be a mainstay of the Providence jewelry trade's spatial organization, eventually supplemented, as at Worcester and Philadelphia, by speculative construction of factories for rental subdivision.[24]

Jabez retired in 1848, and John doubled and redoubled sales to $91,000 by 1854. Rather than retain his father's focus on spoons and such, the second-generation proprietor diversified into hollowware[25] while working to mechanize flatware processes. He hired salesmen to travel his styles and bought advertising liberally to support their efforts, claiming to purvey "only the best wares, in the latest fashion." To add capital, John Gorham brought two more of his cousins into the firm. To build its style sensitivity, he journeyed to Europe for three months in 1852, during which he spent a "hard, dirty" three weeks learning silver casting from a skilled British specialist who later joined Gorham in Providence.[26]

By 1859 Gorham's sales neared $400,000; and innovative metal spinning[27] and drop forge techniques quickened first-stage production steps for hollow-and flatware, respectively.[28] The first war years were wretched, just as they were for the district's jewelers, but Gorham had at least the satisfaction of fashioning a presentation silver coffee service for the Lincoln White House. Orders in 1863 topped 1859's showing and were four times 1861's slack figures. In the final war year, sales reached three-quarters of a million, though this increase must be assessed within its inflationary context, and Gorham incorporated as a closely held family firm, three-quarters of the shares divided between John and his cousin Gorham Thurber.[29]

The company soon opened a New York showroom, whose energetic director would become Gorham's second nonfamily president in the 1890s, and added similar facilities in Chicago, agency relations in other cities, and a set of ten full-time travelers by 1890.[30] Under John Gorham's leadership, the firm proliferated styles at a rapid pace, adding nearly 100 lines of flatware to its 1865 roster of 18 patterns, and ventured into virtually every imaginable subdivision of fine metalworking: custom silver sets for warships[31] and corporate moguls, elaborate trophies, Japanese tea sets, copper lamps and Turkish coffeepots, absurd silver camp kettles ($160 each), ecclesiastical ware (including cast bronze pulpits), architectural ornament, extravagant oxidized-silver dessert knives with medievally inspired pictorial handles, huge bronze statues, and ornate silver tureens. In silver plate, Gorham offered the public 88 *classes* of goods in 1881, and 26 *classes* of copper work the next year, ranging from silver bells, bread trays, pickle knives, and pudding dishes to copper "love cups," pen trays, umbrella stands, and ale mugs. Gorham also operated its own photography department, which captured its products on glass plates to provide travelers, retail outlets, and agents with illustrations for their selling efforts. In 1889 alone, the department made over 80,000 prints for promotional use. Gorham had mastered specialty production and marketing with a facility that jewelers could only envy.[32]

Gorham's selling tactics included acclaimed displays at the Centennial and the Paris Exposition of 1889, establishment of West Coast agencies, and publication of fat catalogs,[33] but there were bumps in the road as well. The 1873 depression forced layoffs and precipitated John Gorham's personal bankruptcy, ending family control. Ever ambitious, Gorham had pledged his shares in the company as security to a set of failed investments far outside silverware. Their collapse forced his departure from the company in 1878. Providence businessman William Crims replaced Gorham; but effective control soon rested with

9. A skilled woman engraver doing custom decorating work at Gorham Silver, 1892. Note the electric lighting to supplement or replace sunlight on cloudy or stormy days. Courtesy of Hagley Museum and Library.

Edward Holbrook, the firm's New York office director, who bought up loose shares and assumed "absolute control" by 1888 from his treasurer's post. Conflicts with workers were few. The Knights of Labor struck several silverware producers in its fevered 1886 surge, but not Gorham, which supported the besieged Whiting shops by refusing to hire strikers seeking alternative employment as a means to sustain their brothers' struggle in New York City. On the other hand, difficulties with designers were endemic. Crims repeatedly complained (to his diary) that the firm's lead designer, who boasted about his centrality to Gorham's stylistic prominence, was often absent. Such annoyances were trivial, however, when compared with the work-space problem that market success created. In the 1880s, Gorham sales soared from a half-million to two million and more yearly. This surge stressed the company's antique facilities, for appreciably more goods were flowing through its downtown workshops than at any point in the firm's history. Holbrook determined to relocate to a more open site.[34]

On thirteen acres in the outlying Elmwood section, Gorham erected and opened in 1890 the largest silverware plant in the United States, encompassing six acres of floor space, with its own electric power generator (driven, of course, by a Corliss steam engine), a private fire department, and ample space for production workers, designers, photographers, and office employees. Surveying the new works for a company-sponsored pamphlet, Alexander Farnum rhapsodized thus:

> In the city of Providence I have seen under one roof an entire block of buildings filled with . . . foundries for casting in iron, brass, silver, gold, and all other metals required; machine shops for every metal, and also for wood work; blacksmiths' shops, rolling mills, lathes, drills, milling and planing machines; shearing, punching, shaping and embossing machines; lofty shops and ponderous machines for die stamping; large rooms devoted to melting and refining furnaces; . . . to electroplating and gilding; to photography; to metal spinning; to finishing by hand and machinery, in more stages, modes, and apartments than could be carried away in memory . . . in short, after walking for half a day and to complete exhaustion, I was congratulated on having seen full[y] half of the Gorham Company's establishment![35]

Gorham's facilities accorded with Farnum's swooning description, for the Elmwood plant represented state-of-the-art design for diversified metalworking. Though the founding clan had been displaced, Holbrook appreciated the factory culture they had fostered. He would coddle designers and recruit skilled labor from all quarters, while relentlessly advertising the company's technical and style prowess.[36]

Specialty items could be priced for high unit profits, but sterling silver flatware could readily slip into price rivalries among various leading producers unless collective action were undertaken, for makers sold it by weight, much as foundries and machine builders priced their products. Holbrook's appreciation of this hazard led to the Silversmiths Company, a thinly disguised price-setting compact among major flatware producers that obstructed retailers' efforts to purchase styles on price comparisons. Members agreed to sell their sterling goods at a standard rate per ounce of weight, making style preferences the only relevant criterion for buyers' selection. As in other trades, there were slips and defalcations, but most producers understood the collective benefits that accrued from defending price and pressing style and quality as the criteria for buyers' decisions.[37] With a network for direct sales in place, Gorham kept many flat- and tableware lines alive for decades and built buffer inventories of their components—a strategy some fine furniture firms, but few textile or jewelry companies, could duplicate. Even so, as bar silver prices trended downward from $1.33 per ounce in 1870 to $0.88 in 1892, cutting inventory levels proved critical to prevent losses on styles made from silver bought above current selling price levels.[38]

Several points of similarity between Providence's jewelry firms and its prominent silverware company stand out. Both made ornamental specialties, traded in national markets, and booked above-average profits. Achieving throughput was not an issue; in both sectors, skill and diversity remained paramount and economies of scope far outpaced those of scale.[39] Both were technologically current and faced seasonally erratic and limited demand. Yet jewelry was chaotic and Gorham serene in its success.

Entry costs surely account for a substantial portion of this contrast, but other factors were also at work. That it required only a few hundred dollars to commence a jewelry business was itself a historically and spatially conditioned circumstance. Silverware firms were scattered across the northeastern states, with no more than one to four firms active in each of the cities supporting the industry: Hartford, Meriden, Wallingford, Waterbury, Norwich (Connecticut), Taunton (Massachusetts), New York, and Providence. Their relative isolation encouraged each firm to build a full complement of productive facilities, steps that set the industry pattern by the 1850s[40] and obviated the kind of specialist networks that cheapened start-up expenses in the concentrated jewelry districts. Both sectors' structures in the 1880s were the product of path-dependent lines of development that diverged dramatically.

Other elements in the trajectories silverware and jewelry firms traced reinforced this linkage between spatial and organizational patterns. The concentration of jewelers at New York and Providence, together with New York's status as the nation's great market nexus, facilitated the rise of wholesalers who redistributed seasonally styled, but anonymous, goods and published the sector's first trade catalogs. The silverware manufacture, however, had no center at all, and early style ranges were relatively narrow and durable. Hence firms sought to sell on quality guaranteed by brand names (for silversmiths had long trademarked their goods), at first using wholesalers, then agents, and finally direct sales through travelers and retail outlets. As design diversity expanded, they published their own catalogs, extending the identification of product with producer and building trade reputations that helped leading firms guide style trends rather than chase them.[41]

Spatial deployments were no more singly determinate of sectoral outcomes than were entry costs, of course, but they did represent both media for and outcomes of the distinctive tracks carved out by sectoral enterprisers. Other features of this dynamic might also be cited. The leading cluster of silverware companies by the 1880s featured the convergence of firms achieving full product ranges from two different directions, which I will delineate in a somewhat simplified sketch. The Connecticut group's roots lay in nonprecious alloys (pewter, German silver, tin). They moved into silver first with plating and later took up sterling lines, dropping most of their earlier base metals. Providence, New York, and Massachusetts companies, by contrast, started in coin silver and later added plated ware and work in other metals.[42] This convergence aligned the interests of the trade's pacesetters and enhanced the effectiveness of their common actions in pricing, establishing quality standards, and disciplining troublesome merchants (by withdrawing a firm's entire line and alerting other makers). Meanwhile, jewelry diverged into up- and down-market districts (New York/Newark versus Providence) with few shared interests to facilitate interregional cooperation. At Providence, further segmentation arose as veteran and novice firms warred to the benefit of distributors and auxiliaries. Hence trade institutions at best could only nibble at the corners of key problems that those differently placed defined differently.[43] Extending these paths would yield, for silverware, consolidation into an oligopolistic core, and for down-market jewelry, intensified rivalry leading toward sweatshop conditions by 1910. In the terminology suggested at the outset of Part II, silverware firms constituted a set of integrated anchors and outliers, whereas jewelry production proceeded from three nodes of networked specialists.

## New York: Printing and Publishing

Prevailing historical images of lower Manhattan's financiers, stock exchanges, wholesaling districts, and contentious garment trades have to a degree obscured New York City's significance as a multifaceted manufacturing center. One probing study of antebellum New York labor called attention to the city's intricate artisanal productive networks, but no comparable analysis of the later nineteenth century

10. R. Hoe and Company's sheet-fed rotary printing press, 1866. This massive machine was succeeded by Hoe's 1871 roll-fed press, which economized on labor and offered increased production speed. Courtesy of Hagley Museum and Library.

has yet surfaced.[44] By 1860 the nation's largest city had surpassed Philadelphia in the size of its industrial workforce and closely matched it in the wide range of trades practiced. New York's American Institute, founded shortly after Philadelphia's Franklin Institute, replicated the latter's technical expositions and forums for information exchange. Key machinery firms dotted Manhattan and adjoining precincts: R. Hoe and Co. in printing presses and saws, John Roach's Novelty Iron Works, Worthington in pumps, and scores of others. By the Centennial, the city hosted the nation's premier manufacturing journals, *Scientific American* and *American Machinist*, sent forth its largest-circulation newspapers, and became the hub for American book publishing.[45]

Among industrial sectors, printing and publishing has suffered a relative neglect comparable to that of New York among industrial cities. The "romance" of publishing indeed has generated dozens of affectionate company histories and one compendious survey,[46] but close scholarly analysis of the sort lavished on steel, autos, or chemicals has been absent.[47] Printing too has had its devotees, but until recently[48] much of the literature was narrowly focused, antiquarian, and/or confined to the era before 1830.[49] Reasons for this historiographical lacuna are not hard to find. The industry was structurally intricate, strongly oriented to local and regional clienteles, and rested virtually unaffected by the waves of business concentration that recurred periodically from the 1880s through the 1920s. In essence, its course did not resonate with the thematic concerns that have animated the last generation of business, economic, and labor historians.[50] The discussion of New York's Gilded Age printing trades that follows will only sketch a few issues and patterns germane to the wider deployment of specialty manufacturing, but further in-depth research on the industry is surely worth undertaking in the present "information age."

The descriptor "printing and publishing" is the first item for explication. As in machinery, from the 1840s through the 1880s, a long process of subsectoral specialization unfolded unevenly across American wordsmithing. Antebellum town and city printers had been generalists for the most part, often publishing a newspaper, doing job work (posters, broadsides, stationery, and trade cards), and printing pamphlets or books to order. By the 1870s, particularly in the larger cities, the linkage between printing and publishing could no longer be presumed.[51] The roots of the divergence may be traced to interactions between changes in technology and labor process at one side and shifts in population and markets at another, but a separation of printing production and publishing was becoming evident.

The mid-1870s have represented a benchmark date at which "printing and publishing were established as separate industries."[52] This was clearly the case for books, but an overview of the trade's components illuminates a more elaborate scene. As products, books, magazines and journals, newspapers, and music were *published* goods, whereas brochures, letterheads, broadsides, labels, calendars, circulars, engravings, railway schedules, and even playing cards represented *printed* matter. The number of possible product configurations was simply extraordinary. Book publishers like Harper's ventured before the Civil War into editing magazines (both a monthly and a weekly), whereas newspapers like the New York *Tribune* commenced issuing paperback book series in the 1870s, and others sought their advertisers' job printing business or the campaign printing of the political parties they backed.[53] By the 1880s, the trade's structure had evolved into four rough configurations.

Book publishers became agencies for soliciting, editing, and marketing. With few exceptions, they contracted with separate firms for printing, binding, illustrations, and their engraving, as did most magazines and journals.[54] Urban newspapers, by contrast, became more integrated and time-sensitive, investing in ever-larger and swifter Hoe presses, facilities for on-site production of illustrations, and, after 1884, Mergenthaler Linotypes that formed and set column-length slugs of type in one operation.[55] Though newspapers' content and layout changed in every issue, their operations became more routinized as their machinery facilitated rapid quantity production.[56] Commercial printers increasingly served institutional and individual needs of every character, including contract work for book and journal publishers. Finally, an array of independent auxiliaries (steel-, wood-, and photoengravers, binders and blank book makers, lithographers, stereo- and electrotypers, type founders) supplemented the efforts of all classes of printers.

Theodore deVinne, one of New York's leading late-nineteenth-century printers, affirmed the sector's specialty manufacturing character. In his view, book and "job" printing was "by its very nature, a custom trade; while certain economies of scale are possible, printing does not mass-produce an article to be warehoused until satisfactory distribution can be made. [For books,] the publisher, not the printer, does much of the risk-taking. . . ." Moreover, printing firms displayed features common to proprietary capitalism in other trades: "participation of the owner in management, . . . absence of stock-market financing, and [a] personal relationship between proprietor and customer."[57] As in jewelry and styled textiles, shop owners overwhelmingly came from the ranks of skilled journeymen schooled in trade customs. DeVinne captured the fragility of these enterprises: "The journeyman of this year may be the employer of the next, in which position he may give wages and work to his old master. . . . There is scarcely [an employer] who feels so rich as to scout the possibility of a reverse of fortune that would compel him to ask work in future years, of the apprentice he is now instructing."[58]

The nation's premier printing center, New York, accounted for a third of U.S. output and half the book division's production in 1860. Lower Manhattan held a thriving printing district wherein labored the bulk of the trade's 7,500 workers, a spatial concentration that persisted into the 1920s.[59] Though the city's share of total capacity would erode (national printing employment expanded more than twice as fast as population 1860–90), it remained at roughly one-quarter well into the new century, feeding on complementary increases in book and periodical publishing and on the city's centrality to paper-intensive financial, advertising, and corporate headquarters activity.[60] Three interrelated aspects of sectoral development through the early 1890s are worth exploring: marketing and profits, technological change and labor relations, and institution creation.

Book publishers found the 1870s and 1880s harrowing. In the absence of effective copyright statutes and the presence of minuscule, federally subsidized bookpost rates, respected and fly-by-night houses competed to vend European classics and popular potboilers, reproduced in cheap papercover editions selling for a quarter or less. In consequence, "regular" publishers felt compelled to cut list prices on new American titles lest they be seen as narrow-market luxuries. Houses sold millions of wretchedly printed and bound volumes during the craze, glutting markets for fiction and travelogues, undermining profits, and setting the stage for political action. Five years after its 1886 formation, the American Publishers' Copyright League secured congressional assent to an international copyright statute that protected publishers' and authors' intellectual property rights, helping to end the fevered stealing of texts. By 1893, this specter had receded, but newly published titles soon had to face the shrinking demand that economic depression entailed.[61]

Before the copyright convention, book publishers had raced to print purloined English and Continental volumes, but their time frames were leisurely when compared with those of newspapers. Daily papers rapidly became waste matter; and over two dozen of them competed for New Yorkers' pennies in 1880, their infinitesimal prices supported by advertising revenues keyed to circulation.[62] Fast and flashy news, briskly gathered, speedily composed and printed, fueled sales and attracted advertisers. By 1890 display advertising increasingly depended on advance contracting for block spaces by department stores, specialist retailers, and national market purveyors of consumer goods (patent medicines, hair tonics, etc.). Selling papers necessitated not only rapid printing but also a tight distribution system through which wholesalers delivered bundles of issues in a few hours to thousands of key street corners and transit stops.[63]

Though separate morning and evening papers might share a common printing facility, all faced real bottlenecks caused by the unevenness of presswork and composition mechanization. Hoe and its rivals had advanced high-volume printing by the 1880s to such a point that output capacities had multiplied tenfold in two decades. However, setting type and framing display matter remained handwork. Hence, before the Mergenthaler, the need for speed brought rising employment, subdivision of labor, and the hiring of women compositors and incompletely trained apprentices to achieve cost savings in shop tasks.[64]

During the 1880s New York's dailies jumped from 29 to 50 competitors; circulation more than doubled to 1.7 million papers six days a week.[65] Together with weeklies and diverse journals, the newspaper and periodicals subsector encompassed nearly 500 enterprises at the decade's close with an annual output valued at $34 million. Overall sectoral growth was equally stunning. Whereas in 1880 the trade's three core divisions counted 445 printers and publishers producing $23 million in goods, by 1890 nearly 1,200 local companies reported sales above $54 million (tables 7 and 8). Auxiliary specialists kept pace. Outside the established arena of binding and blank books (ledgers, diaries, et al.), workforces trebled and sales expanded ninefold (to nearly $10 million), as new capacities in photolithography, photoengraving, and electrotyping supplemented a dramatic growth spurt in regular lithography.[66]

Profits hardly suffered, despite complaints by employing printers and their allies. The 1880 census tables do not permit estimations of profitability, but the published 1890 reports are far more extensive.[67] As table 7 suggests, auxiliary firms coined twenty cents gross margin on every sales dollar, book and job printers over thirty cents, and news and journal firms about twenty-seven cents. Part of this aggregate $15.6 million margin covered the $6 million in salaries for principals and clerks, but two-thirds represented gross profit, a splashy 20 percent annual return on declared capital investment.

Several other elements in the 1890 census compilations illustrate the significance of New York's dis-integrated network. Although the city accounted for only 13 percent of national value added in newspapers and periodicals and 23 percent in book and job work, its printing and publishing auxiliaries contributed over a third of U.S. totals in seven subsectors (table 7).[68] Further, the bulk of the city's $12 million in printing and publishing "Expenses" represented spending for illustrations, presswork, and binding that knitted together separate specialist establishments, conserved capital, and generated profits recycled into purchases of new machinery and fresh marketing efforts.

For commercial printers, marketing was less routinized but no less fevered than for books and newspapers. Rapid turnaround figured as a key selling point, for the number of firms was appreciably larger in

TABLE 7
Printing and Publishing Industry, New York City, 1880

| | Nf[a] | Workers | Wages | Mats. | PVal. | VAdd. |
|---|---|---|---|---|---|---|
| | | | (in millions) | | | |
| Auxiliaries: | | | | | | |
| Bookbinding and blank books | 114 | 4,068 | 1.55 | 2.29 | 4.93 | 2.64 |
| Engraving, steel and wood | 79 | 846 | 1.26 | 0.40 | 2.01 | 1.61 |
| Lithography | 48 | 1,139 | 0.67 | 0.60 | 1.74 | 1.14 |
| Totals | 241 | 6,053 | 3.48 | 3.29 | 8.68 | 5.39 |
| Core printing and publishing: | | | | | | |
| Pocket books[b] | 33 | 619 | 0.28 | 0.53 | 1.03 | 0.50 |
| Printing and publishing | 412 | 9,578 | 5.88 | 7.36 | 21.70 | 14.34 |
| Totals | 445 | 10,197 | 6.16 | 7.89 | 22.73 | 14.84 |
| Grand totals | 686 | 16,250 | 9.64 | 11.18 | 31.41 | 20.23 |

*Source*: *Report on the Manufactures of the United States . . . Tenth Census*, 417–19.
[a]Nf = number of firms; Mats. = materials; PVal. = product value; VAdd. = value added in manufacturing.
[b]Cheap paperbacks.

job work and potential clients were diffused spatially, rather than concentrated at bookstores or transit nodes. Few printers could afford to wait in their shops for individuals needing their services. Instead, proprietors (and, by 1890, salesmen) sought out business—touring offices, stores, theaters, and factories with samples of their work, ready to quote instant estimates. As every potential sale involved an essay in problem solving that could be handled in a variety of ways, and as few buyers of printing understood the full range of possibilities, competition was intense. Only on repeat orders of forms or labels would price comparisons be critical; even there, given the variety of paper qualities, the use of lower-grade stock could compensate for a shaved price to yield decent profits. The information advantage generally lay with printers, and their sales techniques focused on quick delivery, attention to detail, and product quality, which when effectively managed could ensure much-prized, enduring relations with clients.[69]

Technological change may be treated briefly under three headings: materials, equipment, and power. The introduction and improvement of wood pulp techniques fundamentally transformed the paper indus-

TABLE 8
Printing and Publishing Industry, New York City, 1890

| | *Nf* | *Workers* | *Wages* | *Exp.*[a] | *Mats.* | *PVal.* | *VAdd.* | *Estimated Profit*[b] |
|---|---|---|---|---|---|---|---|---|
| | | | *(in millions)* | | | | | |
| Auxiliary Enterprises: | | | | | | | | |
| Bookbinding and blank books | 192 | 4,197 | 2.07 | 0.47 | 1.97 | 5.83 | 3.86 | 1.32 |
| Engraving, steel | 42 | 673 | 0.43 | 0.07 | 0.32 | 1.30 | 0.98 | 0.48 |
| Engraving, wood | 83 | 186 | 0.14 | 0.02 | 0.03 | 0.36 | 0.33 | 0.17 |
| Labels and tags | 18 | 198 | 0.11 | 0.03 | 0.12 | 0.32 | 0.20 | 0.06 |
| Lithography | 79 | 3,036 | 2.21 | 0.34 | 2.34 | 6.28 | 3.94 | 1.39 |
| Photolithography & photoengraving | 17 | 471 | 0.33 | 0.05 | 0.18 | 0.83 | 0.65 | 0.27 |
| Stereo and electrotyping | 23 | 396 | 0.28 | 0.04 | 0.18 | 0.74 | 0.56 | 0.24 |
| Totals | 454 | 9,157 | 5.57 | 1.02 | 5.14 | 15.66 | 10.52 | 3.93 |
| Core printing and publishing: | | | | | | | | |
| Book and job | 658 | 10,228 | 6.45 | 2.14 | 5.10 | 19.89 | 14.79 | 6.20 |
| Music | 11 | 141 | 0.07 | 0.05 | 0.06 | 0.34 | 0.28 | 0.16 |
| Newspapers and periodicals | 497 | 7,345 | 7.64 | 9.80 | 7.48 | 34.25 | 26.79 | 9.35 |
| Totals | 1,166 | 17,714 | 14.16 | 11.99 | 12.64 | 54.48 | 41.86 | 15.71 |
| Grand totals | 1,620 | 26,871 | 19.73 | 13.01 | 17.78 | 70.14 | 52.38 | 19.64 |

*Source*: *Report on Manufacturing Industries . . . Eleventh Census: 1890*, pt. 2, 390–409.

*Note*: For newspapers and periodicals, "Wages" includes both final production and payments to editors, subeditors, reporters, and correspondents, and for books, editors and similar nonclerical staff.

[a]Exp. = expenses, including rents, interest, sundries, and for newspapers and periodicals, commissions on advertising and contracting for composition and presswork, etc. For other abbreviations, see n. a to table 7.

[b]Imputed by deducting wages, materials, and expenses from product value.

try in the two decades after the Civil War, dramatically increasing supplies of newsprint, book, and specialty papers at falling prices while building machine designer Thomas Savery's assets. Newsprint dropped from nearly fifteen to below five cents a pound, 1865–90, as pulp mills increased from eight in 1870 to eighty-two two decades later. As a result, though the printing trade's total output and wages paid tripled, 1880–90, expenses for materials only doubled, thus helping boost sectoral profitability.[70]

Printing proprietors invested a portion of these returns in a host of new machines and devices. William Bullock's web press (1863), which printed both sides of sheets cut from large rolls of paper, soon fell before the capabilities of R. Hoe's "roll-fed printing rotary press" (1871) and dozens of subsequent variants and improvements for folding, collating, and binding. By the later 1880s, Hoe's giant "quadruple" press, built for Joseph Pulitzer's New York *World*, could spit out 48,000 eight-page newspapers hourly, cut and folded. The Mergenthaler Linotype machine was a late arrival (there were only 300 in place by 1891), as was the Lanston Monotype, which found favor in book and job work; but they clearly augured the end of hand typesetting for all but custom orders. Comparable technical novelty was evident in the rapid advance of steel engraving (displacing woodcuts), the elaboration of multicolor lithography, and the adaptation of photographic technologies to printers' tasks.[71]

On the power front, printers led all manufacturing sectors in the speed with which they adopted electric drive systems, following experiments in the early 1880s and the wide publicity given the 1884 production of "the regular editions of *Electric World* . . . from electrotype plates on a motor-driven cylinder press in the [Franklin Institute's] International Electrical Exhibition in Philadelphia." By 1891, the trade press heralded the virtues of abandoning shafts and belting and installing direct drive motors. Printing proprietors converted about 10 percent of their equipment to electricity before the 1893 crash (40 percent by 1899), a shift that proceeded far more slowly in manufacturing generally (electric drive's overall 1899 share being about 5 percent).[72] Printers thus were triply engaged in technical change during the Gilded Age. They benefited materially from the revolution in papermaking, greeted the profusion of new machines with open checkbooks, and appreciated quite swiftly the significance of Edison's first central station in lower Manhattan.[73]

New York's centrality to the nation's printing and publishing industry both made the city an employment magnet (jobs more than tripling, 1860–90) and created a platform for workers' self-organization. After at least three early efforts, New York's typesetters inaugurated a sustainable organization in 1850, supported by the redoubtable Horace Greeley, whose *Tribune* had only recently been joined by James Bennett's *Herald* in paying the workers' desired rate, thirty-two cents per thousand "ems."[74] Greeley, a publisher who was elected the union's first president, noted disparagingly that "[t]here are almost as

128 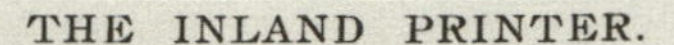 THE INLAND PRINTER.

One Dozen
Improved Perry Quoins
and Key

sent, express paid, on receipt of $3.00.

*Address:*

**THE PERRY QUOIN CO.**

**507 Pontiac Building, CHICAGO.**

---

F. L. MONTAGUE & CO.

**Improved Bookbinders' and Printers' Machinery.**

| | |
|---|---|
| The Dexter Folders, | Signature Presses, |
| The New Wire Stitcher, | Index Cutters, |
| Elliott Thread Stitcher, | Punch and Eyelet Machines, |
| The Acme Paper Cutters, | Patent Gold-Saving Machine, |
| Ellis Roller Backer, | Spooner's Mailing Machine, |
| Ellis Book Trimmer, | Ruling Machines, |
| Ellis Embosser, | Tape, Wire, Etc. |

**F. L. MONTAGUE & CO.**

| | |
|---|---|
| 315 Dearborn Street, | 17 Astor Place, |
| Manhattan Bldg., Room 617, | 140 East 8th Street, |
| CHICAGO. | NEW YORK. |

---

**MANHATTAN TYPE FOUNDRY**

EXCLUSIVE AGENTS FOR THE

Improved ⋄ New Style ⋄ Noiseless

**LIBERTY**

Send for Price List and Catalogue to

**54 Frankfort Street,**

**New York, U. S. A.**

---

**THE EMMERICH**

IMPROVED

**Bronzing and Dusting Machine.**

SIZES:

12x20, 14x25, 16x30, 25x40, 28x44, 34x50, 36x54.

*Write for Prices and Particulars.*

**EMMERICH & VONDERLEHR,**

OVER 800 IN USE. **191 & 193 Worth Street, NEW YORK.**

**SPECIAL MACHINES for PHOTOGRAPH MOUNTS and CARDS.**

EMBOSSING MACHINES

11. Display advertising in the 1890s for printing equipment by trade suppliers of machinery, type, and accessories. Courtesy of Hagley Museum and Library.

many scales as there are offices," and urged his fellow employers to adopt a common payment scheme that properly compensated those whose skills were essential to the timely dissemination of information.[75] Over the next twenty years, however, Greeley and other employing printers found themselves increasingly distanced from the organization they had fostered, for the International Typographical Union (ITU) repeatedly challenged working hours and labor price schedules.

However, the union had its own internal difficulties. Separate organizations for pressmen, stereotypers, and compositors developed, further subdivided by trade branch and language, necessitating a long process of boundary negotiation and coalition building that at last erected "cooperative unionism" by the early 1890s.[76] As the New York labor network solidified, organized printers readily integrated technological changes into their workshop routines and made certain that union members learned the new skills. In 1883 the job control implications of labor's collaborative tactics brought employing printers to revive an earlier organization, the Typothetae, both to deal collectively with the ITU and related groups and to share social contacts and information on effective business practices. Four years later, the New York Typothetae joined employing printers in other cities to form a national association and weathered a two-week strike by local book and job workers for the nine-hour day and a closed shop.[77] Thereafter, walkouts in the New York printing trades were scattered and relatively infrequent through the mid-1890s, being "one-sixth the number waged by garment workers and one-fifteenth the number involving building tradesmen." Cooperative unionism offered employers "the prospect of industrial stability," including rate schedules that stressed competition on product and performance rather than wage shaving to pare costs.[78] In consequence, New York printers' earnings stood 15 to 50 percent above national averages for the various trade divisions in 1890, and 12 to 34 percent above payments to their colleagues employed at Chicago, the second city in printing and publishing.[79] High wages that recognized skill and the power of organization sustained the peace of the shop and little diminished employing printers' returns.[80]

Like other specialists, New York printing firms handled varied demand, created and sustained sectoral institutions, and operated within a spatially concentrated production network of complementary enterprises. As elsewhere, most proprietors had stepped up from the shop floor to ownership and relied on skilled workers. Unlike most others, however, they accommodated to labor's institutions rather than battling to eradicate them. Later union spokesmen facing owners in nonoligopolistic sectors would claim their organizations' salience to regulating trade conditions, arguing that no single employer could establish responsible competition instead of the cutthroat tactics that punished both workers and profits.[81] Perhaps implicitly, New York's employing printers realized the force of this argument and settled in to cooperate with cooperative unionism. The depression of the mid-1890s tested this relationship, but it would endure.

# Chapter 6

## MIDWESTERN SPECIALISTS: CINCINNATI TOOLS AND GRAND RAPIDS FURNITURE

THE INDUSTRIALIZATION of the American Midwest was once thought to have replicated an eastern sequence that moved from extensive agriculture through commerce-based urbanization and transport development to regional manufacturing deployment. Recent work in historical and economic geography, however, suggests the need to modify this scheme along at least two dimensions. First, midwestern farms and factories multiplied in tandem: "industry did not follow the plow, it built the plow." Husbandry and manufacturing interacted to create a "territorial production complex of great geographical breadth and internal richness" in the second half of the nineteenth century. Second, the region's "agro-industrial revolution" was also distinctive both culturally and technically, not modeled on northeastern templates. It featured constant revisions of "agro-processing" procedures, the rapidly increasing scale of enterprises involved, a dispersed network of cities, and a greater emphasis on quantity production of staple goods (boots and shoes, farm tools, flour) than on the specialties and fashions that distinguished Atlantic coast sectors. The region's emergence as the spatial hub of mass production systems ("Fordism") "did not come out of thin air" but derived from the convergence of capacities and practices (ranging from materials handling to bulk marketing) that matured across the preceding three generations.[1]

If specialty manufacturing was not a hallmark of midwestern industrialization, it was neither absent nor insignificant, and in time would become a regional engine of growth on its own account. At Cincinnati, alongside the bulk processing of meat and the transformation of by-products to leather, soap, and candles, a pacesetting cluster of machine tool builders rooted and expanded in the 1880s. At Grand Rapids, furniture works rose to overshadow lumber mills, giving this small Michigan city a durable national identity as a center for quality and style. With less fanfare, machinery and metalworking built a notable presence in Cleveland and Chicago, as did furniture and publishing sectors at the latter site. Whether linked to the tooling needs of agro-industrialists and mining/forest extractors or committed to reworking their intermediate products, midwestern specialists achieved supple growth and profitability through the early 1890s, even as other sectors cemented the foundations for mass production.[2]

Postbellum Cincinnati was widely known as "Porkopolis," a rude testament to its full participation in the agro-industrialization process. In 1880 its six leading bulk-processing sectors[3] all worked on rural yields. Slaughtering and leather converted livestock to commodities, brewing and distilling transformed grains, tobacco drew on southern leaf, and lumber thinned the region's remaining woodlands. Together with baking, these trades accounted for a third of the city's output, more than double the share of batch specialists in furniture, saddlery, printing, and metalworking (table 9). By contrast, Philadelphia's 1880 proportions were quite the reverse, with the five largest bulk trades contributing 15 percent, and leading batch sectors 38 percent, to overall product values.[4] Not only did midwestern urban industrial structures differ sharply from those eastward, their initial batch capabilities chiefly derived from bulk activity. Saddlery depended on the tanneries, furniture on the sawmills, and the metal trades originated with midcentury riverboat building, providing marine engines and fittings for steamers that moved Cincinnati meat and whiskey up and down the Ohio.

The city's industrialization attracted a steady flow of eastern, native-born migrants and immigrant British and German workers, so much so that a key working-class neighborhood was long known as "Over the Rhine." Workers' rising militance during the Knights of Labor era met manufacturers' intransigence in the mid-1880s, generating conflicts of an intensity and violence scarcely matched elsewhere and, after labor's defeat, not reproduced during the ensuing half-century. The post-1887 collapse of workers' organizations marked the Queen City as a trade union desert, for into the 1930s area manufacturers rebuffed virtually every industrial thrust labor activists attempted.[5] Despite the labor turmoil, Cincinnati industries soared during the eighties. In an environment of gradual price deflation, local production increased nearly 70 percent, value added doubled, while the workforce expanded only by half (table 10).[6] Again in 1890, the leading bulk sectors' production nearly doubled that of specialists. Befitting the expansion, new trades joined both leading clusters, but the two formats for manufacturing developed in tandem during the 1880s, nei-

TABLE 9
Leading Industrial Sectors, Cincinnati, 1880, by Production Format
(dollar figures in millions)

| | *Nf* | *Capital* | *Workers* | *Wages* | *Mats.* | *PVal.* | *VAdd.* |
|---|---|---|---|---|---|---|---|
| All mfg. | 2,991 | 49.92 | 52,796 | 18.87 | 61.15 | 102.73 | 41.54 |
| Bulk production and processing: | | | | | | | |
| Bread and bakery | 232 | 0.39 | 699 | 0.25 | 1.16 | 1.83 | 0.67 |
| Leather | 50 | 1.30 | 530 | 0.25 | 3.02 | 3.79 | 0.77 |
| Liquor, distilled | 10 | 3.14 | 750 | 0.31 | 3.60 | 5.29 | 1.69 |
| Liquor, malt | 19 | 4.14 | 1,373 | 0.60 | 2.57 | 4.58 | 2.01 |
| Lumber | 20 | 0.75 | 658 | 0.28 | 1.37 | 1.90 | 0.53 |
| Slaughtering | 49 | 4.07 | 1,143 | 0.34 | 10.45 | 11.61 | 1.16 |
| Tobacco products | 263 | 1.24 | 3,341 | 1.00 | 2.18 | 4.27 | 2.09 |
| Totals | 643 | 15.03 | 8,494 | 3.03 | 24.35 | 33.27 | 8.92 |
| Mixed output: | | | | | | | |
| Boots and shoes[a] | 333 | 1.07 | 3,214 | 1.09 | 2.31 | 4.13 | 1.82 |
| Carriages and wagons | 50 | 1.25 | 3,297 | 1.21 | 3.06 | 5.29 | 2.23 |
| Clothing, men's[a] | 237 | 6.28 | 9,275 | 2.67 | 8.63 | 13.89 | 5.26 |
| Totals | 620 | 8.60 | 15,786 | 4.97 | 14.00 | 23.31 | 9.31 |
| Batch/specialty: | | | | | | | |
| Foundry and machine shop | 90 | 4.01 | 3,634 | 1.75 | 2.59 | 5.72 | 3.13 |
| Furniture | 119 | 2.64 | 3,461 | 1.34 | 1.64 | 4.37 | 2.67 |
| Printing and publishing | 89 | 2.53 | 3,089 | 1.17 | 1.40 | 4.01 | 2.61 |
| Saddlery/harness | 51 | 0.41 | 592 | 0.24 | 0.51 | 0.16 | 0.65 |
| Safes and vaults | 4 | 0.74 | 875 | 0.50 | 0.42 | 1.34 | 0.92 |
| Totals | 353 | 10.33 | 11,651 | 5.00 | 6.56 | 16.60 | 9.98 |

*Source*: Department of the Interior, Census Office, *Report on the Manufactures of the United States at the Tenth Census* (Washington, DC, 1883), 393–95.

*Note*: For abbreviations in column heads, see n. a to table 7.

[a] Both factory and custom-made combined.

ther squeezing out the other. Midwestern agro-industrial processing propelled a parallel advance in specialty production.

In metalworking, apart from Mosler Safe Company and its imitators, that impetus came as much from responses to a specific decline as from general growth. Postbellum riverboat building in Cincinnati tailed off gradually; after 1869 production never again matched the

TABLE 10

Leading Industrial Sectors, Cincinnati, 1890, by Production Format (dollar figures in millions)

| | *Capital* | *Workers* | *Wages* | *Exp.* | *Mats.* | *PVal.* | *VAdd.* | *Gr.Pr.* |
|---|---|---|---|---|---|---|---|---|
| All mfg. | 99.97 | 76,367 | 31.83 | 19.19 | 85.04 | 178.23 | 94.19 | 42.17[b] |
| Bulk production and processing: | | | | | | | | |
| Boots and shoes, factory | 2.11 | 3,984 | 1.70 | 0.16 | 3.15 | 6.02 | 2.87 | |
| Bread and bakery | 1.25 | 1,048 | 0.44 | 0.27 | 1.56 | 3.17 | 1.61 | |
| Clothing, men's, factory | 12.30 | 13,543 | 2.95 | 0.95 | 8.46 | 17.95 | 9.49 | |
| Coffee/spices | 0.64 | 129 | 0.06 | 0.03 | 1.79 | 2.43 | 0.64 | |
| Leather | 3.24 | 878 | 0.46 | 0.15 | 3.14 | 4.14 | 1.00 | |
| Liquor, distilled | 1.86 | 375 | 0.21 | 8.34 | 2.42 | 11.47 | 9.05[a] | |
| Liquor, malt | 11.95 | 1,576 | 1.18 | 1.87 | 2.40 | 7.45 | 5.05[a] | |
| Slaughtering | 2.22 | 675 | 0.39 | 0.16 | 7.87 | 9.51 | 1.64 | |
| Soap/candles | 1.70 | 672 | 0.24 | 0.11 | 2.83 | 3.83 | 1.00 | |
| Tobacco products | 1.35 | 2,335 | 0.91 | 0.48 | 1.13 | 3.29 | 2.16 | |
| Totals | 38.62 | 25,215 | 8.54 | 12.52 | 34.75 | 69.26 | 34.51 | 13.45 |
| Mixed output: | | | | | | | | |
| Carriages and wagons | 4.54 | 3,536 | 1.70 | 0.37 | 4.83 | 8.67 | 3.84 | |
| Lumber and mill products | 2.49 | 1,841 | 1.03 | 0.15 | 2.50 | 4.45 | 1.95 | |
| Paints | 2.10 | 270 | 0.15 | 0.20 | 1.18 | 2.06 | 0.88 | |
| Totals | 9.13 | 5,647 | 2.88 | 0.72 | 8.51 | 15.18 | 6.67 | 3.07 |
| Batch/specialty: | | | | | | | | |
| Clothing, men's, custom | 1.38 | 2,287 | 1.04 | 0.24 | 1.43 | 4.02 | 2.59 | |
| Clothing, women's | 1.74 | 3,344 | 1.03 | 0.22 | 1.95 | 4.75 | 2.80 | |
| Foundry and machine shop | 7.21 | 5,240 | 2.90 | 0.56 | 4.14 | 10.09 | 5.95 | |
| Furniture | 3.96 | 2,978 | 1.50 | 0.26 | 1.45 | 4.21 | 2.76 | |
| Printing and publishing | 2.88 | 2,546 | 1.37 | 0.92 | 2.57 | 6.88 | 4.31 | |
| Saddlery/harness | 1.01 | 992 | 0.53 | 0.10 | 1.60 | 3.64 | 2.04 | |
| Safes | 1.88 | 1,846 | 1.03 | 0.35 | 1.23 | 3.32 | 2.09 | |
| Totals | 20.06 | 19,233 | 9.40 | 2.65 | 14.37 | 36.91 | 22.54 | 10.49 |

*Source*: Department of the Interior, Census Office, *Report on Manufacturing Industries in the United States at the Eleventh Census* (Washington, DC, 1895), 144–55.

*Note*: Gr.Pr. = gross profits; for other abbreviations in column heads, see n. a to table 7.

[a]$9.30 million of the liquor sectors' $10.21 million in expenses represented federal alcoholic beverage taxes, inflating product value and value added artificially.

[b]Gross profits include $10.96 million paid to principals and office staff.

thirty-four vessels completed in that year. This decay pressed engine fabricators toward product novelty—at the Niles Tool Works, prime movers to drive local factories and southern sugar mills, plus machine tools for the metal industries. It was this shift that sent John Richards to Brown and Sharpe and encouraged John Steptoe and William Lodge in their initial tool-making sallies. The results were dramatic. By 1880 metalworking ranked third behind meatpacking and men's clothing in industrial output, second in value added, workforce, and wages paid. A decade later, if the production of alcoholic beverages is excluded, tool building stood second on all counts and paid nearly as much in total wages to fewer than half as many workers as were employed in the bulk menswear sector (tables 9 and 10).[7] Once slaughtering, tobacco, and leather stalled, metalworking and its machine tools subdivision would solidify as venues for reshaping Porkopolis's industrial base.

Eleven of the dozen men credited with initiating the Cincinnati machine tool trade[8] were experienced metalworkers, one of whom took on the twelfth as a bookkeeping and sales partner to form an inside/outside pair. John Steptoe's works, the city's largest general machine operation (employing three hundred in the 1870s), served as a training ground for novice entrepreneurs like William Lodge. By contrast, George Gray moved from machinist to designer to partner at Niles Tool before 1870 and relocated with the firm to nearby Hamilton, Ohio, in the Centennial year. He returned to Cincinnati in 1880 to initiate a series of partnerships focused first on drilling machines, then metal planers. Other enterprises soon moved into specialized tool building, for it proved a profitable departure from general machine-making practice.

For example, machinist R. K. LeBlond, locally trained in a type foundry, took technical classes at the Ohio Mechanics Institute before journeying to Providence for an instructive year with Brown and Sharpe. In 1887, he cycled back to Cincinnati and opened a workshop producing tools and gauges for type founders, filling in dull periods with the manufacture of "dental rolls, tag printing presses, cash registers, small steam engines, and routing machines." Four years later, LeBlond secured a subcontract from William Lodge to build fifty lathes and their accessories and turned his full attention to improving lathe designs, as did Lodge himself. Henry Bickford (drills), Frederick Geier (milling machines), and a dozen other new local tool makers adopted related specialties. Before the shocks of 1893, Cincinnati had already become the hub for "western" machine tool development, a city with technically adroit companies each refining its capacity and performance in one class of tools rather than spanning the entire spectrum.[9]

Profiling the remarkable William Lodge is essential, for Lodge succeeded Steptoe in the 1880s as the regional trade's central figure. His early success, expanding from under ten to seventy-five employees in little over a year, encouraged other "ambitious mechanics" to attempt the move to proprietorship. However, Lodge did far more than serve as an example or a cheerleader. Transferring the shop-based fraternalism of his union days to an entrepreneurial milieu, Lodge routinely contracted out "certain portions" of his products to new firms, helped "the right kind of young men to secure an equipment for the shop work," and "guarantee[d] a steady sale" of their goods.[10] When after 1886 his new partner Charles Davis wanted a "complete line" of tools to enhance his sales efforts, though Lodge wished to "specialize upon lathes," Lodge resolved the conflict by creating a collective marketing scheme that provided Davis ample scope and fledgling firms ample business. Davis sought orders for tools of all descriptions; and Lodge farmed out to others those falling outside the lathe class, along with lathe orders whose delivery dates (or other characteristics) would stress his firm's capacity or deflect concentration on technical improvements, his special interest. An early machine tool historian summarized the impact of Lodge's strategy: "The effect of [this policy] was to build up a number of small tool building enterprises, independent of each other but not competing."[11] Though Lodge and Davis parted company in 1892, supportive contracting continued at the successor Lodge and Shipley firm, assisting both newcomers and sectoral expansion as a whole and in the process creating a "high trust" environment conducive to collegiality and collective initiatives.[12]

While organizing Cincinnati machine tools' entry into a national market long dominated by eastern works, Lodge also promoted the proprietary ambitions of his own workmen, sponsoring their start-ups with subcontracts and on occasion recruiting partners for them from other cities. Area tool builders prominent in subsequent decades included former Lodge draftsmen, superintendents, foremen, vice-foremen, and office staff. A 1907 observer counted fourteen local machine works that "branched directly or indirectly from the Lodge and Davis company," and a later student constructed an elaborate "genealogical chart" of the Cincinnati tool trade that illustrated similar progeny from other firms.[13] Finally, inside his own plant, Lodge strove to install design and technical changes that would reorient lathe manufacturing from custom work to batch production. The key was systematically reorganizing the labor process so that parts fabrication and sequential assembly replaced the customary format in which several men "buil[t] the entire machine." This effort duplicated Brown and Sharpe's approach in the 1880s, though what Lodge knew of Providence practice is not evident.[14] Ultimately, Lodge led Cincinnati tool specialists into the complexities of components inventories, work-in-process tickets, larger batch sizes, and using common parts in a range of models, just as he had pioneered in stimulating new enterprises and coordinated marketing.

Lodge's initiatives established contacts among local metalworks that materialized in and reinforced their spatial concentration west of the city center, along the Mill Creek valley stretching northward from the Ohio River. The importance of this propinquity and its agglomeration effects deepened as machine tool demand and the number of firms rose. Lodge's centrality and efficacy may also help explain the paucity of institution formation in Cincinnati's early machine tool trades. To be sure, city enterprises supported the Ohio Mechanics Institute, which dated to antebellum days, and a series of annual industrial expositions (1870–88). However, machine tool companies stood on the periphery of these activities, which predated their appearance or expired just as their momentum grew.[15] Local trade associations or educational initiatives like those of Philadelphia or Worcester did not precede, but followed, the nineteenth century's last great depression. This delay stemmed from the precise benefits William Lodge's sectoral fraternalism provided local manufacturers. Once depression heightened interregional competition and expansion drew down the supply of skilled workers and shop-schooled supervisors, Cincinnati machine tool firms speedily fashioned both local and national institutions to respond to the threats of price wars and labor bidding, defending and extending their specialty production complex.[16]

Gilded Age Cincinnati was a diverse industrial city and a site for nearly 75,000 manufacturing jobs. Central Michigan's Grand Rapids, a much smaller "city built on wood," articulated a different mode of midwestern industrialization. Unlike many towns spawned by timber harvesting, it avoided the ghostly fate that exhausting local resources entailed for most such settlements, achieving preeminence in American furniture production. Each year more devoted to stylish furniture, Grand Rapids departed from industrial patterns at second-echelon eastern cities committed to bulk manufacturing (Lowell and Fall River, Massachusetts; Johnstown, Pennsylvania), which relied on staple textiles or tonnage steel. Instead, it mirrored the experience of those centers where specialty sectors were the spur to growth (hosiery and textile machinery at Reading, Pennsylvania; silks at Paterson, New Jersey; electrical machinery and locomotives at Schenectady, New York; and furniture at Jamestown, New York, and Gardiner, Massachusetts). In fine furniture works, not sawmills, lay its route to prominence among the many "Middletowns" whose significance for American industrial growth has long been underappreciated.[17]

Grand Rapids was a classic "one-industry" town in the decades after the Centennial (tables 11 and 12). Furniture, lumber, and woodworking accounted for 46 percent of manufacturing employment in 1880 and 49 percent ten years later (plus 44 and 46 percent of all value added at the two dates).[18] Area metal trade firms making woodworking machines, spring sets for bedsteads, or factory equipment depended largely on furniture demand.[19] The predominance of specialty sectors distinguished Grand Rapids from Cincinnati, Cleveland, and Chicago, where bulk outputs overmatched batch enterprise in this era.[20] Yet local manufacturers held a tiny share of national furniture output (3 percent in 1880, 6 in 1890). Unlike printing and publishing or specialty textiles, which New York, Chicago, and Philadelphia anchored, furniture making was spatially diffused. Over a dozen districts from Boston to Chicago shared the market with scores of smaller manufacturing clusters scattered from Maine to the Carolinas to Iowa. In 1890, though its output passed Cincinnati's faltering furniture operations, Grand Rapids' production was hardly half that of Chicago, where an unparalleled rail transport nexus, the immigration of German woodworkers, and a maturing wholesale distribution system spawned the nation's largest furniture concentration by the early 1890s.[21] Thus the heart of the Grand Rapids story concerns how this mouse roared.

As elsewhere, the city's manufacturers were a routine mix of native-born enterprisers and British or German immigrants—shop-experienced, self-educated in designing, and rudimentary in figuring costs. Like other furniture men, they strove in the 1880s to master the integration of machine-based parts production with craftsmanship in ornament and assembly.[22] They invested heavily in plant and equipment, in real 1890 dollars quintupling 1880 capital as workforces rose less than half as much; but other districts responded comparably to

TABLE 11
Leading Industrial Sectors, Grand Rapids, 1880, by Production Format
(dollar figures in millions)

| | *Nf* | *Capital* | *Workers* | *Wages* | *Mats.* | *PVal.* | *VAdd.* |
|---|---|---|---|---|---|---|---|
| All mfg. | 318 | 4.79 | 4,923 | 1.79 | 3.77 | 7.07 | 3.30 |
| Bulk production and processing: | | | | | | | |
| Bread and bakery | 14 | 0.08 | 64 | 0.03 | 0.20 | 0.25 | 0.05 |
| Confectionery | 4 | 0.05 | 95 | 0.03 | 0.12 | 0.17 | 0.05 |
| Liquor, malt | 7 | 0.28 | 65 | 0.03 | 0.13 | 0.23 | 0.10 |
| Lumber | 8 | 1.01 | 252 | 0.08 | 0.46 | 0.76 | 0.30 |
| Tobacco product | 18 | 0.04 | 76 | 0.03 | 0.07 | 0.12 | 0.05 |
| Totals | 51 | 1.46 | 552 | 0.20 | 0.98 | 1.53 | 0.55 |
| Mixed output: | | | | | | | |
| Agricultural implements | 7 | 0.26 | 215 | 0.07 | 0.16 | 0.30 | 0.14 |
| Boots and shoes | 18 | 0.03 | 48 | 0.03 | 0.05 | 0.10 | 0.05 |
| Carriages and wagons | 9 | 0.09 | 93 | 0.04 | 0.08 | 0.14 | 0.06 |
| Clothing, men's | 9 | 0.12 | 195 | 0.06 | 0.17 | 0.32 | 0.15 |
| Totals | 43 | 0.50 | 551 | 0.20 | 0.46 | 0.86 | 0.40 |
| Batch/specialty: | | | | | | | |
| Foundry and machine shop | 13 | 0.27 | 276 | 0.13 | 0.19 | 0.38 | 0.19 |
| Furniture | 11 | 1.31 | 1,807 | 0.65 | 0.82 | 1.83 | 1.01 |
| Printing and publishing | 20 | 0.09 | 210 | 0.08 | 0.05 | 0.17 | 0.12 |
| Builder's millwork | 6 | 0.07 | 97 | 0.04 | 0.09 | 0.16 | 0.07 |
| Tin, copper, and sheet iron work | 10 | 0.03 | 53 | 0.03 | 0.05 | 0.09 | 0.04 |
| Totals | 60 | 1.77 | 2,443 | 0.93 | 1.20 | 2.63 | 1.43 |

*Source*: Same as for table 9, 403–4.
*Note*: For abbreviations in column heads, see n. a to table 7.

rising demand.[23] Along none of these lines were Grand Rapids' firms critically different from their rivals; but in marketing they devised a powerful collective initiative—the seasonal furniture style exposition.

Organizing a biannual wholesale exhibition helped overcome the double dispersion of makers and retailers, drawing hundreds, eventually thousands, of dealers to Grand Rapids each January and July. Expositions enabled firms to consolidate orders for their various styles so

TABLE 12
Leading Industrial Sectors, Grand Rapids, 1890, by Production Format
(dollar figures in millions)

| | *Capital* | *Workers* | *Wages* | *Exp.* | *Mats.* | *PVal.* | *VAdd.* | *Gr.Pr.* |
|---|---|---|---|---|---|---|---|---|
| All mfg. | 15.62 | 10,842 | 4.76 | 1.26 | 8.75 | 18.37 | 9.62 | 3.60[a] |
| Bulk production and processing: | | | | | | | | |
| Bread and bakery | 0.18 | 128 | 0.06 | 0.02 | 0.25 | 0.42 | 0.17 | |
| Confectionery | 0.07 | 119 | 0.04 | 0.01 | 0.13 | 0.24 | 0.11 | |
| Flour milling | 0.74 | 110 | 0.06 | 0.06 | 1.39 | 1.56 | 0.17 | |
| Liquor, malt | 0.41 | 85 | 0.05 | 0.06 | 0.09 | 0.25 | 0.16 | |
| Lumber | 1.93 | 599 | 0.21 | 0.09 | 0.74 | 1.31 | 0.57 | |
| Tobacco products | 0.11 | 129 | 0.06 | 0.02 | 0.08 | 0.21 | 0.13 | |
| Totals | 3.44 | 1,170 | 0.48 | 0.26 | 2.68 | 3.99 | 1.31 | 0.57 |
| Mixed output: | | | | | | | | |
| Carriages and wagons | 0.25 | 134 | 0.07 | 0.01 | 0.16 | 0.31 | 0.15 | |
| Patent medicines | 0.25 | 65 | 0.03 | 0.01 | 0.18 | 0.28 | 0.10 | |
| Totals | 0.50 | 199 | 0.10 | 0.02 | 0.34 | 0.59 | 0.25 | 0.13 |
| Batch/specialty: | | | | | | | | |
| Builder's millwork | 0.60 | 413 | 0.22 | 0.04 | 0.66 | 1.01 | 0.35 | |
| Clothing, men's, custom | 0.15 | 217 | 0.10 | 0.02 | 0.14 | 0.35 | 0.21 | |
| Clothing, women's | 0.05 | 335 | 0.07 | 0.01 | 0.19 | 0.37 | 0.18 | |
| Foundry and machine shop | 0.89 | 500 | 0.26 | 0.04 | 0.21 | 0.75 | 0.54 | |
| Furniture | 5.43 | 4,190 | 1.94 | 0.46 | 2.27 | 5.64 | 3.37 | |
| Printing and publishing | 0.22 | 302 | 0.14 | 0.05 | 0.11 | 0.40 | 0.29 | |
| Totals | 7.34 | 5,957 | 2.73 | 0.62 | 3.58 | 8.52 | 4.94 | 1.59 |

*Source*: Same as for table 10, 230–37.
*Note*: For abbreviations in column heads, see n. a to table 7.
[a] Gross profits include payments of $1.21 million to principals and office staff.

that they could devise economical lot sizes for production. As companies offered new case goods lines every season, each made in a variety of woods and finishes with differing levels of detail and ornament, a firm's full array of style options easily ran from three hundred to six hundred variants.[24] Roadmen could not travel with furniture samples, so they used engravings, then photos, of styles to entice retailers one by one. Even when successful, they forwarded a trickle of orders—three of this, five of that—which either had to be filled from advance inventory (always a hazard in a style sector) or constructed in tiny lots, with profits reduced by salesmen's commissions and expenses. As in jewelry, retailers wearied of travelers' relentless visits; but they soon discovered that the benefits of viewing hundreds of full-scale samples in Grand Rapids far outweighed the trip's costs. Local producers' initiative to organize the market created a "win-win" situation for makers and shop owners alike.[25]

The expositions' origins can be traced to the aftermaths of both the 1873 economic contraction and the Centennial. Faced with heavy cancellations of goods made for clients, three companies (Phoenix, Berkey and Gay, Nelson-Matter) advertised through midwestern newspapers a ten-day inventory liquidation sale in December 1873. The local *Daily Eagle* urged that this become an annual event, and returns proved satisfactory enough that repetitions followed during the next two winters. After the Philadelphia displays, the three firms opened New York retail outlets; but dealers' protests against direct sales to consumers mounted quickly, precipitating the first advance style showing in Grand Rapids at the close of 1878. As attendance at the exhibitions grew, the leaders closed their retail stores and concentrated on drawing buyers to their shows. In the mid-1880s, out-of-state manufacturers began bringing *their* samples to the city and leased empty stores, lofts, clubrooms, and hotel suites (often at punishing fees) to seek a share of the season's sales. Before 1892, local developers erected several exhibition buildings, the largest of which encompassed eight acres, providing a spatial market focus that replaced the scatter of factory showrooms and visitors' temporary quarters. In part to coordinate the exhibitions, but also to deal with exasperating railway rates and insolvent clients, owners formed the Grand Rapids Furniture Manufacturers Association (GRFMA) in 1881, the first of several institutions designed to represent, promote, and regulate the regional furniture trade. Local styles sold heavily, a result of the expositions and area firms' employment of eastern and immigrant designers; and tiny Grand Rapids surged toward sectoral prominence.[26]

All went smoothly until the spring of 1886, when labor conflicts fractured the association. Furniture workers, two-thirds immigrants chiefly from Holland, Germany, and Scandinavia, organized and affiliated with the Knights of Labor, then called for eight hours' work with no change in the day rate of pay (a modest $1.75 on average). Manufacturers widely accepted the eight-hours demand, which went into effect May 3, but only the larger, older firms pledged no reduction in earnings.[27] New, smaller companies refused the second part of this "double-barreled request." On May 4, the Haymarket disaster in Chicago shifted the political ground; and soon two plants announced a return to ten hours. Many manufacturers and the union leaders reaffirmed the eight-hour commitment at a May 30 meeting, but in early June, Phoenix and Berkey and Gay defected, squashed worker resistance, and reestablished ten hours, reversing the Knights' gains and splitting the association. This "falling out" rendered GRFMA "moribund." One faction led its remaining adherents into the new, New York–based National Furniture Manufacturers' Association, while another group created the Grand Rapids Furniture Freight Bureau in December to coordinate shipments and secure "favorable freight rates," rebates, and routings, presumably ignoring all other contentious matters. By 1889, common interests eclipsed hard feelings and a revitalized GRFMA both hosted a national gathering of case goods manufacturers at the January opening and linked arms with the Freight Bureau. Effective labor organizing did not revive for nearly twenty years.[28]

Local furniture capitalists also united in 1886 for direct commercial gain, incorporating the Furniture Caster Association to manufacture and distribute spring-loaded, removable, wheeled feet for beds, bureaus, chairs, and the like—a device patented by Julius Berkey and W. R. Fox. Previously, inferior casters had been hammered or screwed into pieces by dealers. Now the FCA would provide them to Grand Rapids firms and other makers for installation at the factory or to dealers for use on goods whose producers shipped flat-foot pieces. Priced for high profits and energetically defended at law against infringements, the Berkey/Fox caster swept the trade, returning over $250,000 in dividends to FCA principals before its patents expired. Though the company's paper capital was $90,000, it actually started out with only a $5,000 loan, soon repaid from early profits, and expanded by plowing back its growing returns. Next to nothing was ventured, but the gains were huge. Allied specialists could, when the opportunity arose, do the bulk production dance. Millions of patented casters were ultimately made and sold under the Grand Rapids patent.[29]

More formally organized and experienced than Cincinnati tool builders, Grand Rapids furniture manufacturers proved adept at collective action. If early associations fractured or foundered, they rebuilt them. Although the furniture men lagged behind Philadelphians in creating trade schools, banks, and insurance companies, they were far ahead in market organization and in milking returns from technical advances. If they were ciphers on the national political scene, they dominated local politics, whether acting from elected positions or exerting influence from the Board of Trade, Board of Public Works, or their factory offices.[30] Most important, individually, they made money. The Widdicombs' closely held family firm paid dividends averaging 10 percent of its $150,000 capital annually, 1885–92, over and above the kin-centered management team's salaries. William A. Berkey and his three partners shared $200,000 in dividends in the late eighties and early nineties, recouping in seven years more than double their invested capital.[31] Younger, smaller firms were more precariously fixed, yet most shared this prosperity, at least until the 1893 crash. After January 1894, neither Widdicomb nor Berkey paid a dividend until 1900, but their ample cushions carried them through the slump.

Workers, however, had reason to grumble. Yes, there were more jobs each year in the city, and work in general became less hazardous, less rushed,[32] and more varied than in sawmills or builder's millwork plants. But the Knights' organizers had tilled fertile ground in Grand Rapids. Hours were long, sixty a week rising toward seventy in the busy season. Pay for skilled operatives rested 15 to 20 percent below metalworkers' rates, and employment was sharply seasonal, with short weeks and layoffs common in slack periods just prior to the twice-yearly expositions. Movement from one employer to another was difficult. Factory owners usually declined to hire other firms' employees, instead advertising outside the region when adding workers; and employers refused to bid up wages above customary local levels on the few occasions when they breached the informal "no poaching" compact. Moreover, the path to proprietorship was far rockier than at Philadelphia or Providence. Grand Rapids had no William Lodge to sustain novice entrepreneurs, and no manufacturing apartments with power for lease. Furniture production was space-demanding, and area real estate developers put their energies and funds into construction of multistory exhibition halls, not facilities for shared industrial tenancy.

These features of local industrial practice constrained competition decisively, creating a highly imperfect "labor market" and limiting entry to those who could finance construction of a factory (or succession to a failed firm's facility), as well as fund the attendant machinery, supplies, exhibition rentals, and sales force. No Providence-like garret competition would emerge at Grand Rapids, nor would the differentiated functional specializations of Philadelphia textiles and metalworking appear. Instead, numbering fewer than twenty in the early 1900s, a core of veteran companies would prevail, welcoming the occasional able newcomer to their circles of familiar contact, to their clubs, associations, and banks, and to their expectations of collaboration amid market competition. Throughout, the "Furniture Men" remained attentive to detail, quality, and the promotion of Grand Rapids as the Paris of American furniture manufacturing. For decades, Grand Rapids would initiate, and others would emulate—a pattern entirely satisfactory to central Michigan's networked specialists.

## Specialty Production and American Manufacturing, 1890

The introduction to this study included a roster of ninety-three industrial sectors that roughly summarized American manufacturing's distribution among specialty, mixed output, and routinized production approaches in 1909. That account showed leading custom and batch trades (those with $25 million or more in products) and their bulk and mass counterparts each responsible for just over 30 percent of value added nationally, plus one-third and one-quarter, respectively, of industrial employment. To conclude this section's empirical presentation, I offer comparable arrays for 1890 (tables 13–15), based on a $10 million output threshold and showing value added, employment, and value added per worker.[33] What is striking when these two groupings are compared is the contrast between substantial across-the-board increases in employment and value added and the relative stability of format shares.

Overall, between 1890 and 1909, factory employment rose just over one-half and output more than doubled (in constant 1890 dollars, 83 percent), but the specialty and bulk proportions on both measures altered hardly at all (tables 16 and 17). Specialty manufacturing's loss of two share points over two decades little dims the robust growth of trades treated extensively here: machines and metalworking, printing, fashion fabrics, furniture. New champions (auto and electrical) and category shifts derived from market and technical changes (silk, iron

TABLE 13
Specialty Production Sectors, $10 Million+ in Output, by Value Added and Employment, 1890

| | *Value Added (millions)* | *Employment (thousands)* | *VA/Wkr. (hundreds)* |
|---|---|---|---|
| Foundry and machine shop products (includes machinery) | 242 | 248 | 10 |
| Printing and publishing | 206 | 165 | 12 |
| Woolen and worsted goods | 80 | 113 | 7 |
| Men's custom clothing | 76 | 98 | 9 |
| Locomotive and railway shop construction | 63 | 109 | 6 |
| Furniture from factories | 56 | 64 | 9 |
| Blacksmithing and wheelwrighting | 40 | 51 | 8 |
| Silk goods | 36 | 51 | 7 |
| Tin, copper, and sheet iron work | 35 | 38 | 9 |
| Women's clothing (dressmakers) | 34 | 68 | 5 |
| Hosiery and knit goods | 31 | 61 | 5 |
| Saddlery and harness | 28 | 30 | 9 |
| Jewelry | 27 | 24 | 11 |
| Custom boots and shoes | 24 | 35 | 7 |
| Shipbuilding | 23 | 26 | 9 |
| Musical instruments | 22 | 19 | 11 |
| Hats and caps | 21 | 27 | 8 |
| Carpets and rugs | 19 | 29 | 7 |
| Architectural and ornamental ironwork | 19 | 19 | 10 |
| Custom millinery | 18 | 24 | 8 |
| Dyeing and finishing textiles | 17 | 20 | 8 |
| Men's furnishing goods | 15 | 22 | 7 |
| Cabinetmaking and upholstering | 14 | 15 | 10 |
| Monuments, stone | 13 | 12 | 11 |
| Heating apparatus | 13 | 12 | 11 |
| Brass castings | 12 | 12 | 10 |
| Lithography | 12 | 11 | 11 |
| Fancy boxes | 11 | 20 | 5 |
| Packing boxes | 11 | 14 | 8 |
| Coffins and trimmings | 11 | 10 | 11 |
| Lace goods | 9 | 12 | 8 |
| Picture frames | 9 | 10 | 9 |
| Fur goods | 9 | 8 | 11 |

TABLE 13 *(cont.)*

| | *Value Added (millions)* | *Employment (thousands)* | *VA/Wkr. (hundreds)* |
|---|---|---|---|
| Cutlery and edge tools | 8 | 9 | 9 |
| Morocco leather | 8 | 8 | 10 |
| Corsets | 7 | 11 | 6 |
| Wood, carved or turned | 7 | 8 | 8 |
| Plated ware | 7 | 7 | 10 |
| Tools, not elsewhere classified | 7 | 7 | 10 |
| Brassware | 6 | 7 | 9 |
| Trunks and luggage | 6 | 7 | 9 |
| Umbrellas and canes | 6 | 7 | 9 |
| Plumber's supplies | 6 | 5 | 12 |
| Gloves and mittens | 5 | 9 | 6 |
| Stamped ware | 5 | 7 | 7 |
| Refined gold and silver, not from ores | 2 | 1 | 20 |
| Totals | 1,336 | 1,570 | Avg. 8.5 |

*Source*: Department of the Interior, Census Office, *Report on Manufacturing Industries in the United States at the Eleventh Census, 1890*, pt. 1, *Tools for States and Industries* (Washington, DC, 1895), 36–45.

*Note*: Variations in VA/Wkr. figures are due to rounding in cols. 1 and 2. Col. 3 figures were derived from full original data.

rolling) swelled the mixed middle, and fiddling with categories might add a bit to the bulk class. Yet these rosters at least suggest that specialty-oriented sectors constituted solid elements of the industrial fabric both before and after the first round of throughput triumphs and mergers, for as a group they continued to develop in synchrony with bulk and mass enterprises.

There was substantial unevenness, of course. Sectorally, foundries, machine shops, and furniture ran well ahead of average employment gains, suggesting limited capital-for-labor substitution, whereas by 1909 printing output soared (up 160 percent in nominal dollars) despite only an average increase in workforces, indicating heightened productivity. Spatially, these tables can say nothing about variations in urban and regional processes, an issue that will be of considerable interest in Part III. First, however, an attempt to conceptualize the diversity of specialty sectors and locales will close Part II.

TABLE 14
Mixed Format Sectors, $10 Million+ in Output, by Value Added and Employment, 1890

| | Value Added (millions) | Employment (thousands) | VA/Wkr. (hundreds) |
|---|---|---|---|
| Men's clothing, factory-made | 122 | 156 | 8 |
| Cotton goods | 113 | 222 | 5 |
| Factory boots and shoes | 102 | 140 | 7 |
| Builder's millwork | 78 | 87 | 9 |
| Carriages and wagons | 72 | 84 | 9 |
| Brick and tile | 55 | 109 | 5 |
| Agricultural implements | 50 | 43 | 12 |
| Factory-made women's clothing | 34 | 42 | 8 |
| Paper | 32 | 30 | 11 |
| Glass | 29 | 46 | 6 |
| Marble and stonework | 26 | 24 | 11 |
| Railway cars, freight and passenger | 25 | 32 | 8 |
| Confectionery | 25 | 27 | 9 |
| Timber products, not elsewhere classified | 23 | 46 | 5 |
| Patent medicines | 22 | 9 | 23 |
| Shirts | 18 | 33 | 5 |
| Cooperage | 18 | 25 | 7 |
| Hardware | 16 | 20 | 8 |
| Clay and pottery | 16 | 20 | 8 |
| Paints | 16 | 9 | 17 |
| Photography and materials | 13 | 12 | 11 |
| Furniture, chairs | 10 | 13 | 7 |
| Electrical apparatus and supplies | 10 | 9 | 11 |
| Cordage and twine | 9 | 13 | 7 |
| Rubber goods | 7 | 10 | 8 |
| Bottling | 7 | 9 | 8 |
| Wire rope and cable | 7 | 8 | 9 |
| Mattresses | 7 | 7 | 10 |
| Rubber boots and shoes | 7 | 4 | 17 |
| Totals | 969 | 1,289 | Avg. 7.5 |

*Source*: Same as for table 13.
*Note*: See note to table 13.

TABLE 15
Routinized Production Sectors, $10 Million+ in Output, by Value Added and Employment, 1890

| | *Value Added (millions)* | *Employment (thousands)* | *VA/Wkr. (hundreds)* |
|---|---|---|---|
| Lumber | 172 | 286 | 6 |
| Iron and steel | 135 | 153 | 9 |
| Tobacco products | 119 | 136 | 8 |
| Malt liquor (brewing) | 119 | 35 | 23 |
| Distilled liquor | 89 | 5 | 167 |
| Slaughtering | 81 | 48 | 17 |
| Flour milling | 80 | 63 | 13 |
| Bread and bakery products | 56 | 53 | 11 |
| Manufactured gas | 43 | 15 | 29 |
| Leather tanning | 38 | 34 | 11 |
| Chemicals | 26 | 17 | 15 |
| Paving materials | 17 | 23 | 7 |
| Petroleum refining | 17 | 12 | 14 |
| Soap and candles | 15 | 9 | 16 |
| Sugar and molasses | 15 | 8 | 19 |
| Fertilizers | 14 | 10 | 14 |
| Wrought iron pipe | 12 | 12 | 10 |
| Canned fruit and vegetables | 11 | 51 | 2 |
| Nails | 11 | 17 | 7 |
| Cheese, butter, and milk | 11 | 14 | 8 |
| Lime and cement | 10 | 14 | 7 |
| Mineral and soda waters | 10 | 8 | 12 |
| Sewing machines | 9 | 9 | 10 |
| Coffee and spices | 9 | 5 | 18 |
| Nonpetroleum oils, not elsewhere classified | 8 | 3 | 26 |
| Brooms | 7 | 11 | 8 |
| Wire | 7 | 8 | 9 |
| Bolts, nuts, etc. | 6 | 7 | 8 |
| Prepared foods | 6 | 4 | 15 |
| Malt | 6 | 4 | 17 |
| Varnish | 6 | 2 | 30 |
| Coke | 5 | 9 | 5 |
| Cottonseed oil | 5 | 6 | 8 |

TABLE 15 (*cont.*)

| | Value Added (millions) | Employment (thousands) | VA/Wkr. (hundreds) |
|---|---|---|---|
| Boot and shoe stock | 5 | 4 | 12 |
| Nonpaper bags | 4 | 4 | 10 |
| Smelting | 3 | 2 | 17 |
| Lard | 3 | 1 | 3 |
| Totals | 1,190 | 1,102 | Avg. 11 |

*Source*: Same as for table 13.
*Note*: See note to table 13.

TABLE 16
Format Summary, 1890

| | Value Added (millions) | % | Employment (thousands) | % | VA/Wkr. (hundreds) |
|---|---|---|---|---|---|
| Custom and batch | 1,336 | 34 | 1,570 | 36 | 8.5 |
| Mixed formats | 971 | 25 | 1,288 | 30 | 7.5 |
| Bulk and mass | 1,190 | 31 | 1,102 | 25 | 11 |
| Other (240 small sectors) | 368 | 10 | 395 | 9 | 9 |
| Totals | 3,865 | 100 | 4,355 | 100 | Avg. 9 |

*Note*: Five construction sectors were removed from "manufacturing" totals so that comparability with later censuses would be feasible. They were carpentering, masonry and brickwork, painting and paperhanging, plastering, and plumbing, totaling 358,000 workers and $345 million in value added.

TABLE 17
Format Summary, 1909

| | Value Added (millions) | % | Employment (thousands) | % | VA/Wkr. (hundreds) |
|---|---|---|---|---|---|
| Custom and batch | 2,745 | 32 | 2,259 | 34 | 12 |
| Mixed formats | 2,473 | 29 | 2,220 | 34 | 11 |
| Bulk and mass | 2,607 | 31 | 1,625 | 25 | 16 |
| Other (145 small sectors) | 704 | 8 | 511 | 8 | 14 |
| Totals | 8,529 | 100 | 6,615 | 101 | Avg. 13 |

*Source*: See table 2.

## Contingent Conclusions and Conceptualizations

The Centennial Exposition announced America's emergence as a sophisticated industrializing nation. Though the bulk of the population still worked the land or resided in small towns dependent on agriculture and extraction, ever larger numbers migrated and immigrated to urban places where manufacturing led development. The seventeen years between the exposition and Chicago's 1893 festivities formed one of the most peculiar periods in American economic history, featuring repeated recessions, substantial deflation (wholesale prices down ca. 20 percent in real terms, 1876–90), two waves of massive labor conflict, persistent political battling over gold and greenbacks, *and* striking, though uneven, industrial growth. Traditional accounts of this industrial surge have displayed memorable images: an extensive railway network feeding carloads of beeves and acres of wheat to Great Lakes cities, provisioning the nation; a basic metals complex, drawing on immense reserves of ores and coal, redefining the contours of organizational and technical capacity; cotton mills erected near cotton fields and hand labor in tobacco-stripping sheds feeding Bonsack cigarette machines, as the South fixed its sights on processing what it cultivated; oil pipelines threading eastward across the Alleghenies to ever larger refineries; and the achievement of parts interchangeability for production of watches, sewing machines, and reapers in the hundreds of thousands. The great names of the era still ring: Armour, Pillsbury, Carnegie, Duke, Rockefeller, McCormick, and of course Morgan, Drexel, Gould, Huntington, and Vanderbilt. Such men were the architects of industrial throughput, the managerial corporation, railway system building, and the first phases of mass production. Yet like antebellum Lowell's cotton magnates or Springfield and Harper's Ferry's armory practice innovators, they represented but one dimension of American industrialization.

Gilded Age businesses followed at least three *other* tracks toward profit and accumulation. Many pursued the manufacture of bulk staples in sectors where, for various reasons, achieving throughput and scale economies, much less market dominance, was a chimera. Another cohort energized specialty production, realizing gains from opportunities and practices distinct from those of bulk-oriented industries. The third group, more prominent in later decades, integrated in some fashion the routinized and specialty approaches, a strategy evident by 1893 among the leading electrical corporations and adopted in steel following the merger wave. If this is a plausible distribution of industrial formats at the time, what can be said in summary about batch specialty enterprise? Second, what can be proposed conceptually and specifically about it at the three levels interwoven in the preceding narrative: the sector, the firm, and the city?

Into the 1890s, specialty production was substantial, expanding, and ubiquitous across the American manufacturing belt. Most prominent in the older eastern cities of the first and second rank, batch trades (other than styled textiles) also had a notable presence at midwestern centers of agro-processing. They kept technically current and active in patenting and, with the exceptions of tailoring and dressmaking, were anything but dead-end holdovers from the age of handicraft. Consistently the nation's leading employers of skilled labor, they possessed a magnetic appeal for immigrant craftsmen and entrepreneurs. Furthermore, as a group, these trades developed into impressive engines both for creating jobs and for adding value to raw materials and intermediate products provided them by bulk enterprises. Their efforts were not simply labor-intensive or capital-intensive, but instead featured a differentiated, productive meshing of skill and investment. Finally, whereas throughput leaders offered the managerial corporation as their organizational legacy, specialists built institutions that reached across firm boundaries for collective ventures in marketing, education, banking and insurance, labor relations, politics, and proprietary sociability. Striving toward forms of collaboration within the wider competitive maze, largely denied the oligopoly option, they mapped their own paths toward product diversity and specificity, rejecting price competition as the ultimate determinant of sales and profits.[34]

How might we then conceptualize specialist sectors? If classes of durable and nondurable goods can be taken as a starting point and considered in relation to entry costs and the extent of markets, a rudimentary profile of shared and distinctive characteristics among groups of batch producers can be framed. At one extreme in durable goods lay the production of what Stephen Meyer has called the "big stuff,"[35] heavy machine tools, ore-crushers, cranes, locomotives, steel ships, prime movers, Fourdriniers, and printing presses—the province of Corliss, Sellers, Baldwin, Allis, Hoe, Cramp, Pusey and Jones, and Richards' San Francisco hydraulic works. Here entry costs were high by 1890, in terms of both capital and accumulated knowledge, and the number of firms often fewer than a dozen. Here too one would find, on the consumer market side, silverware manufacturers like Gorham or piano builders like Steinway and the other Baldwin (in Cincinnati). All dealt with a restricted clientele, high expectations of quality and performance, and the likelihood of wide cyclical swings in demand. These businesses are here described as "integrated anchor" enterprises, anchoring both a specialty trade and, at times, an industrial district's auxiliary firms.[36]

The same swings and expectations confronted batch producers of durable goods for whom entry costs were more modest, but who had far more potential customers to offset the larger number of rival firms. Makers of lathes and planers, shop tools and instruments, furniture, lamps, or upholstery and carpets could commence business with capitals ranging from a few hundred to perhaps twenty thousand dollars, depending on the availability of leasable quarters, used machinery, and trade credit from materials suppliers. So too could manufacturers of certain consumer nondurable goods: most printers and publishers, fashion textile weavers and knitters. Few nondurable sectors had very high entry costs (e.g., machine lace), but several had quite low barriers (e.g., saddlery, jewelry, apparel, or job printing). Together, such firms occupied economic spaces substantively different from those filled by the integrated anchors. They created sectorally varied webs of interdependence, contracting, and contact to accomplish the production and marketing of finished goods, along with institutions to provide equally varied collective services and foster solidarity. As their networks helped create and capture the external economies central to industrial districts' vitality, these enterprises constitute the class of "networked specialists."

This division of batch producers entails considering a third group, businesses that supplied diverse inputs and services to networked specialists and in many cases to integrated anchors.[37] These "specialist auxiliaries" included independent foundries and machine shops, tool and die works, pattern makers, versatile yarn mills, textile finishers, Jacquard card cutters, electroplating and enameling shops, furniture hardware makers, lithographers, and stereotypers, among others. Entry here was cheap to modest, yet endurance depended crucially on delivering quality and profitably pricing contracted tasks. Bulk processors might well supply specialists with steel, copper, lumber, or newsprint, but a constellation of auxiliaries was no less significant to the dis-integrated structure of production in shipbuilding, publishing, and fashion fabrics.

One additional concept is necessary to set these sectors into motion—uncertainty. Uncertainty differs from risk in being inestimable. It floats like a specter over enterprises, indifferent to calculation, insurance, and investment. Yet as recent theorists have suggested, uncertainty itself varies with market structure, technology, and other elements of the business context, which suggests that it may well be differentiated by industrial format and sector.[38] Throughput giants initially acted to cope with uncertainty through investments and strategies to control their environment that instead intensified competition (adding to uncertainty) and helped precipitate the great merger movement that finally brought greater stability to oligopolized sectors.[39] Specialists moved individually and collectively to *adapt to* uncertainty, rather than attempting to control markets, labor, or technology directly. Many firms sought to limit inventories of finished goods, to fund growth from retained earnings, and to utilize varied configurations of personalism to foster trust-centered relations.

Yet there were also class-specific options. Integrated anchors employed at least three specific tactics. Facing both seasonal and long-cycle swings in demand, they tended to prosper at sites where laid-off workers could readily find alternative employment until recalled and where additional skilled operatives could be recruited in boom periods. Such enterprises also cultivated direct and intensive working relationships with clients, at times in effect codesigning products with them, and regularly provided after-sale advice, service, and repairs or replacement parts, so that continued satisfaction would ensure later orders. In high-end consumer durables, this involved custom work as well as the early creation of firm-owned or licensed dealerships. Third, though evidence is spotty, the small number of firms made active collusion on bids, price lists, and contract terms attractive and informal defense of trade practices simple. Rarely were formal sectoral institutions deemed necessary.[40]

By contrast, networked specialists defined the heartland of trade-organizing activity. Though the very largest and smallest firms opted out at times, it was in these sectors that clusters of entrepreneurs widely fashioned instruments for sectoral representation and regulation. Whereas anchors routinely sponsored their own apprenticeship programs, for example, networked firms more often established schools and, later, labor exchanges. Inside the firm, it was in this class that early and sustained interest in restructuring production, chasing product costs, and retooling selling methods materialized, for such enterprises had fewer chances at the cost-plus-profit or customer-negotiated contracts integrated anchors favored.[41] However, networkers at the lowest end of the entry cost spectrum struggled with more severe uncertainties. As mushroom rivals could start up at any time in jewelry and apparel, firms tried to preserve style secrecy, pushed workforces to accept staggering overtime hours "in the season" to complete orders and deliveries before knockoffs appeared, and moved toward increased divisions of labor and skill dilution. These tactics, however, entailed greater dependence on auxiliaries for skilled services (in jewelry—enameling, plating; in clothing—contract buttonhole and cutting shops), and on specialist suppliers for components (findings, drilled stones, buttons, trim).[42] Auxiliaries and suppliers, of course, served garret competitors as well as contracting jobbers who adopted putting-out roles. In this context, institution-formation generally failed to realize its objectives. One implication of these trends led toward the sweatshop and, in apparel, toward a near-total displacement of custom tailoring, dressmaking, and millinery. Another vector drew clothing firms away from styled and toward staple goods,[43] and a third featured various forms of niche hunting (in jewelry, religious goods and lines for fraternal orders). Clearly low-end networked specialists held minimal capabilities to adapt successfully to uncertainty.

Auxiliaries may be separated into two divisions: those firms working on goods delivered to them by others and those making specialty intermediates or components on contract or for stock. Some were service providers (dyers, electroplaters) and the rest producers (machine shops, lithographers), but all confronted a unique uncertainty—that their specialty could readily be added to the capacities of successful clients who passed the scale threshold at which a captive facility could be utilized sufficiently to offset its capital and operating expenses.[44] Spreading a firm's contracting among many customers was the obvious response. Even if some were slow payers and others obsessive about quality results, that was far less hazardous than becoming largely dependent on orders from two or three companies. This perception had a locational correlate; auxiliaries had the best chance of success when sited in proximity to large clusters of networked enterprises.[45] Where entry was relatively inexpensive, price shaving and rotten estimating by rivals represented another uncertainty, to which one response was the circulation of customary rate schedules that new starts might adopt. Figuring metal casting by the pound, machining or woodcarving by hourly charges, or dyeing by the square yard refocused clients' attentions on quality, though pricing issues leaped forth in depressions.[46]

Auxiliaries depended on their clients' good fortunes. If the foundations of their primary sector were imperiled, ruin promptly loomed. Hence, where feasible, auxiliaries sought to cross trade boundaries, a hedging strategy that mandated learning on the firm's part yet could provide indirect assurance of longer-term vitality. In metalworking, machine shops could serve an enormous range of trades, including bulk plants needing quick replacements of machine parts. Brass founders mixed marine work with plumbing supplies, whereas some iron foundries sought flexibility by bidding on brass work. Electroplaters bridged jewelry and hardware; dyers stretched from yarn packages to piece goods made of varied fibers and blends, knitted and woven, each of which demanded different handling. Operating in a climate of doubly derived growth,[47] durable specialist auxiliaries sought to maximize scope as a defense against uncertainty.

Following the courses of diverse batch and custom manufacturers from the second Cleveland administration into the 1920s will be a principal task ahead. Yet these firms and sectors were urban creatures as well, and their relation to the industrial configurations of American cities cannot be ignored. Grand Rapids and New York had few anchor companies, Philadelphia and Providence substantial presences among all three groups, and other centers different mixes. Moreover, sectoral concentrations of batch manufacturing and the local significance of specialty versus routinized production varied dramatically. How might we think usefully about these patterns?

Maverick urbanist Jane Jacobs provides important clues. Over twenty years ago, in a provocative comparison of England's Manchester and Birmingham, she highlighted the surface efficiencies of the former's cotton goods production system and the putative waste of the latter's congeries of manufacturing "oddments." Then, noting Birmingham's early mix of leather, hardware, and tool trades, she observed:

> In the seventeenth and eighteenth centuries, the city had enjoyed a large trade in shoe buckles, but the shoelace put an end to that. A rising button industry had more than compensated for the loss. Some of the button makers used glass decoratively and this afforded opportunity to makers of bits and pieces of colored glass who, working from this foothold, had managed to build up a considerable glass industry. In the nineteenth cen-

tury, Birmingham was also making, among other things, guns, jewelry, cheap trinkets and papier-mâché trays. The work of making guns afforded opportunities for making rifling machines and other machine tools. . . . At the time of all the intellectual excitement about Manchester, nobody was nominating Birmingham as the city of the future. But as it turned out Manchester was not the city of the future and Birmingham was.[48]

Why? From Jacobs' perspective, in its sloppy, uncoordinated way, Birmingham was accomplishing an immense range of "development work," that "messy, time- and energy-consuming business of trial, error, and failure" that creates new possibilities, new goods, fresh urban "exports" whose returns reinvigorate the creative city. "In effect, [Birmingham] contained a great collection of development laboratories. This fact was not obvious because the 'laboratories' were also doing production work. Viewing the city's economy as a whole, one can think of it as a great, confused economic laboratory, supporting itself by its own production."[49] Unlike Manchester, the center of productive routinization, Birmingham was a mess, but a fermenting, portentous mess nonetheless, in which serial interactions among sectors produced seemingly endless novelty.

Picking up this theme in a later work, Jacobs added that "these tight-packed bunches of symbiotic enterprises . . . have always been the strengths and wonders of creative cities . . . [T]he huge collections of little firms, the symbiosis, the ease of breakaways, the flexibility, the economies, efficiencies and adaptiveness are precisely the realities that, among other things, have always made successful and significant import-replacing a process realizable only in cities and their nearby hinterlands."[50] Jacobs here offered a vision of supple urban economies generating momentum toward a structured versatility that could repeatedly rebore regional industrial engines so that technological and market shifts energized rather than debilitated local capacities. For such cities and city-regions, rich webs of connectedness and associated institutions were as critical as individual innovations, and conversely, placeless abstracted decision making could be as disastrous as dour commitments to resolutely refining routine. From this angle, America's Birminghams were those places where industrial flexibility either reigned or intersected creatively with bulk capacities. Yet these urban centers, like specialty sectors, were not all cut from the same templates. What sort of conceptual net can be cast that will enclose and sort them sensibly?

Largely to provoke discussion, a four-part distribution follows. Consider these process labels for cities featuring notable activity in specialty manufacturing trades: interactive, parallel, derivative, and narrow focus. Interactive cities are Jacobs' ideal, true Birminghams at which intra- and cross-sectoral relations engendered novelty both technically and in institutional forms. Such cities were incubators for wide-ranging product innovation; they fostered contacts among specialists in adjacent and distantly related trades and thus yielded unexpected complementarities. Here, American candidates for tentative inclusion are Philadelphia, Worcester, and Providence, plus perhaps Chicago and, later, Los Angeles. By contrast, parallel process cities possessed several prominent specialty trades, but their intersections were thin or nonexistent, limiting their generative potentials. Here the deep separations between New York's clothing, printing, and jewelry industries are exemplary. Each sailed along on its own, but they rarely had useful points of connection. A similar case might be made for jewelry, leather goods, and electrical products at Newark or specialty tanning, machinery, and shipbuilding at Wilmington, Delaware. In each case, although individual sectors prospered, their intersections were barely noteworthy and serendipitous creativity (as at Birmingham) rarely materialized.

Third, there were cities in which such specialty production as appeared was derivative of or secondary to dominant bulk manufacturing sectors. Though William Lodge managed to sponsor machine tool entrepreneurship and innovation at Cincinnati and complementary metal-forging machinery builders gathered around Cleveland's iron and steel complex, in most such cities specialists attended to either local demand (printing) or bulk processing's needs (foundries and machine shops). They were present but not potent, unless the leading bulk sectors faltered, as at Cincinnati. One might add St. Louis, Pittsburgh, Buffalo, Milwaukee, and Minneapolis to this fraternity. Each had nontrivial elements of specialty manufacturing by 1890, but none of them, with the exceptions of Pittsburgh glass and St. Louis stove makers, were then major city-region forces.

Last, "narrow focus" refers to those smaller cities for which one or two lines of batch manufacturing provided significant growth vectors. Whatever their prowess or potential, by the 1890s these cities were off the track that seemed to lead toward major metropolitan status, yet they were neither company towns nor simply extractive, commercial, or administrative centers. Instead, as at Grand Rapids, their manufacturers fashioned and deepened a flexible capacity in furniture, metalworking (Connecticut's silver and brass centers), or textiles (knitting towns in southeastern Pennsylvania). Their relations with major cities have been barely explored, much less the networks of contact among them (Grand Rapids, Jamestown, Rockford, and Evansville in furniture, for example). Both matters deserve closer scrutiny.

If this distribution of centers linked to late-nineteenth-century batch production is at all interesting, these cities' subsequent interactions with the three sectoral classes may provide a lens through which to view the radically different urban and industrial trajectories (toward fertile prosperity or stunning collapse) they traced across the twentieth-century American economy. For now, several basic city-centered observations must be appended. The categorical designations suggested above cannot be taken as static. With the decay of its metalworking sectors, Paterson (New Jersey) descended from a parallel-sectors city to a narrow-focus place where only silk meant money. By contrast, after 1910 Grand Rapids developed sizable metalworking capacities and regional outliers that reoriented its wooden furniture firms' propulsive thrusts.[51] Metamorphoses were possible in industrial capitalism, in addition to the familiar rises and falls.

Second, with Jacobs, it is essential to underscore the point that city boundaries were not stable guidelines for the analysis of urban industrial activity. In large urban areas multiple neighborhood-sized industrial districts coexisted with regionwide networks among producers (as textiles and metalworking in Philadelphia districts overlapped with shipbuilding from Kensington south to Wilmington along the Delaware). Last, American state and federal policies have thus far been largely background noise, but after 1890 statutes designed to deal with bulk sector accumulators also impinged on specialists, with uncertain implications for batch and custom firms and for the cities they inhabited and invigorated. Tracing the development of the three classes of firms and of sector relations in particular urban contexts, as the national economy slides into depression and through it to renewed prosperity, is the next agenda item. The first stop is Chicago.

# *PART III*

## DEPRESSION AND ADVANCE, 1893–1912

*Chapter 7*

# CHICAGO AND GRAND RAPIDS: PALACE CARS AND FURNITURE

HORDES OF middle-class Americans entrained for Chicago in 1893 to tour the Columbian Exposition's triumphalist exhibits and savor the adjacent Midway's exotic delights, even as gold outflows and shrinking credits signaled another spasm of the economic and social order that the White City celebrated. En route, thousands of travelers ate, relaxed, and snored in rolling domestic spaces—strings of dining, parlor, and sleeping cars constructed at the Pullman works in Chicago's far southeast corner, cars attended by an immense but dispersed African-American workforce managed from company headquarters in the city's central business district. Polar opposites in industrial terms yet powerfully linked to the nation's rail network, the Pullman shops and Chicago's slaughterhouses competed with the exposition as visitor destinations, each representing grittier realities beyond the shimmering exhibit halls or the Midway's Street in Cairo. The stockyards illustrated Chicago's command over perhaps a million square miles of agricultural terrain in the Great West, documented its centrality to feeding the East, and exhibited the Big Four packers' technological and organizational mastery of bulk processing's "disassembly" line. By contrast, George Pullman's modern industrial town exemplified the paternalism of a proprietary capitalist, the productive versatility of an integrated anchor, and the rising significance of custom and batch trades within the regional industrial structure. Indeed, between the 1880s and the second decade of the twentieth century such trades would supplement, though not supplant, Chicago's core agro-processing, steel, and machinery sectors, becoming growth nodes built upon achievements in bulk manufacturing and mass production.[1]

Here, an assessment of two of Chicago's industrial specialties will introduce a gradual eastward trek, which will allow us to reexamine fine furniture making at Grand Rapids and, in the next chapter, the dynamic machine tool district in Cincinnati's Mill Creek valley through 1913. During the two decades after the Columbian Exposition, specialist sectors flourished in the same Midwest whose industrial history had long been tied to the routinized production of meat, steel, and flour. Thereafter, returning to specialty production's bedrock, we will visit the new electrical equipment industry's leaders as well as now-familiar trades in Providence and Philadelphia.

## Chicago and Specialty Manufacturing, 1893–1912

Exposition-bound travelers gliding west, north, or east along steel rails tracking the gentle hills of the upper Midwest may well have encountered Chicago as an "unexpected metropolis," but its lakeside elite had cherished building a great city for generations before 1893's big show. From midcentury, boosters trumpeted the regional resources and transportation vectors that "naturally" made Chicago the crucial transition point between the settled East and the open West. Surmounting its disastrous 1871 fire with frenzied reconstruction, the city swelled past half a million in population by 1880. Chicago's Board of Trade managed grain markets of international significance, while its railway corporations reconfigured the economics of space and agriculture across the West, hauling McCormick reapers and goods from Montgomery Ward to rural destinations, then returning laden with wheat for milling or cattle for slaughter and shipment in refrigerated cars to Memphis or Boston.[2]

Yet even as connections solidified among agriculture, extraction, the railroads, and Chicago's bulk-processing sectors, a second set of firms making specialty goods drew on readily available raw materials, growing urban markets, and the city's omnidirectional rail network. For example, four piano and melodeon builders' shops opened in the late 1850s, and Cincinnati's J. M. Brunswick added a Chicago factory in 1864, in part to supply "the city's many gambling establishments" with ornamented billiard tables. Increasing sales of eastern keyboard instruments encouraged dealer Wallace Kimball to start manufacturing reed organs and pianos on his own account in 1879. Local artisan cabinetmakers likewise took advantage of Chicago's central role in the lumber trade. Fifty modest furniture factories were in place by 1870 (employing over 1,100 men), many specializing in upholstered parlor pieces whose hardwood frames were covered with durable eastern haircloth and Brussels carpet (likely from Philadelphia).[3] The great fire wiped out dozens of them, as well as scores of lumberyards and much of the main business district where leading furniture firms had located their showrooms.

As the "transportation network remained relatively unharmed," materials for rebuilding soon flowed in. Though insurers' defaults hampered many proprietors' efforts to reopen, most proved able to celebrate their own and the city's revival with furniture exhibits at the lakefront Interstate Industrial Exhibition in 1873, shortly providing a new set of downtown hotels (Palmer House, Grand Pacific, Briggs House) with the "ornate and costly furnishings" their commitment to opulence required. By 1880, a map distributed to the trade showed over 125 furniture establishments located in central Chicago, most clustered either along rail lines or within short distances of six major freight depots. As elsewhere, auxiliary firms sprouted, offering springs, curled hair (for stuffing), fringes and tassels, veneers, hardware, and varnishes. This furniture concentration was one element within the West Division "industrial belt," which also held foundry and metalwork specialists and other wood-using firms (sawmills, box and barrel makers). These deployments represented the second phase of Chicago's shift, in Jane Jacobs' terms, from "depot" to "manufacturing city" through "import replacement." In the 1850s, area enterprises had begun making for local consumers "a very large range of the common city-made goods of the time," substituting them for many eastern items. A generation later both the stockyard masters and hundreds of specialists produced with a regional or national clientele in mind, moving into an export-generating phase, which at Chicago yielded "explosive growth."[4]

During the 1880s, Chicago's population doubled to just over a million, a far faster growth rate than older, eastern cities charted. More striking, its industrial workforce rose 133 percent and aggregate earnings tripled during a decade when national price levels dropped roughly 20 percent.[5] Most remarkable, despite deflation, the value of Chicago's manufactures increased 159 percent (to $627 million), value added leaped 260 percent, and reported capital investments quadrupled. Explosive growth, indeed, yet how did bulk and specialty trades share in it?

Consider the four largest sectors in each format for 1880 and 1890. The leading bulk industries—slaughtering, basic iron and steel, tanning, and lumber—delivered 44 percent of Chicago's product value in 1880 and 43 percent in 1890 (by which time brewing and factory men's clothing had replaced the fading lumber and static leather businesses as leaders). By contrast, in 1880 the top specialists—foundries and machine shops, furniture, builders' millwork, and printing—generated only 10 percent of production by value, a share that rose over the decade to 14 percent (by which time railway cars displaced millwork).[6] Chicago's agricultural and extractive processing trades remained dominant, but specialists were growing more rapidly. Indeed, when value-added figures are assessed, the leading bulk and specialty sectors' shares rest close together at both dates (bulk—22 and 24 percent; specialty—18 and 21 percent). Moreover, a disparity in value added per worker evident in 1880 diminished rather than grew (bulk/specialty, 1880: 1.44; 1890: 1.08).[7] Though specialty manufacturing was far more important at Philadelphia, it represented a rising force in Chicago's diversified industrial structure, expanding alongside stockyards and steel mills into the 1890s.[8]

Just as Philadelphia's Baldwin Locomotive and styled fabric mills exemplified anchor and networked specialist firms, at Chicago Pullman and the furniture trades occupied comparable technical and market positions. Both exhibited at the Columbian. Pullman displayed a scale model of its industrial town and several full-size "palace cars" in the transportation hall, and 32 of the city's 250 furniture houses presented samples of their handiwork in the manufacturers' building.[9] George Pullman's company is most often remembered for three things, each of which evokes ambivalence. It provided the funeral car that carried Abraham Lincoln's body to its final resting place in Springfield, Illinois, in 1865, generating a publicity coup for Pullman's fledgling firm. Its paternalistic town was the flashpoint for the 1894 shopmen's strike that brought Eugene Debs and his American Railway Union first to national prominence, then, respectively, to jail and ruin. Its sleeping car service became the largest employer of black workers in America, offering "servile" jobs that also "contributed directly to the creation of a black middle class."[10] Here, however, the Pullman Company's specialty manufacturing capacity is of immediate interest.

In its first quarter-century (1866–90), Pullman revolutionized long-distance railway travel through a series of innovations and a unique marketing strategy. The Pullman shops built palace cars to railways' varied specifications, then negotiated contracts for their carriage, retaining car ownership and responsibility for operations and maintenance. This tactic saved railroads capital investment and repair expense while providing them with attractive accommodations kept to high standards through oversight from Chicago. The roads collected regular fares from palace car occupants and forwarded surcharge fees to Pullman. Following the Lincoln funeral, the company placed forty-eight cars with five railways by the end of 1866. George bought out his partners, incorporated, and quickly sold $1 million in shares to Chicago businessmen and railroad executives, thus funding the 1870 construction of the company's Detroit erecting shops, which employed a thousand men to fabricate over a hundred luxury cars annually. In the interim, with Andrew Carnegie's aid, Pullman struck deals with the Union Pacific and the Pennsylvania, extending his service network to both coasts.[11]

Technical and design innovations soon multiplied. Kitchens appeared as palace car features in the late 1860s; but shortly, functional specialization ruled, as the works diversified to fashion separate sleeping, dining, and parlor cars whose "interiors were made as different as possible." Linked together they became rolling luxury hotels, one of which hauled 130 affluent travelers from Boston to San Francisco in ten days during 1870. Pullman introduced day-coach parlor cars for shorter journeys in 1875, the year it commenced constructing passenger and baggage cars for direct sale to various lines. By 1881, annual earnings matched the firm's capital stock at $3 million, and assets reached $18 million. The extent and diversity of its production moved Pullman to purchase 4,000 acres adjoining Lake Calumet in south Chicago and to invest $5 million in erecting workshops and housing (1880–82). Symbolically, the company acquired the Centennial Corliss engine for $130,000 and installed it as the shops' main driver behind plate glass windows visible to passengers arriving in the city along the Illinois Central's tracks. The new plant employed 2,700 workers by the mid-eighties and 5,500 at its 1893 peak, when their average annual earnings reached $613. Addition of a streetcar division and technical refinements dotted the 1880s: car heating through subfloor coils, gas lighting (electric in the 1890s), and safe and secure vestibule connectors between cars linked by "powerful spiral springs." These cycles of innovation overmatched two of Pullman's three competitors in the field; by 1893 only the Vanderbilt-backed Wagner Company remained as a rival.[12]

In the exposition year, "[f]inancially, the Pullman Corporation was among the soundest institutions in the United States." It had long paid shareholders a solid 8 percent annually on par values, which accounts for stocks quoted at twice par in contemporary share trades. Despite such payouts, residual profits added $2.7 million to company

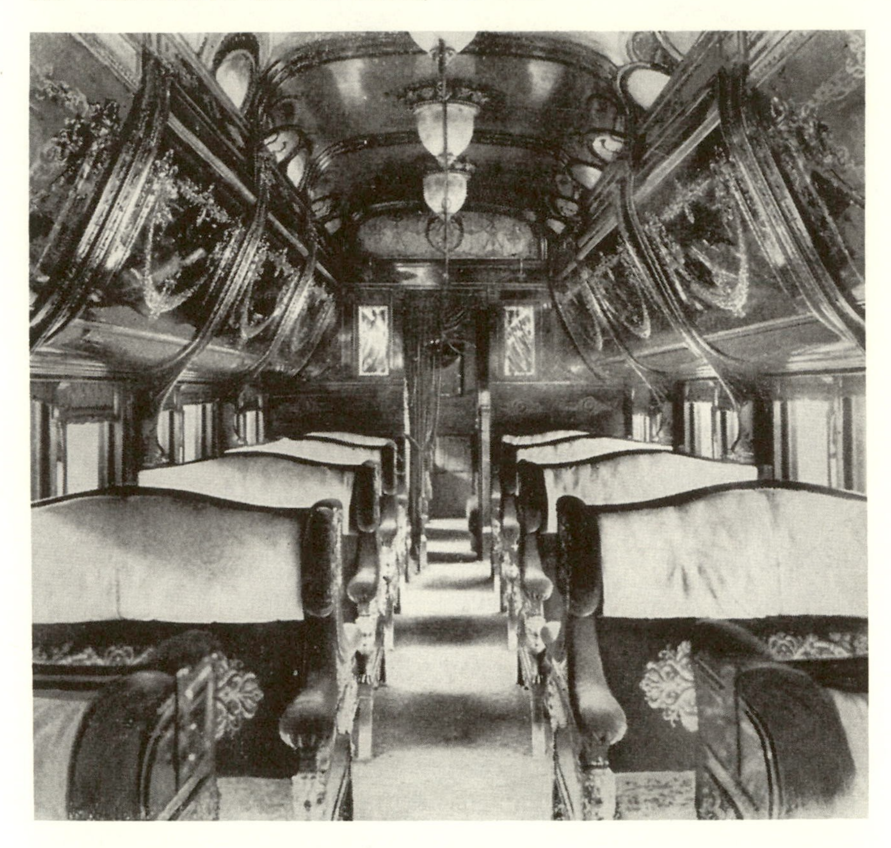

12. Highly ornamented interior of a Pullman car on the Pacific Coast Limited, 1896. Courtesy of Hagley Museum and Library.

reserves in 1881 alone. Thus Pullman could make debt-free plant construction and machinery purchases and "finance large orders which railroads could not pay for outright." Earnings in 1893 mounted to $11 million ($2.1 million to dividends), and corporate reserves passed $25 million. Depression conditions would squeeze car operations as ridership decreased and would hammer the workshops as car orders simply evaporated, but they could hardly threaten the company's long-term viability.

Two specific elements of Pullman's successful market and labor practices accentuated the crisis that blossomed into the famous 1894 strike. First, the shops had become heavily committed to fulfilling outside contracts for batches of passenger and freight cars; only 36 percent of the workforce constructed or repaired Pullman cars in 1891. Second, the industrial town, designed to attract and hold thousands of essential skilled workers, shifted from an asset to a liability when general market orders collapsed. Housing about half of the shops' 5,500 employees in 1893, the town was a component of a labor strategy keyed to trade fluctuations. In "normal" cycles, Pullman residents would form the core labor force; workers from nearby areas could be added or subtracted as work levels varied. However, ownership and management of the town built rigidities into an otherwise flexible stance toward financial and labor requirements. Pullman simply did not envision an 80 percent workforce reduction that slashed the active complement to 1,100 by November 1893. Though the firm called in hundreds of cars for repairs, closed its Detroit plant, and secured $1.4 million in low-bid contracts during the winter, increasing local employment to 3,100 by April 1894, George Pullman foolishly refused to reduce or forgive rent charges for workers in his town. Nor did he revise or discard a recently installed piecework system whose rate-setting practices "assumed a repetition which usually did not exist . . . due to the diversity of work performed in the shops." Key features of a successful expansion strategy, Pullman's direct car sales and housing provision set up a conflict escalated by the founder's inflexibility.[13]

The three-month strike proved a disaster for many of the workers involved and for George Pullman personally,[14] but the Pullman company recovered its momentum. Employment at the shops rebounded to 4,000 by 1897, the year of George Pullman's death. His will listed 58,000 shares of company stock as his largest asset; and one of its provisions set aside $1.2 million to inaugurate a free school of manual training for those living or working in Pullman.[15] By the time the will completed probate in 1903, the value of Pullman's stock had doubled amid continued corporate prosperity. A financial move and a technological initiative had spurred this surge. In 1899, the firm issued $4.4 million in shares to the Wagner Company's owners and absorbed its last competitor. The next year, its shops commenced using steel frames in car building, then broadened into all-steel construction by 1908, investing $5 million in new shops and machinery.[16] While diversifying its lines of passenger, freight, and redesigned sleeping cars, Pullman also introduced its interurban "Red Rockets" in 1907, "the prototype of all later cars."

Ending the last vestiges of the founder's experiment, the company sold off its town's homes and apartment blocks in 1907, some to Pullman workers (who had first choice at moderate prices, with mortgages provided), but most to outsiders and speculators. In 1908, as employment reached 10,000, workers' average earnings of $823 "provided a higher standard of living and greater security" than at any point since the '93 crash. Conservatively managed by Robert Lincoln, its president after 1901, Pullman built its capitalization to $120 million by 1910 and its car construction and repair workforce to 15,000 three years later. Standardization of some components across models and efforts to secure larger, batch orders brought greater system and cost efficiency to the shops, reducing the skilled proportion of the workforce to 40 percent, as Pullman (like Baldwin) anchored one of the era's most demanding specialty trades.[17] Unlike Baldwin, it held an effective monopoly and sustained its profitability into the 1950s.[18]

Progressive Era Chicago held other large-scale, nationally significant metalworking specialists, like the Crane Company in plumbing supplies, the great Fraser and Chalmers machinery shops, and the Lake Michigan yards of the American Shipbuilding Company.[19] However, rather than examining their practices and achievements, their contests with organized labor, their associations, mergers, and missteps, we will move on to consider the furniture trade's array of midsize firms, auxiliaries, and organizations. In structure and process, Chicago woodworking and Philadelphia textiles resonate provocatively. As sizable, highly diversified sectors within major metropolitan manufacturing complexes, each was populated with hundreds of proprietary firms in considerable part initiated by immigrant entrepreneurs and staffed by thousands of skilled workers. Both displayed disaggregated and specialized production networks, spatial concentrations within industrial districts, a profusion of stylistic variations, constant attention to technological and market shifts, and ambivalent, recurrent conflict with craft labor unions. At both sites, manufacturers formed subsectoral and "umbrella" trade associations to address collective problems. Both figured as network specialist concentrations, but they operated within different processes of urban and specialty manufacturing development: interactive at Philadelphia and derivative (from bulk trades) in Chicago. Indeed, the Chicago furniture industry's emergence as the nation's largest complex flowed not only from the city's early and massive lumber trades but also from its stockyards, whose by-products (bones, hair, hides) were locally processed into key raw materials—glue, curled hair for stuffing, and leather for upholstery.[20]

In 1890, the city's furniture sector included 250 companies marketing about $13 million in finished goods (double Grand Rapids' output). A trade journal estimated that two-thirds of the firms were "composed of men of foreign birth and training," adding: "Not only are Germans, Swedes, Norwegians, and Frenchmen at the head of these concerns, but Germans, Frenchmen, and Scandinavians, with a few English cabinetmakers, are the technical heads of the factories. The only industrial relation native Americans have to this vast industry is that of mechanics in the lowest grade of its manual labor."[21] Typical earnings reached $2.50/day in the early 1890s, dropped over 10 percent during the depression years, then rebounded slightly to $2.30 in 1898. Workers had mounted a series of strikes for the eight-hour day in 1886 (defeated in Haymarket's aftermath), but the unions they created endured. During the nineties, the trade exhibited bifurcated labor relations—union shops running nine hours, nonunion firms, ten. Woodworkers remained relatively quiet until a turn-of-the-century strike wave, but upholsterers conducted a three-month walkout in 1896 to replace nine and ten hours at day wages with eight hours and piecework rates. With a fair piece-rate schedule, they expected to accomplish as much work in fewer hours without an earnings loss, but allied employers refused their demands and prevailed.[22]

Proprietors first organized in the 1870s, creating a Furniture Manufacturers' Exchange to monitor retailers' credit ratings, seek favorable insurance quotes, and discuss responses to craft unionism. During the 1880s, FME members drafted agreements "setting minimum prices to ensure a stable market," then initiated the Chicago Furniture Manufacturers' Association (1888) to contest railroad rates and confront unions. Dues based on member firms' employment levels supported a salaried secretary who negotiated with the lines, tracked legislation, and managed the "Labor Fund," a reserve for recruiting strikebreakers. Subsectoral associations soon multiplied, reflecting the trade's differentiated structure. As an 1899 commentator observed, Chicago firms "now confine themselves to the manufacture of a few articles . . . turned out in the greatest variety of styles." Thus emerged specialist associations in upholstered parlor furniture, parlor frames, chamber goods, tables, chairs, and case goods, most dedicated to establishing pricing minima to regulate competition.[23] Most important, in 1895, area firms duplicated on a permanent basis the twice-yearly style openings that anchored Grand Rapids' rise as "rightfully . . . the great western furniture market." The Chicago Furniture Exposition Association (after 1909 the Furniture Market Association) organized and publicized January and July exhibits for retail buyers. Lacking a central hall, firms either exhibited in their own shops and downtown showrooms or rented space in converted warehouse buildings near the Illinois Central station. In addition, larger companies marketed their styles through annual catalogs mailed to retailers. S. Karpen and Brothers, the city's most prominent maker, inaugurated a monthly house organ in 1897, along with direct-to-consumer magazine ads in the *Ladies' Home Journal, Saturday Evening Post,* and *Vogue* by 1900. Other firms bought display space in one or more of Chicago's furniture trade journals, whose publication helped swell the city's expanding printing industry.[24]

Stylistic diversity and product specialization defined Chicago furniture's market and spatial organization. By 1880, discrete districts coalesced: chair makers in the northwest, frame makers in the north, custom craftshops on the "Near North Side" close to concentrations of skilled German workers, chamber set factories to the southwest. Upholstered goods companies were more scattered, however; for them, some activity was recorded "in every ward of the city." The profusion of styles was most evident at larger firms, but all adjusted their lines as home furnishing tastes shifted from Eastlake-inspired designs to Queen Anne or Renaissance revivals and Turkish or French constructions. Ford, Johnson alone offered "3,000 varieties of chairs, rockers, and cradles" in the mid-1880s in various combinations of carving, veneers, or seat forms. The unrelated Johnson Chair Co. "advertised 1,000 patterns of office and dining chairs in more than 400 styles" at the turn of the century and provided Congress with new seating for the House and Senate. Though differences among styles might be minor, most had to be manufactured in batches when called for, as space and expense constraints entailed that only a small number of patterns could be stocked as regular inventory.[25]

Chicago furniture firms functioned within a network of specialist auxiliaries, contracting for lumber preparation, fine carving, die cutting (for embossing), veneers, a portion of designing, and, in upholstered goods, for frame construction.[26] Some devised "bridge" strategies, undertaking large orders for staple goods (for Sears or Ward) to level out seasonal variations in fashion-sensitive lines. Success here could create its own problems. When Ford, Johnson scored a Columbian Exposition contract for 54,000 chairs, tight delivery dates and the array of work in progress at its own shops forced the firm to subcontract much of the order to local and Wisconsin makers. One enterprise, A. H. Andrews, essayed an integrated production format. Its fate surely inscribed a cautionary tale for the rest. Having commenced with standard school desks and seating for theaters and meeting halls, Andrews focused its capacities on business furnishings in the 1880s, picking up on the wave of office building construction and leading corporations' growing white-collar staffs. By 1890, it offered a hundred varieties of office desks and employed 500 workers in its three city factories and an Indiana branch plant. Andrews worked on both custom and staple terrains. The firm created special departments for outfitting executive offices, banks, and churches with furniture, interior woodwork, and stained glass, and supplemented its schools division (which sold 50,000 standard desks yearly) with related items (bulletin boards, globes). To secure its supply lines, Andrews invested in woodlands and constructed its own lumberyards. To hold its bulky inventories, it built substantial warehouses. When depression retarded payments and collapsed the market, overextended A. H. Andrews entered bankruptcy in 1895. Creditors sold off several of its segments in 1896; its warehouse burned a year later. Reorganized as a much smaller specialist in bank furnishings, Andrews started over in the new century.[27]

Integration and expanded scope failed as a business strategy in styled furniture. Only in the 1920s did the Simmons Company, which made standard metal beds and springs, echo Andrews' approach; but in acquiring furniture plants, it sought to achieve repetition production within a narrow style range instead of mixing custom and staple lines.[28] Andrews' decline reinforced Chicago specialists' concentration on design novelty and technological innovation within their individual subsectors. Although the transition to machine woodworking had transformed the wood-shaping segments of furniture production by 1890,[29] Chicago enterprises created or adopted an array of fresh techniques in the following two decades. Greater precision in rotary veneer-cutting, which shaved a continuous sheet from a turning log, brought paper-thin hardwood veneers that seemed to solve the problem of diminishing forest reserves. Machine carving and die-press embossing simplified ornamentation on midprice and cheaper lines. S. Karpen's mechanics invented and patented an "automatic" multiple spindle–carving machine (ca. 1900–1902), using the pantograph principle to make eight duplicates of a decorative master "without the aid of an operator." Its originators consigned this device to a woodworking machinery specialist for production, as was done with an 1899 novelty tufting machine that relieved a crucial bottleneck in upholstering. Other Chicago firms created variations on seating spring sets for upholsterers, who also adopted new, specialized sewing machines designed for their trade. This second generation of technical novelties "substantially reduced the cost of labor while increasing production" yet "also made possible . . . a wide range of styles and products." At Karpen, Chicago's self-proclaimed "Makers of New Patterns," these shifts confirmed a commitment "not to sacrifice quality for speed" and facilitated efforts to "concentrate . . . the skills of their workmen on those processes best done by hand, such as joining, finishing, spring setting, and upholstering."[30]

Technical change also expanded the range of woods that could be used for furniture. High-speed cutters worked ultratough ash and birch; steam-heated dryers held warp-prone aspen and sycamore straight; cheap elm, gum, and oak soaked up new durable stains and were marketed as imitation mahogany and walnut or were used as bases covered by prestige veneers.[31] Fine hardwoods' growing scarcity and rising prices also triggered two other types of "technical" innovations. As hardwood costs doubled between 1900 and 1912, scores of Chicago firms making business, chamber, and novelty furniture experimented with willow, rattan, bamboo, and metals (brass beds, steel chairs), while their associations tried to establish oak as a worthy successor to walnut and a replacement for imported mahogany or teak.[32] Second, together with colleagues in other centers, area furniture men attempted to devise cost-finding systems that could rapidly document rising expenses and inform profitable pricing for their diverse styles.

During the 1890s, manufacturers were more concerned with managing men, machines, and designs than with shop systems and cost accounting, if trade journal articles are any indicator. Though comments on individual job ticket schemes and pleas for closer attention to figuring costs occasionally surfaced, extended series on tracking expenses and framing sensible prices regularly appeared only after 1900.[33] In 1907, a production engineer addressed an "indifferent" furniture manufacturers convention on cost analysis, but by their 1910 Chicago meetings, "'Cost of Production' was about the only matter discussed."[34] Not only had materials prices risen, hundreds of new firms had also entered the industry, swelling capacity, while both workshop errors and trade abuses menaced profitability.

One manufacturer, who usually added "not less than 5 per cent [to cost estimates] to cover things overlooked, unseen, and neglected," explained:

> There seems to me to be innumerable occurrences . . . creating expense that we are not inclined to consider when figuring. A car of lumber not quite dry, causing glue joints to open; a poor lot of glue, causing the same trouble; a lot of poor varnish, or unfavorable weather, [yielding] crazing, checking, sticking to the [packing] paper and otherwise creating cause for damage claims; defective mirror plates; breakage in transit and by carelessness of workmen; furniture refused and returned on account of damage caused by criminal recklessness of transportation companies, or arbitrary actions of unscrupulous dealers, causing us [either] to submit or incur greater expense of a law suit; breakage and damage in transit computed arbitrarily by customer at two or three times more than the true amount, and also arbitrarily deducted from the invoice with equally arbitrary notice that you can accept [reduced payments] or take your goods back.[35]

Each created uncertainties for which his 5 (or 10) percent add-on inadequately compensated.

Little wonder that manufacturers tried to install systems that could give them better financial and work process control, but too often "experts" unfamiliar with the industry applied approaches designed for bulk manufacturing. A furniture trade adviser observed, "When one considers the number of boneheads who are posing as business systematizers, it is not difficult to figure out why manufacturers have had very little to do with them."[36] Given the complexities and contingencies of production itself, the shop management, piece rate, and bonus plans devised by Taylor and his rivals seemed appealing but worked out poorly in styled furniture plants. One company officer, having studied Taylor's principles, commented on his attempts to apply them. "It is only after two years of more or less continued effort that I now give the problem up. . . . Piece work or premium system, either, cannot be applied to work, the *duration of which is short*. It is not worth while to keep track of a short job, because in so doing, *more is lost than saved*. . . . Furniture manufacturers would be glad of any practical method of placing *direct* incentive before the employees, but unless some 'flat rate' method can be invented to apply to all varieties and styles of articles made, miscellaneous to a great extent, I don't believe it can be used at a profit to anyone."[37] Such schemes might work in the steel industry, he noted, but their application to specialty furniture was both expensive and useless, given the constant change of products. Fashioning adequate accounting systems and more effective wage payment plans remained a project in 1912, if not a dream. Although some manufacturers had learned to work depreciation and selling expenses into prices, the rest continued to pay labor by the day and priced goods by calculating "prime cost plus percentages" for general expense and profit.[38]

If such puzzles troubled Chicago's furniture proprietors in 1910,[39] these manufacturers nonetheless had much to be cheerful about. In 1909, two hundred local furniture companies employed over 10,000 workers (97 percent men), a 20 percent increase over 1899, to produce furniture worth $20.5 million, up 43 percent in constant dollars.[40] Moreover, they had begun to resolve a pair of enduring regional difficulties. For years, a half-dozen freight consolidators had controlled the shipment of furniture (and much else) from Chicago to the nation. As it was in their interest to assemble sets of filled cars to each destination, they routinely delayed time-sensitive furniture deliveries for weeks until boxcars for Cleveland or San Francisco had been filled. In 1910, the furniture association cut through this tangle by incorporating the Chicago Furniture Forwarding Co. and opened its services to association nonmembers to build volume and reduce delivery lags. This innovation in coordinating transport was an immediate success and continued to operate into the 1980s. Second, in 1912, completion of the National Furniture Exchange of Chicago, a huge structure permitting tenants "to display their samples year round," remedied the long-rued absence of a central hall for furniture exhibits. An Illinois Central spur into the lower level of the building encouraged distant firms to rent space and ship carloads of samples directly to the exchange, thus establishing Chicago's claim as Grand Rapids' superior in furniture marketing.[41] That rivalry was not so easily settled, however.

## Grand Rapids: Proprietors' Solidarity and Style Leadership

Between 1890 and 1910, Grand Rapids matured as the nation's center of fine furniture manufacturing. Despite depressed sales in the mid-1890s and several poor seasons at the turn of the century, local companies' numbers rose from 31 to 56 in twenty years as employment swelled 80 percent to over 7,800. Attendance at style openings multiplied nearly tenfold during the 1890s, reaching 1,500 registered buyers at the 1900 shows and 2,400 in 1912. Entrepreneur Philip Klingman focused the markets by inducing local investors to erect three central exhibition halls (ca. 1899–1908) for incoming exhibitors, a group that included over 150 noncity firms by 1908. In tandem, James B. Pantlind constructed a vast hotel near the Grand River; others followed his lead in providing accommodations, though on smaller scales. As for auxiliaries, by 1910, in addition to various makers of veneers, furniture hardware, specialty machinery, and factory equipment, Grand Rapids' industries included nearly 50 foundries and machine shops that handled repairs and fashioned the metallic components increasingly designed into office lines. Furniture enterprisers gathered informally and for association meetings at the downtown Peninsular Club, where their noted (or notorious) bonds of affinity materialized in shared policies and practices that became the local industry's customs. As in Philadelphia, company owners created new banks (three between 1905 and 1911), both as sensible investments for earnings and to provide the working capital facilities crucial to a trade in which three to six months typically elapsed between shipment of and payment for goods.[42]

If exhibitions drew buyers and "captive" banks financed production, design novelty made the sales. After their Centennial embarrassments, chastened local firms recruited eastern talent, notably David Kendall, who worked at Phoenix for over two decades and after whom the sectorally sponsored design school (1928) was later named. Julius Berkey brought in A. W. Hompe, who created lines in Sheraton and eighteenth-century French styles, while others trained family members (e.g., Ralph Widdicomb) or promoted talented hand carvers who showed a flair for ornament and proportion. Yet Kendall "more than anyone else was responsible for giving . . . Grand Rapids such an obsession with period styles that, [ca.] 1895–1910, the gamut of styles of all ages was introduced and in large part discarded."[43] Confident of design excellence, local firms welcomed outside companies to the expositions, for the latter's samples either paled by comparison in case goods lines or filled market niches not occupied by Grand Rapids makers (cheaper furniture, curiosa, upholstered goods). Incomers complained at times about competition with their hosts, the expense to ship samples, or the high rental charges and joined some retailers in pressing for one annual opening; but they returned in rising numbers, for expositions were plainly "a good thing for the manufacturer who is not represented in the centres of trade."[44]

13. A classic, heavily carved, period-style sideboard from Grand Rapids' Phoenix Furniture, ca. 1912. Such pieces were made to custom orders. Courtesy of Hagley Museum and Library.

Middle- to high-grade household and commercial furniture constituted Grand Rapids' two main trade divisions, and firms rarely crossed the lines between them. The larger household group and its retail clients dominated the expositions, but office and store furnishings firms were far from a negligible factor. Fred Macey Co. (later Macey-Wernicke), Grand Rapids Show Case, and Gunn Desk echoed Andrews' mixed output strategy in Chicago, but not its grandiose integration. They filled custom orders for corporate offices and specialty store fixtures while selling longer runs of sectional bookshelves, display cases, and filing cabinets to railway, insurance, and industrial corporations and retail chains (e.g., Woolworth and Kresge). By 1910, nine such firms employed just under 20 percent of the city's furniture workers.[45]

Men who had moved "up from the bench" to shop superintendence most often started new firms in Grand Rapids. A reputation that drew

14. A period-style walnut bed from Phoenix, shown in the 1912 Market Exposition. Depending on orders, such models would be made in lots ranging from four units upward. Courtesy of Hagley Museum and Library.

the support of other proprietors proved no small aid to their initial prospects. Records from the F. L. Furbish shop (thirteen–fifteen workers in the late 1880s) indicate that subcontracted orders from major companies like Berkey and Gay or Nelson Matter represented from a quarter to a half of its monthly sales. Other local contacts proved equally valuable until a new start could achieve some impact at the expositions. In Furbish's case, sales of telephone cases to the new local company, cabinets to the First National Bank, and furniture for Pantlind's hotel helped its proprietor maneuver through his first decade. Such relations laid the basis for reciprocity and solidarity among Grand Rapids' furniture capitalists, facilitating the diffusion to lesser firms of trade practices that sectoral leaders initiated.[46]

Reinforcing the market expositions, a set of customs and private compacts regulated competition among area firms, managed their labor market, and assisted their growth. By 1900 Grand Rapids furniture makers had silently agreed not to show new lines before exposition openings, to decline selling samples at the exhibits, to refuse special pricing for large buyers, and, responding to retailers' wishes, to refrain from attaching makers' names to goods (thus rejecting the brand-name strategy that Chicago's Karpen adopted). They also set common day rates for labor, pleged not to poach skilled workers or designers from one another, and moved collectively to shorten work hours in weak seasons (to avert layoffs and workers' dispersion to other sites). In later years, their banks supported home mortgages for furniture plant employees, thus tying them more solidly into the regional production complex. The Employers Association (1905) established the Furniture Workers Free Employment Service to place incoming operatives, monitor grumblers, blacklist unionizers, and find new jobs for those displaced by trade reverses or business failure. Collectively their association not only contested railways' rates but also—by 1900, with the assistance of a Muskegon boxcar builder—persuaded them to carry over five hundred freight wagons specially designed for minimal-damage transit of Grand Rapids goods. In 1905, the "Furniture Men" formed a separate organization to establish a shared trademark identifying "authentic" Grand Rapids–made furniture and to litigate against pretenders and fakers. Given that these customs, agreements, and collective efforts intersected with fashionable designing, product quality, and the market expositions, Grand Rapids' centrality to the national furniture trade is hardly astonishing.[47]

When a cluster of specialty manufacturers strives to focus competition on style, the question of price management arises. At the level of the firm, one approach involved proprietors' calculating production costs, adding overhead and profit percentages to fix a net wholesale price for each style, then bargaining with clients over delivery dates, payment terms, discounts, and the like.[48] Interfirm agreements on discounts and terms could take these elements of a sales contract out of competition, but their effectiveness depended on the manufacturing group's solidarity. Collective attempts to fix the prices of goods were futile, for the diversity of styles afforded nothing like the opportunities to standardize prices that were available in sectors producing generic products (such as basing-point systems to even out transport cost differentials that derived from a spatial scattering of production).[49] Even the minimum pricing tactic for goods of the same general construction involved such complexity that it foundered whenever attempted by various furniture associations. Grand Rapids' innovation in this arena was the perception that collective action on price *movements* was feasible and effective, whereas efforts to manage *setting* individual prices were useless.

In the 1880s, retailers "received printed notices from nearly all the factories in Grand Rapids that, owing to certain conditions there would be an advance of five, six, seven, or eight percent in their entire line of goods." This pattern broke up in the depression years, as collective reductions proved more difficult to establish than advances; but once a measure of prosperity returned, a prominent Chicago retailer wrote John Widdicomb in 1899, urging resumption of the earlier practice. H. E. Scholle explained, "The only way for you Grand Rapids people to advance prices is for you all to get together, and advance one and the same time, and then on a percentage basis on everything you make, not on a basis here and there, as you have been doing, but on *everything*, because that would benefit the dealer." Benefit the dealer—how? With a uniform advance on Grand Rapids' furniture, a retailer "could then go through his stock, and mark it all up accordingly," thus adding to all those goods that he had previously purchased a sum sufficient to cover the price increment for the next season's inventory. Retailers would not be harmed by uniform advances if they could conserve their own capital by drawing the added funds for replenishing stock from consumers of their present inventory. Scholle's advice, however, was ignored until about 1908, by which time rising materials costs pressed local firms toward across-the-board percentage increases, announced a month or so before each exposition and scheduled to take effect January 1 or July 1. Thereafter news from Grand Rapids that prices at openings would (or would not) be elevated was eagerly awaited throughout the fine furniture trade, and producers at other centers frequently echoed Grand Rapids' price movements.[50]

Tracing the course of one of these firms and its proprietor will add specificity and incident to the foregoing overview. In 1893, John Widdicomb, one of three brothers who founded the Widdicomb Furniture Company in 1865, headed an enterprise from which his sibling William had departed for ventures in banking, real estate, and retailing. John had embarked both on factory expansion and technical improvements and on a quest to buy up company shares held by other family members, borrowing "heavily . . . from friends and the banks to make payment." Then "the panic came and brought business to a standstill and knocked values endways." Sales collapsed and WFCo.

stock slid to a sixth of its peak prices, leaving John at least $50,000 in debt as the depression worsened. Creditors intervened and forced him "out of the management" by 1896, replacing him with his banker brother and triggering a family feud that boiled away for two decades. Furious at William's complicity in his displacement, John Widdicomb determined to start over on his own account and to build a business that would vindicate him and erase his debts.[51]

After thirty years in the trade, John Widdicomb possessed a reputation for honest dealing that was his chief remaining asset and a network of associates to whom he turned for assistance. He "scraped together" $5,000, incorporated the John Widdicomb Co. with brother Harry, and negotiated with the Fourth National Bank to buy a bankrupt factory it had seized. His new firm issued $30,000 in bonds, which the bank "took up," providing him with $5,500 in cash and crediting him with $24,500 against the plant's $35,000 purchase price. Fourth National then created a $10,500 mortgage for the balance owing. Given that "it was necessary to make up a large stock before we could go on the market with our goods," his $10,500 working capital was far from sufficient.[52] Thus Widdicomb mobilized his personal creditors, who had loaned him funds for his ill-timed stock purchases. The Berry Brothers (varnish makers in Detroit), J. P. Uptegrove (New York mahogany importer), Jacques Kahn (New York mirror wholesaler), and R. J. Horner and Co. (New York fine furniture dealers) rallied round, forwarding him materials and taking promissory notes in exchange or sending checks in advance for goods not yet constructed.

"Obliged from the beginning to be under the disadvantage of a large floating indebtedness" and pledged to liquidate his 1893 obligations, Widdicomb sought rapid growth, based initially on style excellence and later on an aggressive search for large-scale contracts to fill slack periods and swell total sales. However, those among his old creditors who supplied short-term financing would have to wait years for repayment of their earlier loans. As he explained in 1905, "with no capital to begin with it could only be by building up a large business that I could hope to accumulate enough to liquidate these old claims, and if I had used the first dollar or the first thousand dollars which I made for this purpose it would have been very slow work to make the second dollar or thousand dollars."[53] Moving his family to a rented house "next door to the factory" and joined by son Harry as company secretary, his designer nephew Ralph, and a complement of skilled workers from the old firm, Widdicomb plotted his comeback.[54]

Though his finances were a perennial high-wire act, John Widdicomb's marketing strategy was straightforward. His firm rented exposition space at every opening, showed the highest grade of fashionable furniture, and sought business from upper-crust retailers and department stores in major cities. Moreover, Widdicomb arranged exclusive dealing agreements with shop owners like Horner in New York and Scholle of Chicago, selling to no other local purveyor and expecting (rightly) that they would order heavily from his seasonal lines. His trade terms indicated his tight finances—1 percent discount if invoices were paid in ten days, net wholesale thereafter—and in dozens of letters, Widdicomb informed buyers that there would be no deviations or concessions, no matter how large their purchases or what terms they might secure from other makers. The company did no advertising, published no catalogs, and initially retained only one traveler to follow up exposition clients for "duplicate" orders as the retail season progressed.[55]

Ralph's design skills and shop workers' craftsmanship brought ready sales; employment grew to 175 by 1900. The firm booked $11,500 in "net profits" on shipments of $115,000 in the second half of 1899, but virtually all earnings were plowed back into facilities for producing an expanding variety of styles.[56] The complexity of fine furniture production created cash flow, work process, and inventory problems that Widdicomb shared with his Grand Rapids colleagues. The key difficulty was that in high-grade furniture "it takes an average of four months to get [the] goods ready for shipment," a process that commenced with a cut order for so many units of an item and proceeded through veneering, drying, sanding, and assembly, ending with a series of varnishing, light sanding, and polishing steps before final packing. In middling lines with simpler designs, this sequence could be completed in two months, enabling deliveries sixty days after exposition orders had been taken on the basis of sample sets. To meet buyers' expectations on shipment dates, Widdicomb and other fine furniture makers could not just make samples but had to start manufacturing basic stocks of new styles and recent best-sellers at least two months before the shows opened, just as the last duplicate orders from the previous season were completed. While this necessity yielded more continuous employment for shop workers, it strained finances and forced proprietors to guess how much of each new and old style, in each wood used, might safely be made up as advance inventory. As John Widdicomb put it, "it is a very difficult proposition to calculate from four to six months before the selling season comes around just what patterns will be in demand and in what quantities." Missing the estimate left a firm either with warerooms of completed, but unsold, goods or with bundles of orders that could not be finished and shipped until May or November, when dealers wanted the pieces in March or September.[57] Widdicomb responded to this challenge by juggling his promissory notes and overdrafts like a credit card junkie, writing impassioned letters to accounts receivable, and building inventory cautiously as exposition dates neared.

Preservation of the firm's cut order books for this early period allows a closer look at the inner workings of a diverse-product specialty firm. In the second half of 1899, John Widdicomb authorized assembly of 9,800 pieces of furniture in six wood varieties,[58] encompassing bureaus, chiffoniers, various bedroom dressing and toilet tables, washstands, "splashers," lowboys, "chevals," stools, and beds, covering over 300 style numbers priced from under five to above a hundred dollars. Work began on over 2,600 of these items during November and December for the January opening, roughly half in six new style lines and the rest as reserve stock selected from the previous year's favored patterns.[59] During June 1899, a slack month, the firm cut stock for only 663 pieces, but these spanned 40 pattern numbers in quantities ranging from 5 single items to 130 bureaus in a hot design and 60 chiffoniers from the same set,[60] each made in four woods (which simplified cutting and assembly but mandated separate treatment in finishing). Over the half-year, the firm built 1,004 bureaus in 38 designs and 850 chiffoniers in 39 styles, for which the average cut orders were 18 and 15 pieces, respectively.[61] Thus was specialty production of consumer durables handled profitably, despite the complications of work routing, multiple supply lines, design intricacy, and guessing about styles and optimal inventories.

At the opening of the January 1900 market, Widdicomb's plant held finished or in-process furniture valued at $120,000, most of it unsold, and the owner was in debt up to his earlobes. His letters convey the image of a stolid but witty middle-aged entrepreneur sweating through a bitter Michigan winter, plotting revenge, and seeking means to surpass the first three years' tentative but substantial accomplishments. A $200,000 turnover with 10 percent net returns would not suffice for a man in quest of "large business," so Widdicomb extended the scope of his manufacturing capabilities. Holding a solid position in the finest lines, he added a set of midmarket styles in December 1900.

These could be "rushed through the factory in about one-half the time required on the higher grades." They sold so well at the January openings that Widdicomb arranged to purchase an idle plant at Charlotte, Michigan, and sent son Harry there to supervise production, which would not "interfer[e] with the better grades." Expansion at the dead run was his object.[62]

Widdicomb's next maneuver is more intelligible in this context, for it ill conformed with his experience and Grand Rapids' industrial practice. In 1901, John Widdicomb took a bulk production subcontract, agreeing to build 200,000 sewing machine tables for Singer, shortly after an effort to float preferred stock failed and a brief flirtation with a fine furniture merger scheme ended.[63] With a contract for over $750,000 confirmed, he had little trouble securing bank facilities to buy a second Grand Rapids factory and invest $70,000 in mill improvements and special machinery for these standardized products. Widdicomb had quoted Singer $3.83 per unit for the tables and expected to earn a handsome $1.25 gross profit on each, which, had all gone smoothly, would have provided a margin of $250,000 to cover acquisitions, improvements, and overhead, pay off old debts, and boost the profit account. However, as the Singer contract wound to an end in 1905, he admitted that "many things . . . have conspired to hinder our making even an ordinary profit" on these goods. In 1908, when a major client inquired whether Widdicomb might be interested in some "sewing machine woodwork," he was "rather short" in his reply, "as I certainly felt that . . . I never wanted to hear about sewing machine work again."[64]

The usual disasters that might befall a specialist firm adding standardized goods to its output surfaced during the Singer contract—and others besides. Builders' slow delivery of dedicated machinery for repetition production created initial bottlenecks, and outfitting the plant cost double Widdicomb's estimate. Experienced Grand Rapids woodworkers put in too much care and time in constructing and finishing the tables, wrecking Widdicomb's labor cost figures. When urged to lower their standards and hurry the job, the workers turned out carloads of tables that Singer's Elizabethport, New Jersey, inspectors rejected and returned for repairs or refinishing. Once a balance was struck and an eastern Singer official commented that Widdicomb cases were superior to those made at the corporation's South Bend, Indiana, plant, an order arrived at Grand Rapids to direct subsequent shipments to South Bend. Upon *their* inspections, Indiana managers alleged numerous defects, held up payments for months, and instructed Widdicomb to cut production in half. A 1903 railway car shortage left ten thousand finished cases unshippable; later that year, carloads of required veneers were sidelined in Kentucky for weeks. Another load of veneers sized for Singer tables was cut a half-inch short by the supplier, causing more bottlenecks, delays, expense, and headaches. Hot, sticky summer weather ruined the varnish on several thousand units, and spring floods (the first in twenty years) swirled two feet of water through the factory's first floor and storerooms. Worst, midway through the contract, Widdicomb learned that Singer had expanded its South Bend woodworking complex "and therefore wanted to . . . shut us off or at least cut down our output to a very small quantity, and . . . I was really frightened because if they cut us off from the whole or any great part of the $9,000 weekly receipts which we had counted on I would simply be unable to provide for maturing obligations." Widdicomb did have one backup. Early in 1903 he had sought and secured case and table contracts with National Sewing Machine of Belvidere, Indiana, and he soon solicited trade from Domestic Sewing Machine in Newark, New Jersey. In for a penny . . .[65]

If anything, the National and Domestic contracts yielded more aggravation than Singer's. Domestic proved a notorious "kicker" and a "slow pay." When pushed to cover outstanding receivables late in 1904, it replied with a $3,100 charge-back for "defective" work received at Newark. The next spring, having held $20,000 in unpaid invoices for months, Domestic demanded a 6 percent discount in exchange for a partial payment of $5,000. Refused, it paid nothing till midsummer. National likewise stalled months on bills due. In the fall of 1904 Widdicomb, affirming that "we are the nicest kind of people and always endeavor to avoid disputes with the people with whom we do business," made it clear that unless revenues were forthcoming all work for National would cease. By the time the Singer contract closed out in mid-1905, Widdicomb had determined to get out of the sewing machine cabinet business entirely. He offered Singer $12,500 worth of special machines at 40 percent below their 1902 cost, put the Charlotte works up for sale, and transferred its midprice furniture operations to the Grand Rapids Kent works, the rehabilitated sewing table plant. In 1901–2, before bulk production had begun, Widdicomb recorded profits of $57,000, but earnings dipped to $32,000 in the first "sewing" year, despite nearly $300,000 in shipments to Singer alone.

Once the sewing contracts ended and the firm returned to its focus on styled furniture, profits rebounded to $65,000 in 1906.[66]

Committed to widening his scope, John Widdicomb skirted financial ruin several times during his sewing table venture, earned smaller profits than he anticipated, and learned painfully how different were bulk and specialty manufacturing. But his effort at bridging the two formats did yield gains and satisfactions. His rivalry with his brother's Widdicomb Furniture Co. and his overhanging debts surely had helped propel him into seeking Singer's business in the first place. Scoring a million-dollar contract (he always rounded the total upward) gave him room to crow about the "large business" he was doing, even if he quietly ate a good deal of crow in getting it done. He was delighted that his workforce (624 in the fall of 1903) matched that of the city's largest firms and remained larger than WFCo.'s into 1906.[67] More concretely, through the sewing contracts and seasonal furniture sales during and after that period, Widdicomb paid off his old debts by 1908, some of which had been "written off by creditors as losses sustained." He also possessed two amply equipped factories, the second extensively remodeled and expanded, had redeemed over half his bonds, and yet had a smaller long-term debt in 1908 than in 1897.[68]

Focusing his full attention on fine and middle-grade goods brought Widdicomb $99,000 in net profits on 1907's sales, $652,000. That summer, before beginning a trip to visit his daughter in London, he paid cash for both "quite a little new machinery" and a $10,000 warehouse addition. The flurries of urgent letters begging credit and note extensions ended. Though the firm continued to issue short-term notes, banks now discounted them readily. Widdicomb redeemed them on time, passed the year-end panic with hardly a murmur, then took another vacation abroad in February 1908. Clearly the era of scramble and struggle was over.[69]

Although Ralph's designs drew acclaim for the company's lines and the firm would build no more sewing machine tables, John Widdicomb used his Singer experience to good effect in seeking less harrowing opportunities for large batch sales consistent with the firm's case goods facilities and in finding new outlets for his midprice styles. On the first count, Widdicomb started making four types of kitchen cabinets for Sears, Roebuck in 1907 and soon hived off this trade as the separate Michigan Cabinet Company, hiring Chicago's O. B. Starkwather to have his commercial travelers "do a little missionary work" rounding up orders. Starkwather excelled at the task, finding takers for Michigan Cabinet's goods at Montgomery Ward, the Spiegel, May, and Stern companies, as well as at "premium houses" like Larkin in Buffalo and Crofts and Reed in Chicago. Having diversified the cabinet factory's clientele, Widdicomb was well prepared when Sears pushed for price reductions in summer 1908. Refusing all but one of its buyer's demands, he instructed Sears to "go outside" to place orders for cheap sales "leaders." "We assure you we appreciate your business [but] we cannot furnish you with all the patterns . . . at less than cost, nor do we care to compete with every little jerk water factory which may be hard up and willing to [accept] ridiculously low prices in order to keep running." Whatever the short-term reaction to this tart rebuff, within a year Sears called for midprice bedroom suites in fifty- to one-hundred-unit batches worth $9,000–$10,000 per shipment. In 1909, the retailer advanced Widdicomb four $10,000 accommodation notes to cover materials for orders booked. When Sears head Julius Rosenwald requested a specially designed, solid mahogany chiffonier, Widdicomb had it built, then sent him the staggering bill, adding a "fair profit" to the shop expense. As his upmarket furniture company was clearing $750,000 in yearly shipments, to which the cabinet branch added $200,000, he no longer needed to curry favor, not even with retailing magnates.[70]

In 1910, John Widdicomb completed forty-five years as a partner or proprietor in Grand Rapids furniture making and had climbed fully back from the depths of the mid-1890s. His original network of suppliers and supporters remained intact, its members doubtless relieved at the lack of flying promissory notes. Now he could spend $30,000 without hesitation to install fire-safety sprinkler systems in his plants, take leisurely vacations, and grouse contentedly about the excessive beer and cocktail consumption at employers' association meetings. Like owners elsewhere, Widdicomb griped about the shortage of skilled workers and stressed increased efficiency and an end to idle chatter when hiking wage rates. Yet he also quietly paid the costs of "cures" for the alcoholics among them, negotiated with creditors of the improvident, and penned references for the most talented. A colleague's comment in 1908 summarized his career: "[After starting over,] he gradually made headway, supply houses, lumbermen and the banks who had confidence in him giving him encouragement and his workmen standing by him loyally. . . . His hard work . . . and good furniture sense have won him such success that on a clean-up he would today be pretty near half a million to the good—all made since the dark days of '93 and every cent of it made honestly." During his lifetime, John Widdicomb never faced a workers' walkout, but that lifetime ended in 1910. Harry, Ralph, and a non-kin treasurer took over "active management." The next year, the local furniture trade faced the fiercest labor upheaval in its history, the "Great Strike" of 1911.[71]

Long before that conflict, Grand Rapids furniture workers had organized for common purposes. Though an 1886 Knights of Labor effort failed, as did an 1890 hand carvers' walkout, union advocates mounted another drive six years later and renewed their attempts in 1900. When the small complement of city upholsterers mounted a strike in 1905, the Employers' Association assessed all members for funds to back the affected firms and defeated them. Oriel Furniture rebuffed a 1909 call for wage increases by forty-five cabinetmakers, fired their "spokesman" early the next year, then stood firm during the ensuing strike, rehiring all but the leader once resistance faded. No local furniture maker accepted any "outside" agent's representation of its employees or negotiated with any collective on matters of hours, wages, or work rules, consistent with manufacturers' near-universal hostility to labor unions nationally. Nonetheless, the United Brotherhood of Carpenters and Joiners made one more try in 1910.[72]

Reacting to the Oriel fray, a group of skilled woodworkers from several firms initiated a UBCJ chapter in February and, gathering strength, sent representatives to discuss hours and wages with the manufacturers' association in July. Playing for time, the association urged a delay until the summer market exposition had closed, then in August rejected any colloquies. Terms of employment were matters of individual adjustment between a worker and a proprietor, they averred. As the strike's historian explained, "The issues of managerial autonomy and complete control of the workplace shaped manufacturers' response from the very beginning." Despite sharp criticism from the city's mayor and its three daily newspapers, employers rested intransigent while union leaders secured support from national headquarters during the winter of 1910–11.[73]

On February 9, Grand Rapids workers, advised by UBCJ organizer W. B. MacFarlane, renewed their call for conferences and listed their core demands: a nine-hour day (fifty-four-hour week), a 10 percent increase in the day rate, abolition of piecework, and installation of a minimum wage. Over a month elapsed without a reply; on March 25, more than three thousand workers assembled to approve overwhelmingly an April 1 strike date. That this group accounted for only a third of the city's furniture operatives caused the unionists no disquiet, for the vote at last brought an employers' response. Three days later association secretary Francis Campau issued a statement declining any meetings (which could imply recognition of the union as a bargaining agent) and denying workers' claims that their wages fell short of those paid at other furniture centers and that "no material increase has been made in wages in the past five years." Attempting to avert a full-blown clash, the mayor rapidly appointed a five-man Citizens' Commission to review evidence regarding the conflicting assertions and report its findings to the public. Deferring any strike until the commission completed its task, union members worried that it had no power to arbitrate the dispute. The group's charge, drawn up by the local Catholic bishop, was only to "obtain . . . a statement of [workers'] grievances and proposals," convey this to manufacturers, secure replies, evaluate both positions, and "make such further independent investigation of . . . conditions in the furniture industry here or elsewhere as may be necessary."[74]

The commission held closed hearings at which workers shared their frustrations with the rising cost of living, argued that manufacturers had amassed vast profits, and maintained that only implementation of their four demands would remedy the troubled situation. Proprietors replied with presentations documenting a 16 percent increase in wages since 1906 (yielding earnings comparable with those at other centers) and showing their plants active on 97 percent of possible working days over the previous four years. However, they declined to discuss profits and made no response to workers' observation that "selling prices had been advanced ten per cent on three occasions" in the last five years. When the commission shared this reaction with union leaders on April 18, and before it could commence its investigative work, the woodworkers renewed their strike authorization. The walkout started on April 19. Although four to five thousand employees stood down from their benches and machines, employers urged the commission to move on toward its report phase. Association members had no intention of negotiating a settlement, but they did believe that their calculations were correct and that wage hikes could not be justified. The citizens' group reviewed twenty-two firms' books and records during the next several weeks; its report appeared in mid-May. The commission affirmed both that local woodworkers' earnings were "practically what they are elsewhere," and that manufacturers' contention that "outside competition" regulated wage levels was "substantially correct," as were the makers' "statements as to wages." Yet it did support workers' efforts to secure a nine-hour day, scored piecework as "liable to serious abuses," and called attention to the problems of "low-priced men" less able to manage increasing living costs. This mixed judgment, broadly but not uniformly favorable to the employers, came too late to affect the course of the strike.[75]

After the fact, Royal Furniture owner Robert Irwin dryly commented, "The history of the strike, from this time on, is not vastly different from that of the average industrial struggle of this magnitude, except that it was of much longer duration." With local government and the press backing the strikers, the pace and intensity of rowdy confrontations between pickets and nonunionists accelerated through May, culminating in a May 15 window-smashing spree along factory-crowded Fifth Street, a wild burst sparked by the provocations of William Widdicomb and John's erratic heir, Harry. After boisterous court proceedings, two sets of special marshals undertook to keep the peace in the factory district—a hundred union supporters appointed by the local police and two hundred nonunion men commissioned by the county sheriff. In the first months, several small companies settled, largely on the workers' terms, as did American Seating, a Chicago school furniture corporation's branch plant. American, making standard goods and with 90 percent of its employees on piece rates, gave the UBCJ much publicity but little of substance. It agreed to nine hours only for its forty day-wage workers but rejected any increases or minimum rates. That the union accepted this hollow offer indicated the strike's bleak prospects. Strikers' support payments from the union covered less than a third of most workers' earnings, and the men began drifting back by late June. The association's move to import strikebreakers at the end of May added to the pressure and fueled further violence. Picking up on technical changes in Chicago, Grand Rapids Show Case recruited 150 skilled woodworkers displaced by "the Pullman's Company's shift . . . to the construction of steel and iron cars" and lodged them in "dormitory facilities" at its plant. The Great Strike had failed, though it dragged on until mid-August, when the last thousand holdouts returned to their posts, defeated.[76]

What of the aftermath? Union efforts in area furniture trades floundered until after the Wagner Act (and the CIO organized few plants even then), whereas manufacturers were quietly gleeful. Their clients had come in force to the midstrike July openings and had helpfully ordered goodly quantities of the styles local firms had in inventory, clearing out old samples, aging nonsellers, and such of the new lines as could be made up with short staffs. The strike also vitalized manufacturers' organizations. Soon after, the separate Grand Rapids Guild and Grand Rapids Furniture Association (which, respectively, promoted collective catalogs and the city industry's trademark) merged into the Employers' Association, forming a comprehensive Furniture Manufacturers' Association (FMA). Strike solidarity forged organizational consolidation and confidence.[77]

In 1912, when July market sales exceeded those of the strike summer by 40 percent, the FMA calmly turned toward regional trade governance. Having secured assessments on members of $8.50 per employee to cover the 1911 victory's costs, the FMA agreed that increasing earlier annual levies per worker from under a dollar to two or three would help fund wider collective programs. The association also promoted reshaping the structure of local government to eliminate a strong mayor who posed political counterweights to its interests, and soon provided clear mandates for members' discipline, codifying the customs of earlier days. Ironically, workers' unprecedented struggle for unity provoked a successful welding of specialty manufacturers. This group's subsequent trade leadership and internal cohesion would eventually precipitate a Federal Trade Commission challenge, but by 1912 allied networked specialists ruled narrow-focus Grand Rapids, generating solid profits and cementing their district's centrality to the fine and midprice furniture industry.[78]

As John Widdicomb's bulging income suggested, Grand Rapids companies had hiked prices while increasing their style ranges and adding new technological and work process innovations. The 1911 crisis augmented "Furniture City" manufacturers' solidarity and set the stage for the extension of collective action. Similar opportunities and challenges also confronted localized capital goods producers in these turn-of-the-century decades, notably Cincinnati's machine tool builders, our next focus.

## *Chapter 8*

# FASHIONING THE MACHINE TOOL HUB: CINCINNATI

FOLLOWING a three-year journeyman's trek, in 1887 R. K. LeBlond commenced a general metalworking business in his native Cincinnati. Brother John joined R. K. in 1889, becoming his fifth employee. Two years later, William Lodge ordered a batch of lathe components for use in his works, then, evidently satisfied with LeBlond's output, contracted for fifty complete "12 inch engine lathes." Lodge superintendent Nicholas Chard soon joined LeBlond as a junior partner. Together they designed a novel fourteen-inch lathe that embodied "a great many improvements . . . which made it one of the best machines in the country at that time." Though others would soon testify to its advantages by copying "many features of this design," the 1893 panic "nearly wrecked the business." Bereft of orders for eight months, the two brothers alone staffed the nearly silent shop, for Chard's "strained financial condition" brought him to accept Lodge's offer of his former superintendency. Yet by 1912, R. K. LeBlond and Co. rose from these miseries to become one of the "Big Six" firms leading Cincinnati's machine tool industry.[1]

With Lodge's encouragement and a sound design, LeBlond left the job shop trade and specialized in lathes. Flickers of business in 1894 brought his workforce to ten; then within in a year, "owing to the tremendous expansion in the manufacture of bicycles," the company booked orders for "what at that time seemed large quantities of lathes." Though the craze was short-lived, bicycle makers required dozens of small lathes for shaping parts, nicely intersecting with LeBlond's capacity.[2] Tracing a familiar trajectory, the firm constructed its own plant and incorporated as a closely-held family business (1898), three LeBlond kin holding the controlling interest. Soon after a 1901 reshuffle brought in a fourth family member, the LeBlonds added a second tool line, milling machines and small grinders for sharpening milling cutters, and again expanded their production facilities. However, John LeBlond noted that "after every addition to our plant . . . a period of business depression invariably followed, so that where we had provided for an increase in floor space, we were generally left without orders to utilize it properly." Eventually "business always returned with such an increased volume that we were compelled to start other additions," and the cycle recommenced. A decade of effective designs for larger, faster tools widened LeBlond's markets, and the company built an entirely new plant (in Hyde Park on the east side), completed just as war orders more than doubled employment to over a thousand workers.[3]

In 1910, the Big Six tool builders headed a roster of thirty-two Cincinnati companies, each producing one or two of the five basic types of metal-cutting devices: lathes, drills, shapers, planers, and milling machines.[4] Lathes bring a fixed cutting tool into contact with a spinning workpiece fastened securely along its centerline, the contact edge essentially "peel[ing] the metal away." Multiple lathe heads take several simultaneous cuts at different locations, whereas turret heads bring a series of edge tools sequentially to the same spot. Drills and their cousins, boring machines, drive a rotating cutter into fixed pieces; some drills feature multiple adjustable "spindles" to make several holes at the same time in repetition work. If designed solely for a single mass production component, the spindles are fixed permanently in place, creating "dedicated" tools. Both shapers and planers take a linear cut across the surface of a bolted-down workpiece, but in shapers the cutting edge moves in strokes across the plate or casting, as in hand woodworking, whereas planers' sliding powered tables drive the workpiece into a fixed cutter. Again multiple heads are feasible, planers at times being designed with "side heads" to shave the edges of a piece while other tools work the top surface.

Milling machines combine shaping/planing and drilling, sending a rotating cutter across the surface of a fixed or sliding workpiece. Elaborately designed cutters can carve complex grooves in a single stroke, obviating a series of production steps that would be necessary if the piece had to be shifted among machines. Turret heads, as in lathes, serially bring differently sized milling cutters to bear on the work. The speed and versatility of milling machines, as well as their capacity for dedication to single outputs, resulted in their adoption across the entire metalworking spectrum from customized turbines to auto parts after 1900, propelling Cincinnati Milling Machine to regional leadership in employment and sales.[5] Expanding demand for sophisticated and durable machine tools in all five classes established Cincinnati as the nation's metal-cutting machinery hub, yet the variety of designs,

15. A turn-of-the-century miller from Cincinnati Milling Machine. The rotary cutting tool would be fastened on the rod to the right of the power belt cone, cutting metal as the workpiece, fixed to the moving table at right center, passed underneath it. The nameplate lists the five patents covering this machine tool and improvements to it. Courtesy of Hagley Museum and Library.

the diversity of clients' needs, and the uncertainties of demand linked the city's machinery builders to industrial specialists in Philadelphia textiles or Grand Rapids furniture. Take G. A. Gray's planer factory as a case in point.

Illinois native George Gray (b. 1839) apprenticed locally at Miles Greenwood's antebellum iron works and later relocated to Hamilton (Ohio) to superintend the Niles Tool Works before returning to Cincinnati as an entrepreneur in 1880. Though Gray designed weapons and patented improved drills and lathes before he started his own business, within a decade he concentrated on planers. By 1887, his firm offered potential buyers thirteen varieties of metal planers based on his patented design, stone planers for lithographers, and eight sizes of lathes.[6] In 1889, he accepted Henry Marx, principal agent for machinery dealers Hill, Clarke and Co. of Boston, as a "financially-interested" partner and rapidly expanded the range of planer designs Marx would market. The 49 models proffered in 1889 (before Marx's arrival) nearly doubled to 92 in 1890, when the firm announced that it had adopted as an option the "spiral geared" power train pioneered by Philadelphia's William Sellers. Gray also devised special heavy planers suitable for railroad supply work on switches and "frogs," and ended lathe production. In 1892, the model range reached 125, from modest 22″ planers listed at $455 to 60″ monsters costing $5,300 and weighing over twenty tons.[7]

Key features of planer design included "accuracy, ease of adjustment, regularity of speed, and durability." Gray's machines met all these requirements, owing to both their "massive, yet simple" construction (to achieve rigidity and balance) and the precision of components' preparation and assembly, for which the Gray shop gained international renown.[8] In a 1903 interview, however, Gray "declared, half seriously, that there were times when he doubted the advantage of such a reputation." Some clients purchased a "first-class" Gray planer, casually set it to work expecting perfection, then complained in "surprise and indignation" that its accuracy was deficient. In such situations, firm members or dealers had to show the users that they had not leveled the machine properly, had failed to use uniform, precisely sharpened cutters, or had paid inadequate "attention to [the] fixtures for holding the work" solidly on the table.[9]

Gray first constructed each new design for use in its own shops "as something of an experiment," expecting "that some changes would probably be made after it was started running." Once design modifications stabilized, the drafting room certified final drawings, made and filed blueprints, and had castings order sheets printed, listing each model's array of numbered parts. Small parts common to several models (bolts, cutter holders) were "made up in lots," but inspection of the castings sheets shows that not more than 5 percent of components could be used in planers of different throat size.[10] Thus receipt of individual planer orders triggered foundry orders to cast components for shop machining and assembly, a sequence that lasted roughly four weeks for smaller models with about 60 main components and up to four months for the largest, with over 130.[11]

In 1892 Gray contracted for display advertisements in *American Machinist* to announce electric motor drive options for its planers, but the economic slump dropped 1893 sales to half their 1892 level. Revenues reached only $58,000 in 1894, the lowest total between 1890 and 1920. How many of his ninety workers Gray laid off is unknown, but the firm pushed hard for orders, sending Marx to open a New York City sample room and entering its planers in the 1895 Brussels Exposition to hook European sales. News of a Brussels gold medal cheered the Cincinnati shops and snared customers for Gray's agents, Schuchardt and Schutte of Berlin and Charles Churchill in London. Shipments doubled in 1895, fell back the next year, and at last passed their 1892 showing once the general economy rebounded in 1898. The bicycle boom boosted calls for LeBlond's lathes, but planer demand flowed from heavy industry and specialty metalworking. Thus in 1897 Gray sold planers to machine tool builders and railway shops, to Westinghouse's East Pittsburgh electrical plant and Bullock Electric's Buffalo shops, American Brake in St. Louis, the Brooks Locomotive Works (soon a part of the American Locomotive merger), and others who banked their machinery needs until the business horizon brightened.[12]

Quick shipment of smaller planers that year, some the same week that orders arrived, suggests that Gray kept his workforce together in slack times by building a modest inventory of three smaller models that provided 30 percent of unit sales, though less than a sixth of annual dollar volume. The context for this tactic is worth explicating. Gray's 1897 price list offered 118 base models with one cutting head and a square throat (e.g., 30″ × 30″), all geared for belt drive from line shafting, framing the firm's capacity for building "standard" planers. Yet any of these base models could be specially constructed with up to four heads, a "widened" rectangular throat (e.g., 30″ × 42″), direct electric motor drive, extra-long tables, and other options. In toto,

Gray stood prepared to construct over a thousand variants on its basic designs, fewer than a hundred of which were called for in 1897. Contracts for midsize (30″–38″) and large (42″ and over) planers far more commonly specified diverse combinations of these "extras" than did those for small planers (22″–28″), nine-tenths of which were ordered as base models. As three of the twenty basic small-throat-and-table variants accounted for three-fifths of 1897's smaller planer orders, building stock of these items made sense, whereas making basic midsize and large machines for inventory did not. Overall that year, Gray built 167 planers (worth $128,000) in 41 throat-and-table combinations, of which 59 were "specials" (35 percent), chiefly among the 55 midsize and 13 large tools manufactured. No order specified more than two identical tools.[13]

Much like other machinery firms in and beyond Cincinnati, G. A. Gray & Co. erected its planers singly or in small batches in 1897—a pattern that would continue well past World War I. Its markets, however, showed no such regularity, for they shifted dramatically both in export-domestic proportions and among U.S. regions in the domestic trade. Gray's 1897 sales fall into three categories: domestic agent, foreign agent, and direct to users. Though the firm had selling arrangements with dealers in Boston, New York (having closed its own office), Baltimore, Cincinnati, Cleveland, Chicago, and San Francisco, those seven dealerships amassed orders for only 60 planers (net $44,000), while its two European agents forwarded contracts for 89 tools (net $62,000). Meanwhile, Gray secured direct orders for a dozen planers, mostly specials, that realized another $21,000—almost twice as much per tool as in agency sales, in which base models then predominated. Exports proved crucial to Gray's revival after the 1893 crash, but they were no more consistent or reliable than any regional domestic source of orders.[14]

Over the next fifteen years (table 18), export sales never again reached 1897's 52 percent of net revenues, ranging from a high of 32 percent in 1900 to a low of 7 percent in 1909. From 1900 through 1912, domestic agency sales bounced between three- and four-fifths of revenues, with direct contracts to users generally holding near 10 percent, spiking to 19 in 1909 when foreign sales flagged. Fluctuations were equally evident among the U.S. dealers, who had exclusive regional territories. New York's McCabe and Co. booked $46,000 in contracts for 29 mostly midsize and larger tools in 1900, slumped to $24,000 (for 15 tools) in 1903 and $1,251 (for one) in 1909, then recovered to $45,000

TABLE 18
Gray Planer Orders, by Sales Outlet, 1897–1912, in Percent

| | *Numbers of Tools (%)* | | | | *Revenues (%)* | | | |
|---|---|---|---|---|---|---|---|---|
| | *Dom. Agt* | *Direct* | *Fgn. Agt* | *N (T)* | *Dom. Agt* | *Direct* | *Fgn. Agt* | *N ($,000)* |
| 1897 | 37 | 7 | 56 | 164 | 34 | 13 | 52 | 128 |
| 1900 | 57 | 8 | 35 | 167 | 59 | 9 | 32 | 193 |
| 1903 | 77 | 9 | 14 | 159 | 80 | 8 | 12 | 206 |
| 1906 | 63 | 6 | 31 | 321 | 69 | 11 | 23 | 348 |
| 1909 | 79 | 11 | 10 | 127 | 74 | 19 | 7 | 167 |
| 1912 | 69 | 6 | 25 | 177 | 74 | 11 | 15 | 257 |

*Source*: Gray Papers, Box 21, Order Books, vols. 3–5, Cincinnati Historical Society.
*Note*: Dom. Agt = domestic agent; Fgn. Agt = foreign agent.

TABLE 19
Gray Planer Orders, Domestic Revenues, by Agent, 1897–1912, in Percent

| | *Bost.* | *NY* | *Pgh.* | *Cleve.* | *Cin.* | *Chi.* | *Others*[a] | *Direct* |
|---|---|---|---|---|---|---|---|---|
| 1897 | 7 | 24 | —[b] | 11 | 5 | 15 | 11 | 27 |
| 1900 | 6 | 35 | 6 | 5 | 8 | 17 | 10 | 13 |
| 1903 | 7 | 13 | 24 | 11 | 16 | 14 | 6 | 9 |
| 1906 | 13 | 12 | 11 | 16 | 11 | 11 | 12 | 14 |
| 1909 | 21 | 1 | 9 | 14 | 6 | 15 | 13 | 21 |
| 1912 | 8 | 21 | 6 | 25 | 7 | 11 | 10 | 12 |

*Source*: See table 18.
[a] Includes sales agents at Baltimore, St. Louis, San Francisco, and, briefly, at Denver and New Orleans.
[b] There was no agent in Pittsburgh at this date.

(26 planers) in 1912 (table 19). Less dramatic but noteworthy swings among other regional agencies meant that Gray could never predict where planer demand might come from and with what specifications. Although small and midsize planers' share of total output was fairly steady, demand for large machines varied wildly. Moreover, after 1897, "specials" accounted for nearly half of all orders (table 20), limiting the utility of building base models for stock.[15]

Yet amid these uncertainties, machine tool prices remained stable for long stretches from the later 1880s into the first decade of the

TABLE 20
Gray Planers, by Size and Proportion of Specials, 1887–1912, in Percent

| | *Small (22–28")* | *Medium (30–38")* | *Large (42"+)* | *Specials* |
|---|---|---|---|---|
| 1897 | 53 | 40 | 8 | 35 |
| 1900 | 40 | 38 | 23 | 54 |
| 1903 | 37 | 50 | 13 | 50 |
| 1906 | 47 | 44 | 10 | 47 |
| 1909 | 48 | 41 | 11 | 46 |
| 1912 | 43 | 41 | 16 | 47 |

*Source*: See table 18.

twentieth century, as several historians have noted. How could such producers fail to moderate prices in times of stagnant demand? How would they approach the pricing question when the trade boiled and buyers were eager for new tools? Gray records provide answers that move beyond issues of price inelasticity, i.e., producers understood that price cuts would not generate additional sales, hence maintained their list rates. Crucial here was makers' management of agency discounts. List prices indicated the cost to clients of machine tools but disguised the division of revenues between dealers and makers. If the Gray company's history is any guide, stable list prices concealed a quiet struggle between dealers and producers over their respective shares.

For two decades after the late 1880s, Gray's price lists were stable, but the discounts allotted to U.S. dealers (providing their sales commissions) slid from 33–40 percent in the early 1890s to 15 percent in 1899 and 10 percent in 1902–4. European dealers' discounts thinned from 25–30 percent in the mid-1890s to 15 percent in 1900, rose to 25 percent (1902–6), then dropped to 10 percent in 1907. These shifts open a window onto specialty producers accessing national and international markets. Dealers handling little-known tools well earned their sizable commissions in the late eighties and early nineties, but as Gray's reputation for fine planers spread, the Cincinnati office judged that selling had become less challenging. Dealers' efforts should therefore command a slighter proportion of contract prices, with the makers reaping a larger share. Hence shrinking discounts fueled Gray's profitability without any increase in prices to clients. For example, in 1903 Gray made eight fewer planers than in 1900, and fewer of its large, big-ticket models, yet booked $13,000 more in revenues by shaving the dealer discount 5 percent. Gray's dealers may have muttered, but none dropped the firm.[16]

After a quarter-century as a machine tool builder, George Gray retired in April 1905; Henry Marx became president. Gray's son Wallace had little involvement with the business, though he inherited a two-fifths interest in it when his father died suddenly that summer, aged sixty-six. *American Machinist*'s obituary underscored the elder Gray's care and conservatism. Consistent with Jane Jacobs' notion of specialty plants as informal "development laboratories," Gray "never put out a tool until it had been very carefully studied and tested in his own works." Despite his success, he "resisted all temptation to enlarge . . . simply deferring or declining orders beyond the capacity of the works." Building on that foundation would be Marx's task.[17]

Tensions over dealers' discounts came to a head in 1906, when Marx attempted a second time to lower domestic agents' rates to 10 percent. That level had first been announced in November 1902, but dealers evidently pressed Gray to restore the 15 percent rate, which held for two years after May 1904. When nearly a hundred planer orders flooded in during the first quarter of 1906, continuing 1905's heavy demand, Marx reinstalled the lower discount. By July, Hill, Clark and Co.'s Chicago and Boston offices lodged the first demands for concessions, followed shortly by Strong, Leslie and Co. (Cleveland), Pacific Tool and Supply (San Francisco), and Cincinnati's E. A. Kinsey, who sought southern outlets, principally at railroad supply and repair shops. The company gave ground gradually and creatively, coming to an understanding with agents that took advantage of users' weak information network concerning machine prices. In September, dealers began forwarding orders in which the same planer models carried varied discounts and were quoted at different prices to separate clients. Thus PT&S ordered four basic 24" planers with 6' tables, two of which were invoiced at $594 each, less 10 percent (netting Gray $535), while the other pair went out priced at $855 each, less 30 percent (net $598). A third of the 93 tools sold from September through the year's end reflected advanced prices and discounts. Gray also charged well above list for the 13 tools it sold directly to users (e.g., $644 for a base 24"/6' planer and $931 for a 28"/8' listing at $764, both to Cincinnati Frog and Switch, no discount). At the same time, those dealers who had not "kicked" about discount rates (in New York, Pittsburgh, St. Louis, and abroad) continued to receive shipments at list, less 10 percent.[18]

As in other specialty trades, periods of rising demand created avenues for opportunistic pricing by machine tool builders.[19] The wave of discount reductions that ended with the 1904 reversal had buoyed Gray's profits and compensated for a postdepression rise in materials and labor costs without affecting the list's stability. However, the 1906 effort to boost returns at dealers' expense clearly failed, leading to a crazy quilt of prices and discounts through which both the firm and its agents dipped into buyers' pockets. That purchasers accepted the higher charges for planers late that year emboldened Gray's management team to regularize prices and discounts by fashioning a new list, issued in January 1907. Confirming the 30 percent domestic discount but retaining 10 percent for export orders, the schedule increased base prices for the smallest model ranges by 51 percent, hiked midsize planers 48 percent, and added 31 percent to previous quotes for the largest machines. Thus PT&S's high-end $855 charge for a 24"/6' planer in late 1906 became the new list price, adding 10 percent to Gray's return and nearly $200 to agents' commissions for each one sold. A veteran marketer himself, Henry Marx had sweetened dealers' incentives, augmented his company's per unit income, and surely watched with delight as 1907 sales swept past those of the previous year.[20] Though the 1908 market contraction was an upset, if hardly a surprise after two overheated years, Gray's sales averaged nearly a quarter-million annually, 1910–13, drooped after the Wilson tariff reduction, then broke all previous records once war demand accelerated.[21]

Though George Gray's reluctance to expand set him apart from his Cincinnati machine tool colleagues, in terms of technical innovation his firm was typical. Incremental innovations in design were routine at Gray, as throughout the industry. The firm regularly distributed thousands of flyers to announce improvements and to promote more ambitious novelties, like cast aluminum pulleys or a "speed variator" that provided four rates of table movement. In 1906, Marx also took the opportunity to caution machine tool users about illusions that "high speed tool steels" had brought to the metalworking trades. At the time, nothing could cut metal but other metal,[22] and alloy processes had made possible "tool steel" cutters that could shave iron or steel at higher turning or planing speeds than had previously been feasible. Such cutters withstood the heat the work generated without losing their sharpness. As this advance lessened the labor time necessary for production and as labor costs were a major segment of total costs in high value-added metalworking, a rage for speed penetrated nearly every corner of machine-tool-using trades.[23]

Along with others, Gray objected that quicker was not necessarily more efficient. In planing, moving tables faster against cutters had two drawbacks that "counterbalance ... the benefits derived from the excessive speeds so glibly recommended." First, a planer tool "receives a sudden strain—almost a blow—every time it enters the work ... increased in direct proportion to the speed of the machine." These shocks soon chipped the new cutters' edges, which then so roughened the workpiece's surface that an extra "finishing cut" was mandatory, "sacrificing much of the time gained" through advanced speeds. Second, when tables reversed at the end of a stroke to return to their initial position for the next cut, planer workers had to shift the drive belts to "start the table in the other direction." At fast speeds, bringing the moving table and "revolving" driver parts to a stop, so that the next stroke could be taken without damage to the gearing, virtually erased the time "gain" derived from speeding the machine. Marx urged Gray's clients to avoid "the present 'high-speed craze,'" for in planing, its advantages were "in actual practice ... offset" by the "practical limitations" of precision work. As others would affirm at a 1910 mechanical engineering symposium, the speed and economy gains promoted by Frederick Taylor and his allies represented "theoretical conditions, which entirely ignore adverse facts" pertaining to planing.[24]

In general Gray built its planers individually; but other makers achieved batch production levels, fabricating Cincinnati Shaper's tools in dozen lots or Bickford's upright drills in groups of sixty or eighty, roughly a three-months' stock supply of base models. All sought efficient production strategies that would stretch standard components across a number of designs. Most discussion on this theme centered on the "unit system" often credited to Cincinnati Milling Machine, but also used by Bickford and others. In redesigning its millers "for high power work" in 1907, CMM determined to "produce a line of machines to meet the possible wants of a customer ... and capable of being changed by him at any time." By creating two frame options, three driving formats, and two speed control choices for each machine size, the company could boast that "no less than 12 distinct machines are made" through various combinations of the three features. Moreover, should a client shift its output mix or power transmission system after installation, the miller could be reconfigured simply through substitution of a motor driving unit for a belt drive unit or a vertical frame and cutting tool spindle for a horizontal one, eliminating the need to purchase an entirely new machine. These substitutable units added flexibility to CMM's tools in an era of uneven technological change and simultaneously made it possible for CMM to build stocks of standard components for assembly on demand into a variety of millers matching clients' needs, while minimizing inventories of finished models. Furthermore, CMM could incorporate incremental improvements in, say, gear boxes, without necessarily disturbing core designs or devaluing stocks of other unit components. As in other trades, a degree of parts standardization could sustain or enhance flexibility in production and use. Perhaps most important, the success of the unit system led CMM to take the next step in flexible specialization, setting up in 1913 "a new engineering service to custom-design machines that would do specific jobs" for clients. "Sales engineers" visited "a customer's plant, analyze[d] his needs and then tailor[ed] the Mill's products to meet them." CMM "no longer simply made milling machines; it sold production performance as well."[25]

While defining the cutting edge of technological innovation, Cincinnati tool specialists also addressed other key technical issues and shared their findings through journals and at regional and national association meetings. Three firm-level concerns were prevalent: shopwork systems, schemes for cost finding, and approaches to wage payment. Lodge and Shipley managers devised a parts storage facility that could rapidly provide from stock many components for various "engine lathes." Under the older plan of building lathes in "complete lots," work floors had been "filled with one size of lathes and only orders for this particular size" could be shipped rapidly. Clients ordering others had to await the relevant model's turn, unless finished stock lathes were on hand. With an extensive parts wareroom, "all sizes may be erected at one time, enabling complete mixed orders to be promptly filled," without increase in inventory of types not yet called for. L&S also recognized the endemic problem of idle machines and "operators . . . waiting for work." Hence by 1911, it reorganized shop scheduling so that "each operator shall be supplied with work for one day in advance." "One day" here underscores the variability of tasks that skilled workers had to accomplish. With dozens of lathe models, each having hundreds of parts, workers rarely repeated the same operations for days at a time. Instead, they finished fifty of one part, ten of another, then took on new tasks. If designs were fairly stable and the firm retained its skilled workers, the collective proficiency level would rise, but tool building remained far from routinized production.[26]

Cost finding for firms with diversified output was an enduring misery. Shop ticket systems recorded materials and labor expenses for parts preparation and assembly and accumulated "direct" costs for varied goods, but they left proprietors with the task of assigning other (indirect or nonproductive) costs to each item in a more or less arbitrary fashion. The complexity and uncertainty of this process made "rule of thumb" add-ons (flat percentages for overhead or "burden") a durable feature of costing practice in specialty production. However, applying a standard overhead rate illogically assigned the same cost rate of power, office work, selling expense, and so forth to basic and special tools alike, although managers knew that specials demanded greater attention. Standard costing, through which calculations of "proper" expenses would generate base figures against which actual practice could be measured and managed, did not address specialty manufacturers' needs. Tonnage steel mills might have standard outputs for which these estimates were relevant over long periods, but tool builders could not use a homogenizing cost system. In the selection of an accounting method, one observer commented, "The question of quantity of product is, in one shape or another, at the bottom of all questions of choice of action."[27]

Lodge and Shipley led Cincinnati on the cost-chasing front. As accounting theorist Alexander Church argued, the heart of costing for specialty producers was connecting direct and overhead expenses to particular products (1) to determine which were profitable and whether pricing was appropriate, and (2) to locate those production areas where cost inefficiencies ("leaks") merited attention.[28] L&S used shop "tickets" for batches of work in process, but attaching overhead figures was a vexing challenge. Rejecting a general percentage rate for overhead, L&S worked on a "machine hour" basis. Here depreciation, repairs, power, and space "rent" were figured for each tool and charged to shop orders on an hourly use basis, giving a fuller portrait of production-related expenditures to which office staff added a small overhead percentage for "nonproductive" costs (administration, maintenance, etc.).

Lodge's "traveler books" were bound packets of triplicate order slips for each step in manufacturing lathe components. The "stores" slip identified materials issued for the job, while the "shop" copies followed the work and were inscribed with the hourly charges for labor and machine time. After each stage, the third copy went to the "tracing department," which tracked the general flow of components against stock levels and orders for assembly. However, a blizzard of tickets arriving at the office was the core problem with such systems. "As the men do not know what the next job is to be" and handled many different tasks each pay period, tickets worked well for calculating their earnings and premiums, but their volume overwhelmed attempts to use them for further analysis, unless "a small army of accounting clerks" was engaged to sort them repeatedly to develop cost accounts, general expense figures, and "statistics of labor efficiency and material turn-over."[29]

To overcome this barrier, by 1909 Lodge and Shipley installed a state-of-the-art, machine-readable data card system, with "an automatic card sorter, a tabulating machine, a gang punch (or dater) and three hand punches" leased from Herman Hollerith's Tabulating Machine Company. Using three versions of Hollerith's thirty-seven-column cards (one each for labor, material, and accounting classifications), office staff could rapidly calculate costs per piece or per lathe for labor, machine time, repairs, patterns, or sales, and could track expenses for work spoiled in the shops or analyze the earnings of operatives over time. Clerks could punch hundreds of cards each hour, and the tabulator processed 250 cards per minute for computations. As Hollerith's early accounts were chiefly with railroads and insurance companies, it is not evident how Lodge and Shipley came to adopt the system, but a 1906 *Engineering Magazine* article advocating tabulating's "utmost flexibility" may have sparked an interest. In it, Morell Gaines wrote: "Without the necessity of transcribing, [card] records permit the application of re-classification after re-classification to the same set of facts. . . . The revolution in method admits now for the first time absolutely free and close application of searching analysis to cost accounting."[30] Tabulation provided tool builders with a versatile, speedy means to manage complex information without incurring large clerical expenses that offset any savings derived from cost analysis.[31]

On the wage payment issue, Cincinnati machine tool firms reflected the trade's national concern about effective means to manage the wage-effort relation. From the mid-1880s, manufacturers had debated numerous plans for superseding customary day or hour wage rates to heighten output without reducing quality. Piece rates or task and bonus systems were widely adopted, but employers who cut rates after workers improved productivity and earnings undermined shop morale, promoted restriction of output, and encouraged craft union organization. A practitioner of the premium approach, William Lodge was firm in avoiding predatory rate slashing. At the National Metal Trades Association's 1909 convention he "insisted [that] prices [of work] once set should not be changed," and "held that the manufacturer must abide by the agreement not to reduce the rate of premium, even if the workman does show remarkable results," emphasizing practices that were routine in Cincinnati metalworking. Before 1900, Lodge, Bickford Drill's H. M. Norris, and other builders adopted versions of the premium model that enhanced output and workers' earnings and, by restraining proprietors' cupidity, also minimized the appeal of unions. Using historical data "on the cost of lathes of various sizes and of different sized lots," Lodge's managers set standard times for each component's manufacture. "We knew that these times would be bettered in nearly every instance, but we . . . wanted the men to earn more to make it an object to strive for an increased output."[32]

Here is how Lodge's system worked by 1912. A shop order issued for grinding operations on 20 units of a lathe part carried two time standards—an upper limit, 4 minutes for each, and a lower, 2 minutes. Any total time for the 20 parts between the two earned the worker his regular hourly pay (30 cents/hour)—thus 40 cents for a slow job that took the full 80 minutes or 20 cents for a faster one that consumed only 40 minutes. However, were the worker to best the lower, 40-minute barrier, he earned regular wages for the actual time spent, plus half his regular rate for the difference between that time and the full 80-minute limit. Thus, in this case, when completing grinding in half an hour, an employee booked 15 cents in regular wages, a 13-cent premium (50 minutes at 15 cents/hour of saved time), for a 28-cent total—nearly double his hourly rate—then moved on to the next task. In accounting terms, the labor cost of the item rested at the 80-minute standard rate (40 cents), providing the company with 12 leftover cents from this operation. Lodge and Shipley credited 2 cents from each hour's savings to the shop foreman (after complaints that workers were taking home more than supervisors), divided a half-cent between the factory superintendent and his assistant, added 2.5 cents to the company pension fund, and put the balance into the general accounts to underwrite product development.[33]

This "profit sharing" of time saved was viable because (1) Lodge remained reliably committed not to shave inflated "standard" times though they were routinely more than halved; and (2) machine tool prices were themselves firm and, when shifted, moved upward. A "premium" wage payment system such as Lodge and Shipley devised that shared efficiency gains could best be sustained in a specialty trade where product quality and capacity were more crucial to buyers than "first cost," that is, purchase price.[34] Absent records of other area machine tool builders' wage plans, it is unclear whether Lodge and Shipley's approach was typical, but the firm's prominence and the publicity its practices drew suggest that its approaches to costing and worker payments were benchmarks against which others measured their efforts.

Of course, many specialty manufacturing issues stretched beyond the boundaries of individual firms, invoking either their networks of interaction or broader efforts at fashioning collective institutions. On the first count, as in fashion textiles and jewelry, newly started machine tool companies regularly located in factories, built by older enterprises, that were larger than the original firm immediately required. Thus Chambers Machine Co. occupied the third floor of the Cincinnati Machine Tool Co.'s works in 1905, the year that the Oesterlein Machine Tool Co. removed from rented space at Lodge and Shipley to its own, just-finished factory. Five years later, the new Cincinnati Grinder Co. continued the pattern, taking up the second floor of M. L. Andrews' new plant for manufacturing borers.[35] Though no one matched Lodge in spawning new entrepreneurs and coordinating their early marketing efforts, R. K. LeBlond's first foreman and at least three veterans from Cincinnati Milling Machine started their own firms; and in marketing, Fred Holz Jr., son of a CMM founder, worked for a decade after 1907 as a European sales agent, offering a full line of Cincinnati tools.[36] Network connections were also involved when Gray shipped one of the largest working planers from its own shop floor to another firm burnt out of its works, or when LeBlond aided Champion Tool's 1905 start-up by making available its plans for smaller lathes, echoing Lodge's aid to LeBlond's fledgling company a decade earlier.[37] Other local customs endured: shops open to visitors, common movements in wage rates, and separate establishments for foundry casting and tool building. When the city's tool and machinery specialties drew "large foreign buyers ... direct to Cincinnati," proprietors guided them through the network. As Thomas Egan noted: "This is a great advantage as we pass them around and this encourages the smaller factories and enables them to do a larger export business than they could command in any other location.... In shipping for export we [also] all work together. . . ."[38]

By 1900, area foundries and metalworking firms had collectively established regional casting price schedules, which included a sliding scale based on bulk pig iron quotations. Representatives from each sector gathered irregularly to resolve endemic points of friction: charge-backs for faulty castings, damage to patterns that tool builders supplied founders, delivery delays, rush orders, pass-throughs of railway rates for pig iron shipments, and the like. As the utility of steel castings for certain components was gradually recognized, tool builders urged their foundry colleagues to add this capacity, but the pricing convocations were the wrong forum for fostering capital investment in facilities; no local iron foundry would initiate the new process. Tool builders and their colleagues in other machinery specialties[39] turned to one of the city's many collective business institutions, the Cincinnati Industrial Bureau (CIB). Established at the turn of the century to promote the city's manufacturing interests, the bureau sought relocations and facilitated start-ups. For two years, bureau staffers labored unsuccessfully to lure a steel castings firm to the district, at which point its president, William Lodge, stepped forward to head a capital-raising campaign among bureau board members that underwrote formation of the Steel Foundry Company, opened in 1906. Though the firm floundered after the panic and was reorganized, the path to its creation illustrates the multiple levels of interfirm contact in the Cincinnati metal trades and Lodge's continuing leadership role in fostering cooperation.[40]

Other examples of collective action abound. In busy years after 1900 it was usual for companies loaded with orders to avoid overtime, "as the men do not take to it kindly and it invariably causes more or less friction." Instead, they passed their overflow to others "short of orders" who were "called upon to build tools for their more fortunate neighbors, who have more than they can conveniently handle."[41] This practice clearly necessitated a sharing of design and production details that, like the series of joint Cincinnati machine tool ads in *Iron Age*,[42] underscores the construction of an entrepreneurial community whose members collaborated even as they competed for business. Locational concentration eased these interactions, for by 1902 three-quarters of Cincinnati builders operated in the "tool district" along a two-mile stretch of Mill Creek on the city's west side. However, the rush of trade between 1905 and 1907 brought many firms to the limit of their expansion potential within the district. Four enterprises, led by CMM's Frederick A. Geier, leagued together to purchase a hundred acres of farmland adjoining Baltimore and Ohio lines beyond the city limits. At this Oakely site, their Factory Colony Company erected a common power plant and a jointly owned foundry to serve all parties, then constructed individual production facilities. By 1912 the colony sported five machine tool companies and two other factories.[43] This massive project marked Geier, then in his early forties, as a worthy successor (if not rival) to the aging Lodge and extended the tool community's customary collectivism to a green-field site dotted with modern, low-slung, sawtooth-roof factories.

Cincinnati machine builders constructed institutions as energetically as they fashioned tools, largely ignoring the contemporary merger movement.[44] They joined in creating the Cincinnati branch (CMTA) of the National Metal Trades Association (NMTA) in 1899 and the CIB in 1901. Through the CMTA, they held solid against the demands of unionizing machinists in a 1902 strike that involved three thousand skilled workers, then shared in forming the CMTA's Employment Bureau. Within a year, the bureau amassed files on six thousand men, offered free job placement services to newcomers and those laid off, and of course provided employers details on those considered "disturber[s] or time killer[s]."[45] As one of the national organization's strongest branches, the CMTA also took responsibility for publishing the NMTA's monthly *Bulletin*, routinely hosted its national executive committee sessions, and provided office space and staff to the Associated Foundries of Cincinnati, which it helped organize in 1910.

Despite disappointment in the steel foundry episode, the CIB's firm recruitment projects brought sixty-five new plants to Cincinnati by 1906 at an annual bureau operating cost of only five thousand dollars. It also sponsored excursion tours of both operating factories and sites for new plants and worked with immigrant placement agencies in New York to secure skilled workers for area manufacturers. In 1907, Lodge and the CIB secretary traveled to Philadelphia to "inspect the system of building homes for the working men" there, with an eye toward "solv[ing] some vexing local problems," for the local housing supply lagged behind the influx of industrial workers. With the panic, that flow reversed as displaced workers sought jobs in midwestern auto centers. Hence the CIB again focused on labor recruiting. Renamed the Cincinnati Commercial Association in 1911, it took on a fresh task, finding occupants for factories vacated by those departing for the Oakley colony.[46]

To address broader sectoral concerns, local tool firms assembled in William Lodge's office to initiate the National Machine Tool Builders Association (NMTBA) in 1902. Its founders were chiefly CMTA and CIB men who earlier organized the Manufacturers' Association of Cincinnati and Hamilton County, which had earnestly supported the new National Association of Manufacturers and sponsored its first convention at Cincinnati in 1895.[47] It made sense to organize local institutions engaging issues of labor supply, augmenting the industrial base, and technical education on a cross-sectoral basis. However, Cincinnati tool proprietors realized from the start that handling their sectoral problems entailed creating a national institution. Unlike furniture, sprawled across the landscape in its divisions, segmented markets, and diverse associations, or jewelry and fashion textiles, concentrated at a few eastern sites, the machine tool industry had two spatial nodes (Cincinnati and Worcester), a scattering of outlying firms in Yankee, mid-Atlantic, and midwestern cities, as well as a shared set of clients and technological challenges. In this context, local associations could have but little influence.

At initial 1902 meetings in Niagara Falls and Cleveland, twenty-eight NMTBA "charter members" agreed that "cooperative efforts would stabilize business conditions and raise prices to a higher level." Immediately, the association announced a 5 percent price list increase, echoing the tactic used in furniture circles. With Lodge in the presidency during a disappointing 1903, members defended the advance, taking "a strong position . . . in favor of maintaining prices and reducing production" rather than shaving prices to compete for such business as presented itself. A later analyst commented, "NMTBA leaders clearly intended to control price competition or perhaps to eliminate it entirely. . . ."[48] When demand rose in 1905, association president Lodge circulated a letter opening the question of another 5 to 10 percent hike; within months, the majority of tool builders "raised the schedule" accordingly. Though there was neither uniformity nor unanimity (the Gray planer company charted its own course), the NMTBA in three years had become the trade's pivotal organization and Cincinnati its hub. As NMTBA adherents neared a hundred firms in 1909, the year Geier became president, its Cincinnati contingent represented over a quarter of all members, more than double second-place Worcester's 10 percent.[49]

In its first decade, the association addressed price questions repeatedly, stressing collective well-being as the result of joint action. In the hard times of 1908, association leaders argued strenuously for holding the price line, and if weekly trade reports are any indication, few firms failed to comply. It also worked toward coordinating the timing of list changes. Thus when lathe builders announced that new lists would be issued January 1, 1910, buyers filed "heavy" orders during December 1909 "to avoid advances made through the standardization plan." Moreover, before World War I, NMTBA members began exchanging their confidential price lists and routinely sending along updates when changes were made, an early example of the "open price systems" more widely adopted in the postwar years.[50] These steps all contributed to regulating competition and emphasizing product characteristics and capacities in marketing tools. Together with the NMTBA's related concerns for refining cost accounting, extending foreign trade, and training skilled workers, they reflected the generalization of Cincinnati's collaborative relationships to the industry as a whole.[51]

Technical education, a major issue for the Cincinnati tool sector, was deeply entwined with labor supply problems. Though firms tried to shorten hours and spread available work in slack times, layoffs and outmigration of noncore workers often followed. When the machinery cycle again surged, firms urgently sought rehires, lest they be short-staffed facing jammed order books. For example, in the 1899 trade revival, Cincinnati tool building employment jumped 50 percent in a year, forcing employers to recruit throughout the region and beyond. Even in better times Cincinnati pay rates, lower than those in eastern cities (a 10–15 percent differential lasting until the war), pushed experienced men toward sweeter prospects elsewhere.[52] Seeking immigrant skilled workers was a haphazard measure, as was CIB advertising in the eastern press. To augment the local skills pool, Cincinnati machinery builders thus forged a technical education strategy early in the new century, addressing apprenticeship, adult continuing education, and the training of shop-experienced engineers.

If, as contemporaries and later scholars argued, apprenticeship in America was rapidly decaying, it remained vital in the machine tool industry. Replying to an 1896 *American Machinist* survey, Lodge asserted: "Relative to the assumption that apprenticeship is a thing of the past I wish most emphatically to state that this is not so. Every boy we take into our employment is taken on the apprenticeship plan." Apprenticeships were nearly universal in the Cincinnati metalworking district; indeed, the Gray planer company printed its standard apprentice contracts in both English and German, acknowledging the importance of immigrants' children to replenishing the skills base. By 1905 the CMTA circulated leaflets through the public schools boosting four-year indentures that provided $4.40 weekly to start, a $0.55 raise every six months, $100 "in gold" upon completion, and the prospect of upward mobility—"The apprentice of yesterday . . . is the salesman, foreman, superintendent, and shop owner of today." Moreover, local builders redesigned the apprentice course. In 1906, machinist apprentices could choose a two-year term that schooled them in operating "one or two machines" among the range of tools (thus offering them quicker access to full earnings) or a four-year "complete course" that would give them greater experience with all aspects of metalworking practice (and hence longer-term potential for upward mobility). In 1909, following the lead of the electrical trades, they included a classroom component at the city's new Woodward technical high school, in response to "the deficient education of boys coming to them for employment." With school district cooperation secured by Geier and Lodge, each year over two hundred apprentices spent forty-eight paid half-days in technical studies and other paid half-days visiting factories for "practical lessons" later reviewed in classes. Working in groups of about twenty-five with two "continuation school" instructors, they learned blueprint reading, shop mathematics and spelling, mechanical drawing, and, through "a collection of work actually spoiled by apprentices in the shops," a good deal about the sources of error in machining. In 1910, the continuation school added a "foremen's class" and weekly evening lectures by factory superintendents from the metal trades.[53]

Tool builders vigorously supported creation of Cincinnati's technical high schools, Woodward and Hughes, at which class instruction included chemistry, math, physics, and sixteen hours of weekly shop work. In 1910, local firms outfitted the "unusually well equipped" Woodward shops with eighteen of their machine tools so that a basic familiarity with metalworking techniques might be achieved on state-of-the-art equipment. This effort, and inclusion of manual training in elementary curricula, resonated with the goals and approaches of the national vocational/industrial education movement then flourishing, though at Cincinnati, metalworking firms largely initiated the business-government partnership. Geier's prominence in these projects brought him the presidency of the National Society for Promoting Industrial Education in 1911, when he traveled extensively explaining Cincinnati's programs.[54]

Continuing education in the metal trades had long been a Cincinnati tradition; the freestanding Ohio Mechanics' Institute, dating to 1828, was "the Alma Mater of the [local] tool builders." Its low-fee, employer-subsidized evening courses included metallurgy, chemistry, drafting, design, and foundry analysis, and drew scores of workers aspiring to foremanships or superintendencies. Following a half-million-dollar donation by a local capitalist's widow in 1908, the institute (headed by Cincinnati Shaper's Perrin March) commenced fund-raising for a much larger facility. Completed in 1911, the new five-story institute contained, beside classrooms and laboratories, its own machine shops, forge room, foundry, a 4,500-square-foot gallery to display Cincinnati products, an auditorium, a library, and a gymnasium. Mary Emery's gift financed the structure; and CMTA members provided funds for "furnishings" and donated all the necessary machinery, as well as its power plant, boilers, and small tools. Adult technical education was thus expanded and enhanced at private expense. A metal trade foremen's association also discussed technical matters at monthly dinners, blending higher-level continuing education with socializing. In 1911, the Oakley colony cloned a duplicate "Superintendents and Foremen's Club," for distance had inhibited supervisors' attendance at the regular group's downtown meetings.[55]

The city's most celebrated innovation in technical education, the University of Cincinnati's cooperative engineering course, drew instant support from machinery builders but was devised by Herman Schneider, a Lehigh incomer to the university's engineering faculty. Schneider sought a way out of the shops-versus-schools controversy that then bedeviled machinery manufacturers and engineering educators. Collegiate expectations compelled engineering schools to "adopt an academically respectable approach . . . with an emphasis upon scientific theory rather than industrial practice," but graduates often proved "ill-suited for disciplined industrial work and poorly trained in the practical application of their theories."[56] Schneider proposed extending the university course to six years, alternating classroom instruction and shop work in machinery and electrical firms; but when he presented his first outline to Cincinnati employers early in 1906, it appeared to be another approach to the apprenticeship problem.

Suggesting "a plan whereby apprentices might be placed on a higher level by the aid of a cooperative arrangement between the Cincinnati University and the employers," Schneider secured agreements from thirty-one firms by May 1906 (sixteen of them machine tool builders) to select two to ten apprentices, high school graduates only, for "the special course at the university." Chosen in pairs, the "apprentices will [weekly] alternate with each other at the plant and at the university. . . . Thus no machine at the plant will be idle and no loss will have to be borne by the employer." In the context of the CMTA's recruiting efforts and concerns about thinly educated new apprentices, the plan offered a special potential benefit: "that better prepared and more capable apprentices, high school graduates, will be secured" by participating firms.[57]

Despite general enthusiasm, a serious problem soon surfaced: cooperating companies did not have school-finished apprentices in their works. The first class had to be quickly recruited from the 1906 crop of regional high school graduates, enrolled as apprentices and students, then sent to the shops by July for two months' work experience before courses commenced in September. They would earn ten cents an hour for the first six months (25 percent more than regular apprentices to start) plus semiannual penny increments, but this yielded less than $200 in the first year, given unpaid classroom weeks. Over the six-year course, student apprentices would pay the university $400 in tuition charges, leaving, by Schneider's estimate, $1,000 in earnings for other expenses. Only those living at home, with parents furnishing food and lodgings until their sons neared twenty-three years of age, could expect to move surefootedly through this maze.[58]

Still, sixty candidates arrived at local machinery and tool plants in the summer of 1906. Fifteen resigned after the first day; another fifteen quit before classes began in the autumn. Schneider's reprise of their reasons for dropping out sketched the sensibilities of middle-class high school graduates:[59] "I had to get up too early in the morning"; "The work was too greasy"; "I'd rather be a lawyer"; "I want to complete my education in four years instead of six"; "My mother was afraid I'd get killed"; "The boss spoke gruffly to me." Of the thirty who persisted, only three were dismissed at year's end for "poor scholarship." Press attention the co-op engineering course received yielded four hundred applications for the class of '07, which numbered seventy-five. By 1910, the university fielded three thousand inquiries annually for fewer than eighty places. Over three hundred student apprentices mixed shop work and school labor in 1911, even as the Cincinnati co-op format was being copied across the nation.[60]

The efficacy of these diverse technical education projects is poorly documented but is perhaps less significant than the energy that tool builders and their machinery allies displayed in fashioning them.[61] Working together, Cincinnati metalworking firms created, sponsored, or supported training programs that commanded national attention because in specialty production there was no substitute for increasingly scarce skilled labor, there was a growing need for technically versatile supervisors, designers, and engineers, and there were few avenues for single enterprises to develop able and creative employees. Cincinnati's few bulk processors (e.g., Procter and Gamble) ignored these coalitions, for they little needed either skilled workers or engineers attuned to the vagaries of specialty design and labor processes. As in Philadelphia, Providence, and Grand Rapids, the collective building of sectoral institutions and involvement in wider organizational dynamics indicated specialist manufacturers' commitment to sustaining and reproducing production relations essential to their collective prosperity and individual survival.

The capacity to act on that commitment, of course, rested on profitability, which seems to have been substantial. One investigation of Cincinnati machine tool profits showed that Cincinnati Shaper earned twenty cents on every dollar of sales, 1900–1904, or $75,000 on a gross of $372,000. More generally, tool prices and workers' earnings each rose 30 percent in real terms between 1892 and 1912. In the same decades, raw materials expenses increased less than 10 percent while total output value more than tripled (up 210 percent in constant dollars), indicating "a highly profitable situation for the Cincinnati machine tool community." Rising levels of output accompanied by "relatively low prices for a key factor input, raw materials, provide the key to profits," which swelled a bit more through the reduction of agents' commission rates. The sizable expansion of production facilities that commenced after 1898 was accomplished with "no . . . substantial support from the financial community," no marketing of shares or floating of bonds, indicating that tool builders regularly invested "retained earnings accruing from profits" in building industrial capacity, technical innovation, and the jobs base. This last more than doubled between 1900 and 1912, by which time CMM alone employed over 1,600 workers. Machine tool proprietors erected fine mansions and took European vacations but also sensibly rechanneled a substantial portion of profits into enlarging their businesses.[62]

The state, which would have a major role in the industry's course during the next generation, seems to have little impinged on machine tool manufacturers before the Wilson era. Reliably Republican Ohio had few taxes that annoyed industrial interests, and Ohioans McKinley and Taft served in one or another of the nation's top political posts for much of the period. Local government responded to industrial problems, supplying "police protection" on 451 occasions during labor troubles in 1905, none of which arose at tool firms, while supporting educational initiatives and cautiously approaching other contentious issues, like smoke abatement. Yet two federal matters did agitate the local machinery sector. The first was tariff reciprocity, which tool builders heartily supported, for they occupied one of the few specialty sectors largely committed to exports and felt the force of French, German, and Russian tariff hikes on American products in the new century. Characteristically, William Lodge was the sectoral spokesman at reciprocity conferences and congressional hearings. The second was the federal corporation tax, a Taft administration exaction that machinery makers heartily opposed, then paid only "under protest." Little remarked then, but perhaps important later, was the fact that "for tax purposes it was necessary to select and adhere to a[n accounting] method for calculating depreciation considered acceptable by the taxing authorities." Soon, more detailed and officially mandated accounting practices would force specialist firms either to replace existing product-costing practices with standardized accounting systems or to increase their staffs to handle both. This yielded "the virtual disappearance of managerial product costing in manufacturing firms" by the 1930s, most likely owing to the expense of the double accounting burden, propelling a return to overhead averaging.[63]

Cincinnati's machine tool complex developed in the two decades after 1890 from a minor set of promising enterprises into the axis of America's fundamental specialty production sector, fundamental because machine tools shaped components for virtually all other machinery, as well as parts for consumer goods from watches to automobiles. William Lodge's visions of specialization within a specialty trade, and his practical efforts to support new enterprises focused on lathes, drills, or millers, launched the Cincinnati tool district, but technical ingenuity, collaborative and reciprocal relations, and a variety of collective organizations built its national prominence and international reputation. It was from Cincinnati that the machine tool sector derived its national institution (the NMTBA) and through Cincinnati that regional metal trades associations and national manufacturing alliances solidified their positions and perspectives. Local proprietors prized novelty yet shared their technical insights willingly and collaborated to expand the regional industrial base and the skilled workforce necessary for its vitality. With Geier's Cincinnati Milling Machine as the point firm, they forged a durable connection with emergent mass production corporations, while supplying versatile tools to manufacturing specialists like Baldwin Locomotive.

By 1912–13, Cincinnati's tool builders, more than firms in any other American industrial district, had stretched their innovative and flexible capacities, their design units, and their marketing arms to meet the needs of mass producers for customized, dedicated tools while shipping general-purpose lathes, planers, and millers to their specialty metalworking clients. No tool firm had collapsed amid the economic swings of the preceding decades, and more than a dozen newcomers had established themselves. Moreover, tool builders had collectively resisted pressures for price-cutting in slack years and had enhanced profitability, appreciating the trade's status as a supplier of basic technical resources for which capacity and durability were more significant than first cost. However modest their output when gauged against national industrial totals, Cincinnati tool builders understood their strategic significance to metalworking in the national and global economy and acted collectively to sustain and expand it.

## Conclusion

The Midwest, with its slaughterhouses and steel, flour, and lumber mills, has routinely been characterized as a natural base for mass production, which Detroit (with Ford) or Chicago (with Swift or Western Electric) exemplify. Yet, as the preceding cases suggest, both individual giant firms like Pullman and networks of specialists engaged in style-oriented or technically demanding trades built capacities for novelty, reaped sustaining profits from these efforts, and poised themselves for further expansion, drawing on materials delivered by bulk producers and contributing, with machine tools, to the erection of the mass production system. Facing depressions and other trade crises, these companies devised strategies to sustain their trajectories for growth and innovation, strategies that drew minimally on mergers and stock sales and recoded the gospel of efficiency into terms appropriate for manufacturing specialists. Their successes duplicated those of their eastern colleagues in specialty trades, but there were differences. Though Pullman's experience resonates with that of giant electrical producers like Westinghouse and General Electric, the furniture and machine tool networks inscribed quite different patterns from those articulated by the Providence jewelry trades and the apparel and textile manufactures at Philadelphia. Specialty producers would encounter serious pitfalls in the East, as we shall see.

## *Chapter 9*

# BACK EAST: THE ELECTRICAL EQUIPMENT INDUSTRY

AS SPECIALTY production grew from embryo to maturity in the Midwest, its significance at eastern industrial sites deepened. In older centers specialty trades expanded substantially on foundations laid before the nineteenth century's last depression. New York City's fashion apparel, fine jewelry, and printing and publishing sectors remained unrivaled, though often chaotic; and Philadelphia solidified its position as the nation's center of batch manufacturing, adding industrial instruments, fine chemicals, lace, and silk hosiery to its earlier competencies. Specialty metalworking in Connecticut's brass and silverplate centers (Waterbury, Meriden, New Britain, Bridgeport)[1] expanded apace, as did the Hartford machinery manufacture. Worcester continued to diversify and innovate in tools, machines, elevators, and fine metalworking; but whereas Providence's styled textiles and machinery enterprises strode forward, its jewelry sector stumbled in the last prewar years. Meanwhile, Newark and nearby towns extended their capacities in machinery and electrical apparatus, welcomed Tiffany's precious-metals fabrication plant, and revitalized their long-standing involvement in fine leatherworking, much of whose materials originated in Wilmington's fancy morocco tanning district. At the same time, important novelties proceeded from once-peripheral cities: braiding and full-fashioned hosiery machinery from Reading, Pennsylvania; one-piece forged I-beams from South Bethlehem, Pennsylvania; and the core components for electrification from Lynn and Pittsfield, Massachusetts, Schenectady, New York, and Pittsburgh.[2]

Among the many stories of eastern specialty manufacturing in this era, three will be examined in some detail. Discussing the deployment of electrical production in this chapter will help elaborate the notion of "bridge" firms that blended making custom and batch goods (generators, motors, switch gears) with mass outputs (lamps and bulbs), stretching the integrated anchors category. It will also underscore the role of "narrow focus" cities, similar to Grand Rapids, Jamestown,

New Britain, or Reading, in the extension of specialty capacity and will provide a different angle of approach to the problem of technical education. Zeroing in on the Providence regional jewelry trade in chapter 10 will illuminate the constraints facing networked specialist enterprises sliding into hypercompetitive conditions: struggling with stubborn distributors seeking price advantages, unable to manage competition, and drawn toward managerial tactics that for a time led the trade into an outwork maze like that in New York apparel. Finally, examining Philadelphia (chapter 11) presents an opportunity to explore the initiatives of manufacturers and workers in an "interactive" locale that included all three specialty groups (anchors—e.g., locomotives, shipbuilding, alloy steels; networkers—e.g., machinery, tools, fashion textiles and apparel; and auxiliaries). Philadelphia's diversity allows a wider view of the technical and organizational implications of sectoral practices, the efficacy of proprietary management, and the private ordering of industrial policies in major cities.

Although the long depression tested their resilience, eastern industrial specialists might well regard the turn-of-the-century era as the closest approximation to a "golden age" that could be imagined. Employment and investments soared, profits were hearty, innovations and new starts multiplied. Later, after the world war, Providence jewelry's messy problems would spread to other consumer trades; shipbuilding and railway supply would falter, while many machinery producers scrambled after business. Though tool builders, printers, furniture firms, and others would hold their ground or advance in the 1920s, specialty manufacturing sectors gradually diverged, unevenly and for different reasons in individual trades. For now, those hard years lie quietly ahead.

## Electrical Manufacturing: "Bridge" Firms and Specialty Production

At the Philadelphia Centennial Exposition, electrical pioneer Elihu Thomson remembered, "no buildings were kept open at night; there was no means of lighting [them] with any kind of ease or safety." The few electrical exhibits were "chiefly of telegraph instruments," plus a cluster of telephones (a novelty "which nobody believed . . . was of any account") and two displays of electric lighting, each featuring a single arc lamp.[3] By contrast, in 1893, 90,000 incandescent bulbs and

5,000 arc lamps brilliantly illuminated Chicago's Great White Way, giving the exhibition grounds "more lighting than any city in the country." The Westinghouse Electric and Manufacturing Company provided the incandescents, which were driven by "twenty-four, 500-horsepower . . . alternators" built at its Pittsburgh works. "Visually stunning," like Corliss's engine at the Centennial, the lighting displays dominated the exhibition experience, but the fair also included a huge Electricity Building, as well as an array of electrically powered constructions: fountains, elevators, pleasure boats, a moving sidewalk, a railway, and overhead cranes in the machinery hall, which visitors rode for "a bird's eye view of the mechanical exhibits."[4]

Much of the Chicago fair may have reflected utopian visions, but whether fanciful or formidable, its electrical displays signaled the practical dynamism of an industry for which 1893 was a banner year. George Westinghouse's company, founded in 1886 after his earlier successes with railway innovations, used the fair to demonstrate the superiority of alternating current (ac) over direct current (dc) for illumination, as well as ac's easy transformation into dc when desired.[5] That same year Westinghouse completed a California installation that transmitted ac power over a 35-mile, 10,000-volt line, an accomplishment which helped convince Niagara Falls' electric power developers to adopt an ac system and to award Westinghouse their initial equipment contracts.[6]

It was also in 1893 that the new General Electric corporation saw its first full year of operation. The corporation was formed through a merger of Edison General (EG) and Thomson-Houston (T-H), Westinghouse's chief rivals, both built on the inventiveness of the men whose names they bore. While addressing difficulties in production and system-development, the leading electrical manufacturers were financially unsteady in the early 1890s "because they had become capital-intensive enterprises prior to the development of capital markets suited to large-scale industrial expansion." Westinghouse had narrowly evaded bankruptcy in 1891, and EG's Samuel Insull managed expansion "only by juggling numerous short-term loans." Thomson-Houston, however, had forged a capital link with Boston financiers Lee, Higginson & Co., with whom T-H president Charles Coffin kept in constant contact. All three firms needed infusions of funds both to build capacity and to cover the costs of absorbing smaller companies, fighting patent infringement lawsuits, and developing new products. Ultimately, despite foot-dragging by Edison and Thomson, bankers for and managers from EG and T-H reached a merger agreement in spring 1892, a compact that deeply involved both Higginson and J. P. Morgan. EG was the weaker party in the new corporation, for Edison's focus on dc had limited its reach and profits, whereas Thomson and Charles Steinmetz at T-H were chasing the winning technical format for electrification—ac. General Electric profited almost immediately from its most celebrated contribution to the fair, a $500,000 five-mile electric railway; for it soon captured a contract to electrify Chicago's Metropolitan Elevated lines, a first step in displacing steam locomotives from urban transit service.[7]

Over the next twenty years, the electrical manufacturing industry assumed a bipolar structure. At one end resided two giant, full-system corporations, GE and Westinghouse, which together produced 40 percent or more of all equipment and supplies. At the other lay several hundred midsize and smaller firms making electrical components, chiefly standard goods like small motors, lamp bulbs, and switches. In crafting the "big stuff"—large motors, generators, turbines, transformers, and switch gears—the two leaders had nominal competition. For example, in 1904, Allis-Chalmers moved sideways from building heavy machinery into manufacturing large-scale electrical products by acquiring Bullock Electric of Cincinnati and Buffalo; but into the 1920s, A-C's total sales (including both electrical and nonelectrical machines) were a tenth of GE's. Moreover, the two key German electrical innovators, Siemens and AEG, declined to battle for American markets after brief experiments; instead they reached patent-sharing and territory-segmenting agreements with the U.S. leaders. Thus, in major system components, GE and Westinghouse consistently accounted for 70 percent or more of national production, although their commitments to universal electrical coverage mixed these specialties with mass-produced "minor" goods. Thomson-Houston's earlier "success was not based on the cost advantages deriving from the use of high-volume, high-speed production machinery, but rather was due to a strong distribution network that . . . enjoyed economies of scope by using the same sales force to sell and install different systems."[8] Much the same might be said of Westinghouse and GE, which supplemented custom and batch manufacture of heavy electrical machinery with expanded capacity to produce millions of "staple" bulbs, switches, and connectors. The two corporate giants bridged very different production formats in providing full electrical systems. How did they manage the straddle?

The first indication that George Westinghouse and GE president Charles Coffin understood their bridge position was that each corporation maintained separate production facilities for core power components and staple goods. Westinghouse retained its New Jersey lamp works, while building its immense Turtle Creek plants for construction of generators and heavy equipment in the mid-1890s. GE, after briefly considering consolidation of all facilities at a New Jersey site, announced in 1894 that it would "make the Schenectady works the main facility of the company, while the Lynn plant [T-H's complex] will be conducted as an annex for the manufacture of such small standard articles as are required in great numbers." GE's Harrison, New Jersey, works would continue the quantity production of lamps.[9] Engineers also recognized the difference in format. Commenting at an ASME convention, Charles Day noted: "There are two distinct classes of machine shops—*First.*—shops that do duplicate work only. *Second*—shops that do work of a general character." Citing Westinghouse as an example of the second class, Day added, "In the machine-works . . . each job, speaking generally, is an experiment."[10] And well it might be, for both corporations were dedicated to product innovation, and, like machine tool builders, to competing for installation contracts chiefly in terms of product capacity, durability, timely delivery, and economy of operation. Yet as the frontier of novelty was rapidly shifting, clients could hardly know the performance characteristics of newly designed or reengineered generators and transformers in advance. Hence each firm's reputation for quality and service and its ability to provide and update entire systems came into play, often overshadowing "first cost" considerations.[11]

Strategic diversification across traction, power generation and transmission, electric drive, and lighting, when combined with constant technical change, yielded a stunning diversity of products and work-in-process, sizable cost- and record-keeping challenges, and strong incentives for cross-use of components across models. Though inventors and executives, corporate structure and finance, research, international marketing, and company education schemes have drawn scholarly attention, actual production and its management at GE and Westinghouse have remained largely a black box.[12] Peering inside these firms, we will begin to outline patterns of technical development, contemporary managerial practices, and the dilemma of standardization. First, however, an overview of the Big Two's expansion through 1912 will frame the context.

Capitalized at $50 million and employing 8,000 workers at its inception, General Electric soon witnessed the evaporation of its central-station market and a withering of its financial base. Both underlying firms had taken their clients' securities in partial payment for electrical installations. In panic conditions, these could not be sold to raise cash. Worse, while new orders slumped, the flow of payments due on equipment delivered also stalled. To avert bankruptcy, Coffin and his bankers raised $4.5 million from key backers by transferring over $12 million of GE's "best utility securities" to a separate trust. As the crisis ebbed, GE shifted to cash-on-delivery sales terms, aggressively wrote down the book values of securities, plant, equipment, and inventories, undertook efficiency measures to cut production costs, and "redesign[ed] products to simplify manufacture." Zero dividends until 1898 plus write-offs totaling $20 million helped restore GE's finances by 1900, the year the corporation recorded $6 million in profits on income of $30 million, only $2 million of which derived from lamp sales. Dividend payouts recycled just 30 percent of the year's profits; GE retained the balance to build its surplus and support investment in new facilities, including the famed research laboratory at Schenectady.[13]

A new foundry and machine shop added 276,000 square feet to Schenectady's manufacturing work space in 1899, and building proceeded rapidly thereafter. By early 1904, GE's work floors reached 3.7 million square feet, double the 1899 figure, notably through extensions at the New Jersey lamp works and new turbine shops in Schenectady and Lynn, and through interim acquisitions of three smaller competitors. In 1900, GE purchased Siemens and Halske's Chicago power equipment plant, ending the German incomer's eight-year effort, then bought Sprague Electric, an innovator in control systems for electric rail, and Stanley Electric, a Pittsfield, Massachusetts, specialist in ac motors and generators. Employment, which had dropped to 3,000 in the depression's trough, reached its 1892 level again by 1899, then more than doubled to 17,000 by the end of 1903, when nearly 8,000 workers filled the Schenectady plants alone. As vice president of manufacturing and engineering, E. W. Rice managed production in a fashion consistent with the differentiation of work among the various sites. GE facilities "had virtually the status of independent units," each handling its own budgets, wage rates, sales, and inventories, though billing of their 15,000 accounts (1903) was centralized. Headquarters conducted an annual performance analysis, and Rice chaired a "manufacturing committee" of factory managers that established and revised "policies and procedures." Such a "highly autonomous . . . form of operation" did yield "some lost motion," but it also "gave boundless scope for the exercise of initiative."[14]

GE's 1903 sales established a new high ($41.7 million, $5 million in lamps), then passed $60 million in 1906 (lamps, $7.5 million) and $70 million in 1907, a mark that would again be reached in 1910 after the panic's effects wore off. The company erected its Erie motor and transformer works at this point, and in 1912 absorbed the National Electric Lamp works and its 26 subsidiaries after its first brush with antitrust regulation. That year sales mounted to $89 million and the directors declared a 30 percent stock dividend to shareholders. By the end of 1913, during which GE broke through the $100 million barrier, the firm had earned over 21 years nearly $150 million in net profits, $90 million of which had been paid out in cash and stock dividends, after $30 million had been set aside for write-offs. With well over 20,000 employees, a new seven-story research building under construction, and production chief E. W. Rice succeeding Coffin in the presidency, GE rested securely atop the American electrical industry.[15]

In the early 1900s, as a half-century later, Westinghouse was about half GE's size, though vastly larger than any other enterprise in the trade.[16] By turns genial and irascible, George Westinghouse was a classic proprietary capitalist, Charles Coffin's antithesis. Reportedly, his determination not to be a member of any organization he could not head led Thomson-Houston's bankers to abandon tentative Westinghouse merger plans and turn to the Edison deal. As one explained, Westinghouse's strength was "in the arrangement & control of a factory & in dealing with the practical problems. He is not a financier & he is not a negotiator. . . . He irritates his rivals beyond endurance." More the factory master than a corporate leader, Westinghouse routinely toured the floors of his plants—checking the progress of work, chatting with master mechanics, and nodding to apprentices. As he personally "made many of the key decisions," his firms "lacked a managerial hierarchy," for Westinghouse emphasized engineering and manufacturing over "marketing, finance, and organization building." Thus, just as Electric and Manufacturing had been created as a company separate from Air Brake and Union Switch and Signal, when Westinghouse initiated gas and steam turbine production in 1895, he chartered the separate Westinghouse Machine Company, rather than adding a division to E&M. Westinghouse sited each of these enormous plants in adjacent towns east of Pittsburgh, readily placed for his direct oversight. Bulb making, initially part of the Pittsburgh complex, was spun off as Westinghouse Lamp with factories in New Jersey and New York.[17]

At the turn of the century, when E&M employed nearly 7,000 workers, Westinghouse joined GE in acquiring enterprises to strengthen total system provision. Relatively weaker in railway motors than in other lines, Westinghouse purchased two Cleveland enterprises, Walker Machine and the electrical section of Lorain Steel. Each had sound experience in traction drive and both had been sued by the two giants for purported patent infringements, a tactic used repeatedly to force a sale. Westinghouse also faced a production squeeze. Growth in orders after 1898 had overtaxed the electrical and turbine plants' foundry facilities, pushing Westinghouse to contract for castings with companies as far distant as Cleveland. The solution to resulting transport expenses and delays was Westinghouse Foundry, a joint subsidiary of E&M and Westinghouse Machine, which built an immense castings shop at Trafford, five miles east of their plants and joined to them by "a private railway." As Trafford had been only a rail junction, George Westinghouse, like other major metalworking proprietors building facilities on "virgin" terrain (e.g., Henry Disston, Washington Roebling, George Pullman), thought it wise to construct several hundred houses for workers and their families, plus a central inn to serve as hotel, "bachelors' quarters," and meeting/recreation center. However, Westinghouse envisioned "no experiments . . . of the kind that met with such dire results at Pullman," seeking rather "to be regarded [by workers] as a friend, but in the manly sense, not as a condescending patron . . . or a doctrinaire looking on them as a class to be 'elevated.'" Instead, by renting new one-and two-family homes at rates comparable to those charged for "vastly inferior quarters . . . in the city," Westinghouse expected to secure and hold the sizable and stable workforce he needed in his proper sphere of activity—production.[18]

These ventures were not cheap; the Walker buyout cost $1 million, and establishing Trafford's plant, housing, and rail links surely matched or exceeded that outlay. Funds for such investments, or loans toward them, derived from or were based upon the Westinghouse corporations' profits, which in part flowed from advantages secured when GE and Westinghouse agreed to pool equipment patent rights in 1896, a sequel to an earlier agreement to share illumination patents.[19] This pact eliminated the litigation expenses of cross-suits between the two leaders and provided each of them with unfettered use of component inventions that underlay full system provision. It also built a common ground from which GE and Westinghouse could assault infringers and acquire the technically most valuable among them. The patent deal ended GE-Westinghouse court engagements, reinforced the reigning duopoly, and confirmed their shared interest in technical advances, even as they fervently competed for business. With its railway motor additions, Westinghouse won the prime ac contracts for electrifying the New York–New Haven railroad, thus challenging GE's dc New York Central installations. The antagonism between the leaders generated sharp exchanges at subsequent engineering conventions, for their agreement to pool patents merely refocused their rivalry on markets and technologies.[20]

In technical terms, an electrical system can most simply be imagined as having four components: (1) a prime mover that supplies power—with a steam engine, mechanical power that drives a rotating shaft; (2) an electrical generator linked to that shaft, which transforms mechanical power into an electric current; (3) a transmission network that carries the current to its point of use and, through associated apparatus, adjusts its voltage, etc., to the receiving device's characteristics; and (4) a receiving device that uses the electricity by converting it to light, heat, or, in a motor, rotary mechanical power again. Technical change in electrical manufacturing involved both "internal" incrementalism in particular machines and components and a "system" dimension, in which advances on specific units overmatched the practical limits of companion devices, pressing engineers to create innovations that could erase the bottleneck and enhance system performance.[21] Ac motors are a good example of the first pattern; steam turbine development well illustrates the second.

In 1890, when the Baldwin Locomotive Works became the first sizable plant "electrically equipped throughout," its cranes and machine tools drew current from dc motors, the largest of which delivered twenty horsepower (h.p.). Within the decade, competitive innovation between GE and Westinghouse would place Baldwin's apparatus several generations behind the technical edge.[22] Both firms had designed and marketed versions of ac induction motors[23] by 1894, and GE brought out a variation on its first model late that year, in part to avoid conflicts with Westinghouse ac patents. Westinghouse soon announced an improvement on its initial design, the promising Model B series. However, the Westinghouse motor had a tendency to spark, creating a fire risk, and its key moving parts "were exposed to injury, dirt and grit." Westinghouse engineer B. G. Lamme, recognizing these disadvantages vis-à-vis the GE models, scurried to craft improvements, mixing echoes of GE designs with novelties of his own devising, steps that underscore the value of the 1896 patent pool agreements. Within a year Lamme's revisions were embodied in the Type C series, "extremely rugged and reliable," and "inexpensive to manufacture." Indeed, when GE engineers performed a "detailed part-by-part cost analysis" of the two companies' 10-h.p. motors, they found their own was 25 percent more expensive to produce. Price data reflected only part of this differential, for in 1897 the Westinghouse 10-h.p. induction motor sold for only 10 percent less than GE's, adding nicely to Pittsburgh profits.[24]

Why, then, did the Westinghouse ac motor not drive the GE product from the marketplace? Essentially, the two motor series served "different purposes," as Harold Passer explained. "The Westinghouse motor was best suited for operation where the power requirement would vary considerably and where efficiency and speed regulation could be profitably sacrificed to secure flexibility. The General Electric motor gave its optimum performance under conditions of steady load. A predictable load of constant amount permitted the selection of a motor of the correct size to deliver power at high efficiency." Thus the two firms targeted different segments of the ac motor market, even as each continually sought to improve designs and to add motors that matched the capacities of its rival's leading lines. As technical features were refined, the weight-to-power ratio of ac motors shrank, making it possible to attach higher-power individual drives to industrial machines. For example, by World War I an ac motor delivering 20 h.p. was no more bulky than a 5-h.p. model from the late 1890s had been, and a downsized 5-h.p. could be fitted to applications impossible for its predecessors. Though the flow of "internal" technical changes entailed frequent revisions of production practices and a challenge to record keeping (for plans, repairs, parts), it also dramatically widened the uses for ac motors.[25]

As the sophistication of ac motors grew, a "system" problem obstructed their sales: in the mid-1890s, "a.c. power circuits were almost non-existent." The Big Two took different, but "complementary," strategic approaches to this difficulty. Westinghouse, with fuller ac system experience, pushed hard to extend the base of ac central stations with capacity to supply inexpensive industrial power that would then cre-

16. A diverse array of steam turbine components in process of construction at Westinghouse's South Philadelphia Machine Shop No. 3, 1921. Note the overhead cranes and the lifting loops bolted to the large components at lower right, for moving them to planers and other tools along the left and far right corridors. In the gallery (top left), machinists using smaller tools worked on less bulky parts. Courtesy of Hagley Museum and Library.

ate markets for its motors. GE, with a safe, constant-speed motor, worked to place individual ac systems at "isolated industrial plants," with notable success among rural textile mills. By 1900, each had built its equipment lines to duplicate the other's specialties, but expanding demand exposed another bottleneck: the speed and power limits of traditional reciprocating steam engines. Given operating rates of 100 r.p.m., large steam engines could not economically drive generators with capacity of more than 10,000 kilowatts. Thus augmenting central-station power capacity entailed adding pairs of prime movers and generators whose size necessitated constructing costly additions to existing power plants, all of which impeded lowering unit costs of electricity so that demand would continue to grow.

The solution to this system impasse was the steam turbine: its fan-like blades could spin at 3,000 r.p.m., and it had no obvious scale limitations. A turbine was lighter, more compact, less costly per kilowatt of power, cheaper to operate and maintain than a traditional steam engine, and it could replace bulky engine-generator sets to expand power capacity on existing floor space. GE and Westinghouse experimented with turbine designs in the late 1890s, installed their first central-station turbine generators in 1900 and 1903, respectively, and passed the 10,000-kilowatt barrier within the decade. Passer summarized, "These large production units reduced the cost of electric energy sufficiently so that it no longer was economical for a manufacturer to produce electricity . . . in his own generating plant." In consequence, turbine-powered central stations achieved "phenomenal growth . . . in the first decade of the twentieth century."[26]

## Producing Electrical Specialties

Manufacturing electrical system components was, if anything, more complex and technically exacting than steam locomotive building, the peak challenge in Gilded Age specialty production. Each of the three turbogenerators Westinghouse supplied to the Pennsylvania Railroad's Long Island division in 1905 weighed 175 tons, with a 24-ton "revolving element" that generated 7,500 h.p. Each of the six engine-generator sets it provided to New York's Manhattan Railway in 1901 stood 42 feet high and tipped the scales at just under 500 tons. Added to the difficulties of machining immense castings to close tolerances were the demands for precision wiring and winding of generator armatures that stretched to 20-foot diameters, the necessities of careful insulation, and requirements for repeated inspection of components and testing the fully assembled apparatus. To accomplish these tasks, engineers devised new production techniques, most notably the "floor plate" and "portable tools" approach to shaping multiton castings. Before 1900, Schenectady and East Pittsburgh engineers covered machine shop floors with interlocked, slotted steel plates, to which overhead cranes brought castings and a series of heavy tools bolted to these bases to accomplish each step in the metal-cutting process. Crafted singly or in small batches in workshops a quarter-mile long, expected to function effectively for a generation, main elements of the electrical infrastructure could be produced only by exceptionally skilled workers and shop-experienced engineers who struggled to meet performance specifications and delivery deadlines.[27] Comparable levels of skill and complexity were just as fully present at less commanding segments of the production system. As a detailed series of *Electric Journal* articles documented in 1910, the individual bench construction of small or midsize motors necessitated scores of precise steps, much fitting and filing, extensive handwork, and alertness to dozens of points at which errors could be imbedded in a half-built motor. That fault rates on inspection hovered below 2 percent is fair testimony to a craftsmanship created at some distance from the "American system"'s concurrent transition to Fordist manufacturing practices.[28]

Managing these sprawling plants, their swelling workforces and multiple product lines, their intricate finances and thousands of clients, was as challenging as was executing production processes. GE's approach, centralizing finance, billing, and control over major policy decisions while providing considerable autonomy to the manufacturing divisions, became "a standard way of organizing a modern integrated industrial enterprise."[29] The executive-level relations among Westinghouse's E&M, Machine, Foundry, and Lamp companies are unclear, but the organization of apparatus production was similar at Schenectady, Lynn, and East Pittsburgh. Three sorts of orders might be received: *specials* that needed design attention and new drawings prepared, *standards* for products previously made but not routinely manufactured for inventory, and *stocks* for goods regularly assembled in anticipation of demand. The last went directly to the storekeeper, who arranged packing and shipment of finished items in stock. He also scheduled moving inventoried components to an assembly area, once the made-to-order elements of standards or specials had been completed. "Stores" would track supplies and request a stock production order as its inventory reserves diminished or a purchasing order when raw materials levels drooped.

Specials and standards went first to Engineering, which generated designs and drawings for new work or located them for standards, then sent blueprints to the production superintendent, who issued manufacturing orders for castings, motors, and the like, and developed a time-and-parts flowchart ending with a delivery date. Engineering also supplied figures specifying minimum economical batch sizes for stock production requests, a key issue in the convergence of specialty manufacturing's practices with the mathematical skills of twentieth-century engineers. (Obviously, planning optimum batch runs had no interest for staple goods manufacturers, but it would interest flexible mass producers in the auto trades once General Motors' style diversity challenged Ford's standardization in the later 1920s.) As in machine tool building, shop tickets and forms tracked work in process and expenses for materials and labor, but at GE a separate costing department gathered these data from the purchasing and paymaster's offices, then added "the factory expense . . . figured under a logically determined percentage ratio" to tally the manufacturing cost for comparison against contract quotes or wholesale list prices. At Westinghouse by 1911, each of its eight production "units" (e.g., transformers or "medium-sized motors") had "its own percentage of indirect expense or overhead burden, figured on its total productive labor, which percentage is changed from time to time as circumstances seem to warrant."[30]

Though it has been suggested that GE utilized "the work of Frederick Taylor and other practitioners of scientific management" to devise "cost and control procedures based largely on predetermined or 'standard' costing," this does not appear to have been the case before World War I. A 1908 report examined costing at Lynn, where repeat standard or stock orders were more common than at Schenectady. It found the procedures in place fully consistent with the traditional "historical" costing generally used in metalworking, for managers received weekly reports of expenses in each class of motors for comparison with accounts from previous weeks or months. GE's Rice did fashion one cost-keeping novelty appropriate in an environment of persistent technical change and product diversification. The expenses of designs, prototypes, patterns, and so forth for a new motor or transformer were separately budgeted as "development jobs" within each product class. When the article became "marketable," an estimate of sales "for a period of a year or more was made," likely through analogies to related items' market performance; then development expenses were built into its price. GE eventually wrote off against profits any unliquidated expenditures for unsuccessful innovations, rather than lumping them with operating costs or loading them onto apparatus prices in that or other lines.[31] Westinghouse's method for applying overhead was also historical, involving not estimation of optimal costs but periodic revisions, as "circumstances . . . warrant[ed]," of empirically derived percentages. Nor was scientific management dominant in the leaders' premium and bonus payment systems. A third of GE-Lynn's 11,000 employees in 1908 worked on the hour or day rates efficiency promoters despised, "the others on piece work," chiefly without adornments. Together they produced "five-thousand distinct varieties of articles" in manufacturing sequences that involved "over twenty-thousand piece rate prices." Lynn's hourly workers often "turn[ed] in five or six time cards" daily, and pieceworkers "nine or ten each week," attesting to a constant shifting among operations that yielded 200,000 job work "vouchers" weekly for Lynn's Distribution of Labor department to process.[32]

The machinery and apparatus sections of these "bridge" enterprises remained skill-intensive custom and batch production centers well into the new century. They exemplify the integrated anchor format for specialty manufacturing, with operations often located in districts or secondary cities where they dominated the economic landscape and strove to be self-sufficient, rather than becoming enmeshed in a network of related and auxiliary firms as were contemporary shipbuilders.[33] Indeed, though their organization charts were "modern," their techniques for production management and costing differed little from those of other specialty producers in heavy machinery and infrastructure equipment, and may well have been several steps behind those of Cincinnati tool builders.[34] GE's Lynn managers, recognizing the consistent variability of their outputs, "frequently called . . . special meetings" of foremen and, at times, "workmen—selected, perhaps, for their skill on special lines or . . . for their judgment of and influence with their fellow[s]" to assess "questions of shop policy." In an environment of persistent technical change, working with supervisors and shopmen to devise approaches to the jobs at hand laid the *basis* for profitability, rather than being a drag upon an imagined optimal performance.[35]

In this context, the quest for electrical standardization was both understandable and quixotic. System producers like GE and Westinghouse sought to reduce the complexity of their production, repair, and replacement tasks by pressing clients to accept a narrowed set of standard voltages, current cycles, phases, poles, speeds, and horsepower ratings for electrical generators, motors, and associated apparatus. Economies of scale could supplement or supplant economies of scope if the staggering diversity of products could be reduced and long runs of standard goods became feasible (obviously, outside heavy machinery), thus lowering prices and enhancing users' profitability. A GE spokesman appealed to the electrical industry's clientele exactly along these lines in 1901.[36] Though he trumpeted the virtues of standard goods and scored the costs and complications of building special machinery, his plea drew no response. When Westinghouse, shortly thereafter, undertook to create a standards book for its regular outputs, engineers plowed through 120,000 drawings to specify the particulars of 200,000 items, first classified into 350 major categories (later expanded to 700). Electrical goods resisted reduction to a few staple models.[37]

A convergence on a modest number of current cycles and voltages slowly developed, but into the 1920s any broader standardization remained elusive. As a 1923–24 trade journal series revealed, GE continued to contend with "Extreme Variety Versus Standardization." *Industrial Management* editor John Van Deventer argued that the electrical industry's difficulties with "wide diversity, complexity, and frequent changes of design are the problems that are most commonly encountered in the large majority of American shops and factories today. For the average American industrial plant is *not* a one product, mass production manufacturing machine, but rather a plant whose managers . . . must constantly struggle [with] variety, change, special requirements, and small lots." Electric motors, despite all efforts at simplification, remained "comprised of over three hundred different materials" and extravagant in their variations. In the range between 10 and 250 h.p., the region of standard drivers (above 250 h.p., motors were "made to order"), over 16,000 variants "of standard construction" existed, only 600 of which GE built for stock. That figure excluded the lines of motors made for "special service, such as elevator, crane, coal mine, steel mill . . ., etc., so that the number of varieties of ratings is almost unlimited." At the same time, diversely designed large switchboards often included "over a million parts," while in turbine building "scarcely any two orders [were] alike," each taking six months to a year to complete, as in the 1890s. Hence, Van Deventer concluded, "the continuance of this condition of variety is assured for many years to come."[38] In electrical equipment, specialty production and technical versatility long generated the bulk of corporate profits without embodying mass production techniques or adopting scientific management tenets. Work and supervision in lamp and staple goods was of course routinized and repetitive, for operatives, engineers, and managers. However, electrical manufacturing's custom and batch production formats remained central to the industry leaders' success both before and after the Great War.

As in other specialty trades, securing a ready supply of adequately skilled workers and engineers pressed GE and Westinghouse to devise training programs appropriate to their labor process demands and effective in disseminating electrical knowledge not then generally taught in schools or colleges. As at Cincinnati, the Big Two installed apprenticeships for "boys" but handled engineering training initially in-house, not through collaboration with a local university. GE started traditional apprenticeships at the Schenectady works in 1901, but a more ambitious program, developed by Magnus Alexander, commenced in 1902 at Lynn with 12 youths "indentured for four years." Shop classes there involved five paid hours for each of forty weeks during the first three years. Wage rates advanced from a starting level of 8 cents per hour to 16.5 cents in the final year, plus the customary finishing bonus of $100. Alexander explained that GE had taken up the schooling function both because Lynn lacked "proper educational opportunities in the evening" such as were available in Philadelphia, and to provide the "practical knowledge" needed by journeymen and foremen. This included studies of physics, magnetism, and electricity in the second and third years, after apprentices had moved to the shops from the "training room" where they spent their first year. In 1906, the Lynn works employed 190 apprentices, by 1912, "over three hundred." The program's success resulted in its adoption at Schenectady in 1908, and 400 apprentices had enrolled by late 1909. Roughly two-thirds of the "graduates" accepted positions as "skilled mechanics," draftsmen, inspectors, or designers at GE.[39]

Westinghouse copied GE's format in 1909 and embellished it under the direction of Charles Scott, soon after called to lead electrical engineering instruction at Yale. Previously, apprentices had been "scattered through the works," learning in traditional fashion at the heels of

17. An apprentice pattern maker at Westinghouse's Philadelphia Works, 1920. Note the rolled shirtsleeves tucked inside for safety and the work apron covering his suit vest. Courtesy of Hagley Museum and Library.

experienced machinists. The new system involved four paid hours of weekly classes through the full four years. Apprentices spent their first two years in a separate training department, where they encountered, in sequenced two- to six-month blocks, the basics of drill, shaper, planer, lathe, boring mill, and milling machine practice, before being transferred to one of the works' divisions. Pay started at 9 cents per hour, rising to 18 cents in the final year, plus time and a half for overtime and the usual $100 on "completion of the course." Late in 1910, 231 apprentices were on staff and 50 on a waiting list. Both GE and Westinghouse accepted about half of their applicants. There was no shortage of candidates, for, as a labor investigator said of Westinghouse, "this company has more to offer in opportunity than most of its kind in the United States . . . and to learn here is worth a man's while [even] at a wage of $.15 an hour or less." Westinghouse also sponsored the Casino Technical Night School for male employees and apprentices, where senior mechanics and engineers taught advanced mechanical drawing and electrical fundamentals. Those completing the Casino School course were eligible for the "two years apprentice engineering course," where they joined incoming college graduates for advanced electrical studies and a rotation through the shops.[40]

Before World War I, college graduates with practical training in electrical engineering were relatively rare, even as their role in design, inspection, testing, and sales became more critical. At Westinghouse, the "graduate" course regularized an earlier practice of placing "technical graduates [in] the testing rooms," where by 1902 they prepared for "ultimate service in the engineering and erecting departments." Supervisors there readily acknowledged that a lack of shop experience and electrical theory "decidedly handicapped" their charges; hence they designed the two-year course to place training "on a more systematic basis." Engineering apprentices spent their first year in one of the plant's three main divisions (power, railway, or industrial products), working their way through winding, assembling, and testing procedures. The scheme built classroom sessions into the workday and required an additional six hours weekly in "assigned reading, study, and technical meetings." In the second year, supervisors placed them in specialties they deemed best ("pure engineering, salesmanship . . . , manufacturing, erecting, and operating"); and on satisfactory showings, each received "a regular position in the company's service."[41]

Charles Steinmetz and Magnus Alexander shaped General Electric's two-year Test Course as a postgraduate engineering and managerial training experience. Its students rotated among the testing departments within Lynn and Schenectady's many divisions, rather than being assigned to one area and tasked to winding and assembling. Though theoretical studies supplemented testing practice, "emphasis was placed upon the business aspects of engineering, the relationship between design and the dollar." All attended lectures on "the commercial and managerial aspects of the industry," and in 1912 GE introduced a special course for engineering salesmen. The company also encouraged test men to participate in engineering societies, introducing them to "the privileged world of the professional and industrial elite." Westinghouse's Electric Club and GE's Edison Club in Schenectady and Thomson Club in Lynn socialized these novices to the company's "spirit," affording them leisurely contact with both one another and corporate leaders. A 1919 survey showed over half the Test Course graduates to have remained with GE, with the rest holding "prominent positions" in industrial, transport, and utility corporations and in government.[42]

By 1912, Westinghouse's management had taken on the professional style long in place at GE, for "Uncle George," as he was known in the works, stumbled badly in the financial straits of 1907–8. Though his other enterprises remained solvent, the panic caught E&M and Machine overextended, and they entered receivership. A reorganization plan stabilized both firms, in part by converting "merchandise debt" into stock shares; but a new board of bankers and creditors stripped Westinghouse of effective authority. His managerial shortcomings had temporarily damaged the company that his technological enthusiasm and imagination had helped build. When his 1911 effort to regain control failed, Westinghouse departed, demoralized, to his rural New York estate, where he died three years later. Though it also ended an era, Rice's 1912 succession of Coffin in Schenectady was smooth and serene. Neither corporation failed to post profits and pay dividends thereafter until the Great Depression.[43]

Masters of technological novelty and decisive innovations, GE and Westinghouse dominated the electrical industry in and after the 1890s. Their diversification, bridging specialty and mass production, constituted them as two of the industrialized world's four electrical system-builders, yet specialty manufacturing drove their growth and provided the bulk of their sales and profits. Insofar as they were premier integrated anchors, their technical orientations were national and international, like their markets; yet within the firm, each took responsibility for extensive training programs crucial for producing shop-floor skills and engineering capabilities. Neither needed the regional organizations and links to local government and higher education that were necessary to establish trade practices or realize training schemes in Cincinnati's networked machine tool district. Nor, given their commitments to integration and self-sufficiency, did they depend on auxiliary firms or foster potential entrepreneurs among their employees. Instead, each erected a differentiated and specialized corporate apparatus that provided steady, well-paid jobs, avenues for building skills and careers, and, for many, the challenges of precision work on multiton transformers or delicate X-ray tubes. In later decades, the mass production consumer goods divisions of GE and Westinghouse would rise in significance, but into the 1930s the electrical giants far more closely resembled machinery, ship, and locomotive builders, their colleagues in specialty metalworking, than they did American Tobacco, Swift, or Ford, their companions in American big business.

*Chapter 10*

# THE PERILS OF PROVIDENCE: JEWELRY'S ERRATIC COURSE

LIKE THE leading electrical manufacturers, New England jewelry makers relied on metalworking skills to produce an enormous variety of specialty goods, some fairly standard and made for stock but most fashioned once orders had been booked. There the similarity virtually ended. At the turn of the century, the entire workforce at Providence and nearby Attleboro and North Attleboro roughly equaled that of GE's operations but was scattered among over three hundred small and midsize firms. Although the depression had flattened demand and ruined collections in both sectors, jewelers' inexpensive baubles sought ordinary consumers' loose change, not millions from nervous financiers and system builders. In jewelry, there was nothing like the systematic technical change and product refinement, nor anything comparable to the integrated manufacturing and distribution/service formats, that GE and Westinghouse developed. Instead, styles churned chaotically, makers battled with buyers and with one another, while halfhearted efforts to stabilize the trade foundered. Though jewelers joined discussions of—and took action on—many key issues that agitated other specialty manufacturers (technical education, costing systems), they mastered few of them. By 1912, segments of the mid- and low-priced jewelry trade occupied a nexus of design theft, sharp interfirm tensions, and sweatshop/outwork labor that Progressive Era reformers targeted.[1]

In these years Providence's anchor Gorham Manufacturing Company presented a gnawing contrast, over a thousand workers activating a sprawling plant sending forth fine silverware and "art goods" in various metals and sharing with Tiffany the commanding heights of silverwork nationally. In May 1893 Gorham foundry workers constructed an intricate floor mold for a seven-foot-high statue of Christopher Columbus, using plans furnished by Frederic Bartholdi, designer of the Statue of Liberty. They melted a ton of silver (worth $25,000), then deftly poured it into the "sunken pit" to cast the centerpiece of Gorham's elaborate Chicago World's Fair display. Some weeks later, after Gorham announced itself very busy, particularly in "large specimens of silverware," the regional jewelry market report carried the pathetic headline "Business Not Altogether Dead."[2] Gorham would not prove immune to the economic slump (it cut wages 10 percent and shifted to two-thirds time in 1894), and there would be a run of fat years for the jewelers over the next two decades; but try as they might, jewelers could never achieve Gorham's visibility, stability, and profitability. Their trade's mutations will here be reviewed in three phases: depression and recovery (1893–1900); edgy prosperity (1901–7); and troubled prospects (1908–12). Throughout, issues of sectoral structures and networks, technical change, market relations, labor dynamics, and institutional initiatives will be considered.

## Depression and Recovery, 1893–1900

William R. Cobb operated a Providence jewelry "findings" firm in the 1890s, a typical enterprise—modest in scale and sales and immersed in local production networks—yet unusual, for it endured past World War I and left records that have been archivally preserved. Cobb had succeeded Otto Merrill in about 1883 in a business that provided jewelers and other makers of metal novelties with diverse components—clasps, swivels, pin-, brooch-, or button-backs, glazed, gilt, or enameled joints, mountings, bars, and the like. Cobb's few workers made them by the dozen or the gross for roughly two hundred clients, chiefly in the Providence district, in orders that ranged from $5 to $50 and accumulated to $15,000–$20,000 annually. For each item, Cobb arrived at his sales or contract price by summing his materials costs, the labor expense for the half-dozen or more hourly workers involved, and a "shop expense" estimate, then added a quarter of this manufacturing cost figure as profit. Cobb used no piece rates, for the work was too varied and his shop too small to make establishing and monitoring them worthwhile. He figured shop expense at 25 percent of the direct labor bill, ignoring expenses for toolmaking and charges for work sent out to other specialists. Though some manufacturing jewelers included these expenses in their cost bases to yield a larger paper profit, Cobb evidently adhered to the widespread view that the practice generated inflated prices—prices that either would be hammered down in negotiations or would balk repeat orders as buyers sought lower rates from other makers. His network of auxiliary firms included, among others,

J. Briggs and Son and Vennebeck and Co., who straightened and cut bulk wire and rolled it into special shapes or die-cut plate into blanks for shields or bars, and J. P. Bonnett, S. W. Cheever, and W. F. Quarters, electroplaters who gilded and burnished components before or after "making up."[3]

Of course, Cobb was also part of a larger network, that array of findings firms, refiners, die-sinkers, tool builders, and others who supplied final-product jewelry companies. Here some explanation is needed, for what the trade called jewelry making and what we commonly take it to be are somewhat different. In 1895, the regional jewelry complex included 350 companies, roughly three-fifths of which (205) sold finished goods to jobbers, department stores, mail-order houses, and retailers. Most (ca. 140) marketed familiar items—rings, bracelets, necklaces, "ear wires," hair ornaments, and pins for women—but there was also a strong "menswear" trade in watch chains and ornaments, fraternal paraphernalia, patriotic and campaign goods, tiepins, and decorated cuff, shirt, and collar buttons and studs. Sixteen companies specialized in fancy buttons and studs alone, whereas a dozen represented themselves as "badge houses," which included religious, union, and school emblems in their lines. Eighteen others focused on pearl, shell, and stone work for both men and women, and over a score made rings only. Behind these, 68 findings houses provided components, ranging from miles of machine-made brass and plated chain (in 2,000 designs offered by H. W. Wilmarth and S. O. Bigney) to the novelties and settings Cobb and others made on contract. (Cobb worked with firms from every trade division.) Associated with both groups were a dozen platers, 22 die-cutting shops, 6 enamelers, 11 tool and machinery specialists, and 14 refiners (who recovered precious metals by recycling shop sweepings).[4] These auxiliaries worked at the edges of technological change to apply electrical and mechanical advances to the trade's requirements—extending die press work, annealing, and electrochemistry to practical problems. A cluster of printers prospered by specializing in advertising plates, jewelry catalogs, and cards for mounted sales displays, as did a group of jewelry box and sample case builders, and, inevitably, several auctioneers ready to dispose of a season's dead stock or failed entrepreneurs' physical assets.[5]

The auctioneers were fairly busy in the depression's first years; though he survived, Cobb saw his turnover and his workforce halved in 1894. Two-thirds of Attleboro's workers were idle that year, and the town sustained some of the men among them by setting them to repairing public roads. However, the perennial quest for novelty and a falling silver price interacted to create an 1895–96 revival, before the silver goods surge was overdone and trade slumped again until 1898. Noting the steady slide of raw silver quotes, "at least one-third of the jewelers" in Providence commenced making silver specialties, "and many increased their bank accounts thereby." Yet as with earlier and later crazes, the silver balloon deflated once firms rushed to copy successful styles and cut prices to grab jobbers' and department stores' orders, ultimately cheapening the goods to the point that their appeal faded. Other fads followed at decade's end, notably "beauty pins" (twisted wire hair ornaments and brooches with inserted stones), with the same results—huge initial orders, rampant duplications, price slumps, and a collapse of the novelty's desirability.[6]

A later trade observer commented acidly on the jewelry sector's peculiar response to heated demand. In other industries, when buyers were eager for goods, prices stiffened and profits bulged; but in jewelry, followers' eagerness to ape innovators' designs yielded a perverse result—multiplying derivatives of quality low enough to shave prices, wreck profits, and ultimately kill the market. Thus he recorded the process through which jobbers and jewelers helped recode fashions as self-destructing, seasonal commodities. Imitation may be a sincere form of flattery; but in fashion trades, it routinely proved demoralizing.[7]

The "evils of overcompetition" derived from the jewelry sector's own structure and technical capacities. Unlike furniture, where copying styles demanded several months, shop-schooled jewelry proprietors, aided by findings makers, could duplicate current models in a few weeks. Moreover, in jewelry, unlike furniture, jobbers both ruled the bulk of market exchanges[8] and had powerful incentives to promote design theft, especially in cheaper styles or lines that could be made cheaper so as to boom a novelty market. Why pay originators $36/gross for pins wholesaled at $50/gross and retailing at $0.75 each when a copyist might quickly make decent facsimilies for $24/gross that could still wholesale and retail at the same prices? Five hundred gross of the little beasties (bear and frog pins had their vogues) would provide the alert jobber with $6,000 in added revenue, support small enterprises,[9] and prevent the monopolization of styles by rival jobbers with ties to their originators. Manufacturers well knew the game, proclaiming virtuously the necessity of resisting "1/12th dozen" orders likely destined for duplication, yet were unable to resist the lure of large sales that might be reaped.[10] As in styled fabrics and apparel, copying was endemic and annoying, but moralizing appeals proved as weak as patenting, given the sizable expenses of challenging infringers and the brief life span of designs. Jewelry rested far from Corliss's engines or the major electrical corporations' innovations, where the resources for and rewards from battling patent claims were ample.[11]

Marketing practices also shifted decisively during the depression in another way. "Since the hard times of '93–'94–'95, buyers have been more cautious and conservative in making purchases, preferring to give small orders and repeating the same as their wants dictate, instead of placing large ones at the risk of being unable to make clean sales." This imposed costs on manufacturers "on account of the greater expense of making small quantities," but did limit their exposure to "losing heavily by [jobbers'] failures" and reduced the scale of end-of-season returns, an old trade abuse. Jobbers also continued to insist on six-month credits and "dating ahead," which slowed makers' cash flows. These conditions enhanced the attractiveness of sales to department stores or mail-order and "scheme goods" houses (i.e., Buffalo's Larkin Co.), for they paid cash promptly on receipt of goods, and the houses, at least, placed huge orders for cheap chains, bracelets, and collar studs used as premiums. In 1897, Republicans restored to power pushed the jewelry tariff rate to 60 percent ad valorem, nearly doubling the 35 percent rate established in the despised Wilson tariff of 1893 and balking the flow of low-end German imports. By 1899, area jewelers noted that it was no longer necessary to "push goods upon buyers," for at last the market was "drawing the goods from the manufacturers." Though the new "copper trust" had forced brass prices up nearly 20 percent and tales abounded of kickback demands among department store buyers, the worst was over.[12]

One sign of clearing skies was renewed construction of factory "apartment" buildings. Before 1893, five manufacturers had erected multistory plants designed to house their own operations and to provide smaller firms with leasable space that could be reclaimed should the core enterprise expand. In 1893–94, the Kent and Stanley jewelry firm adopted this strategy on a grand scale, erecting the seven-story Manufacturers' Building at a cost of $625,000. They filed bankruptcy within two years, victims of "this monster undertaking," and the property passed at auction to Charles Fletcher, local worsted fabrics magnate. As it filled with new renters in 1897–98, other investors commenced smaller tenant-oriented structures, completing five of them by 1900 in or near the jewelry district, most with retail stores on the ground level and from four to seventeen manufacturers above. Each was sponsored by real estate and banking interests, not by an individual firm on the older pattern. Indeed, when the area's largest ring makers, Ostby and Barton, doubled the size of their factory at this time, they made no provision for tenants, recognizing that market mechanisms had institutionalized the creation of factory spaces for lease.[13]

Outsiders also saw profit potentials in the reviving jewelry industry. In April 1899, drawn by the price increases that the International Silver merger had yielded, New York promoter Seymour Bookman began soliciting leading Providence and Attleboro firms to join a jewelry consolidation that would concentrate about three-fifths of the industry's capacity in one corporation. *Manufacturing Jeweler* mocked the notion, arguing that "it would be utterly impossible to get any considerable number of manufacturers interested, as each one had his own individual opinions and methods of doing business, and would not sacrifice them for a common cause." Better to "organiz[e] against the jewelry credit system that is constantly a loss to them" than to chase this chimera. Manufacturers interviewed allowed that they would be glad to sell their properties to a trust at high valuations and for cash, but admitted that this would hardly limit competition, "for they would immediately go into business again." As one sagely added, "when jewelry is sold like nails, or car-tracks, or any staple commodity, then a jewelry trust might appear feasible." "It would seem as sensible to form a trust of 'artists' brains' as to form a jewelry trust," he continued, because "dealers . . . are ever looking for something new—new creations—and originality." A few staple lines might work out in a merger, "such as collar buttons, plain band rings, etc., but they are a small part of the great whole."[14] Fewer than twenty proprietors responded to Bookman's call for a mid-April meeting to explore his proposal. After findings manufacturer S. O. Bigney vehemently attacked the plan ("Our house will submit to the dictation of no man or set of men . . ."), only one maker spoke favorably about the concept. The merger was interred without ceremony when another meeting two weeks later drew an audience of two.[15]

The jewelry trust idea was plainly deficient, as was its initiator's knowledge of the industry. Yet there was an important insight in J. M. Fisher's dismissive comment: "Jewelry manufacturers could never successfully unite their interests." As outlined above, the trade was structurally and functionally divided among final goods makers, component suppliers, and ancillary specialists. The first group subdivided into companies primarily working gold and gold plate, silver, or brass (or a combination of these), producing tens of thousands of designs for market segments ranging from giveaway premiums to middle-class finery. Given this spread, there were few "interests" or "common causes" around which proprietors could unite, and for each of these they created a separate institution: a Board of Trade for credit checks, a Security Alliance to pursue thieves, a League for life insurance, and a Protective League for theft insurance. At tariff revision times, separate Providence and Attleboro special committees convened to forward manufacturing jewelers' petitions or protests to Congress, though these did have better effect than occasional proposals promoting a public School of Metallic Arts in the Rhode Island capital. The umbrella New England Manufacturing Jewelers' Association was moribund in the late 1890s, reviving as a "rallying point" for the trade only when labor agitation surfaced in 1900. With dues at only fifteen dollars per firm, half of which was literally eaten up at semiannual banquets, it had neither the resources nor the charge to challenge jobbers' market power, establish quality standards, or in any sense regulate competition. This organizational diffusion and incapacity would persist.[16]

## Edgy Prosperity, 1901–1907

The new century's opening years almost uniformly lifted the fortunes if not the spirits of New England jewelry manufacturers. They featured both record sales and the first labor controversies in two decades, plus continued anxieties about fierce competition and prickly relations with wholesalers. The raw numbers were surely impressive. Between 1899 and 1906, Providence jewelry output values increased nearly 60 percent to $21 million, while wage payments rose by half to a workforce only 15 percent larger. Moreover, the increase in women's average earnings (34 percent to $392) far outran men's gains (22 percent) though not their incomes ($668). Trade in the Attleboros jumped more dramatically, from $8.4 million in 1900 to $14.9 million six years later (up 77 percent), and jewelry employment passed 6,000, closing with Providence's 8,150. However, for workers as consumers, 10 to 15 percent increases in the cost of living undercut their earnings advances.[17]

As always, there are stories behind the statistics. A labor upheaval, accompanied by frequent references to rising consumer prices, may have brought wage hikes despite organizers' failure to establish effective unions; but part of workers' income gains likely came just from longer hours in busier rush seasons. There was no change in the trade's severe seasonality—a brief trade flurry after New Year's, flat springs and summers, and a succession of seventy-hour weeks from September until early December.[18] Moreover, swelling output values chiefly reflected "the great increase in the manufacture of gold jewelry" after 1900. Materials expenses thus rose faster than any other cost (80 percent), with the consequence that value added by manufacture expanded by only 37 percent, appreciably short of the wage bill's growth. Put another way, a 60 percent sales increase added just 25 percent to the funds firms could draw on for rent, power, office/selling staff, and other expenses, after paying for materials and labor and before figuring profits. Rising sales and higher-grade goods did not necessarily bring commensurate returns.[19]

The data also conceal the opening phase of a shift in the trade's labor process organization. Between 1899 and 1906, though women workers' compensation rose, their numbers decreased, at least in the factory reports. Given that male employment increased over a thousand, this is odd, for women had long been tasked to jobs (e.g., cleaning and carding jewelry) that should have expanded in tandem. As scattered reports first appearing in 1905 indicate, manufacturers were moving the most routine of these tasks outside their factories and into women's homes, setting in motion an outwork and subcontracting dynamic that saved factory floor space, exploited the labor of married women and their children, and expanded significantly over the next decade. Figures on output and employment also mask the substantial turnover of firms in the jewelry industry. A 1903 analysis of the regional trade since 1893 showed the total number of enterprises to have risen from 327 to 385, but also revealed that 125 of the companies present at the onset of the depression had vanished within the following ten years (38 percent) and that 183 new starts had taken their places. Survivors represented just over half the firms active in 1903. The biggest among them, Ostby and Barton's ring house with 690 employees and T. W. Foster with 305, might exude confidence, but most jewelry entrepreneurs and many workers had reason to greet prosperous years with a caution born of experience.[20]

Though rumors of labor organizing circulated through the region in 1900, a visible union movement did not appear until 1903, when the AFL-affiliated Jeweler's Union and its colleague, the Brotherhood of Silversmiths, attempted to build on recent achievements in the New York district. Manhattan's fine jewelers, who dominated the trade's high gold and precious stones division, staved off a threatened strike in fall 1901 by conceding a nine-hour day to organized workers. The next autumn, in actions nicely timed for the rush season, the Jeweler's Union targeted Newark and the Silversmiths the entire metropolitan area, calling for nine hours' work with ten hours' pay. Newark's leading firms "formed a tacit agreement . . . to resist," and strikes commenced in late October, just as 600 of Tiffany's 1,700 Newark workers presented similar demands. New York silversmiths struck in early November; at least six firms "accepted the men's schedule" within a week, encouraging the Tiffany force to walk out on the tenth. Management "discharged" them, only to discover "a great demand for the Tiffany workmen . . . by New York firms who have granted their request for nine hours. They find they must have more men to finish their orders for the holiday trade." Tiffany and five New York silver companies held out and prevailed by mid-December, but Newark jewelers agreed at year's end to commence nine-hour days on January 1, 1903.[21]

Providence interests, employees and manufacturers alike, keenly watched these events unfold alongside nearer organizing drives among Connecticut brass workers and Massachusetts horn and celluloid ornament makers.[22] Gorham moved first, reducing its workweek to fifty-five hours at sixty hours' pay for four summer months in an effort to preempt unionists, then fired six members of the Silver Finishers' League to emphasize the point. Die-cutters made the workers' first sally, presenting the ten-for-nine demand in May 1903 before conducting a one-day walkout at all the small shops and Gorham. Several small die makers agreed to nine-for-nine and work resumed; but the others refused all propositions, whereas Gorham rejected three different proposals and fired its inside die-cutters. Before the die shops controversy faded out (without gains for the workers), the Jewelers' Union circularized several hundred regional firms on May 15 with its demands: ten-for-nine, time and a third for work beyond nine hours, and a paid half-hour dinner break during rush seasons. Spring meetings had brought 1,700 Providence men onto its rolls, making their

Local 9 the largest in any jewelry center. The local soon laid plans for a women's "auxiliary" and for recruiting north of the border. In tune with the trade's seasonality, the union set September 1 as the deadline for manufacturers' acquiescence or, failing that, a strike vote. Proprietors "through lack of organization, [were] entirely at sea as to what may be done or what should be done," having believed "that it would be impossible to organize the journeymen into an effective union."[23]

The summer's delay afforded manufacturers time to effect a collective response; but far worse, it also provided the occasion for the union to shoot itself in the foot. Even as NEMJA canvassed its members and several hundred nonmember firms, the Jewelers' Union propelled itself into a headlong contest with Manhattan firms that helped wreck its Providence initiative. A New York worker withdrew from the union and stopped paying his dues. His colleagues at an all-union shop demanded his discharge; its owner refused and workers left their benches. The union backed the shopmen's position, but 69 Manhattan jewelry manufacturers supported the owner by locking out 1,400 union workers in early August. NEMJA soon announced that 252 eastern firms had signed a resolution rejecting Local 9's propositions, just as the union called on Providence members for 50-cent weekly contributions to aid those idled in New York. The women's auxiliary failed to ignite measurable support, and earlier enthusiasm for a strike waned. On September 1, the union president temporized, saying, "We are willing to let [owners] bide their time; we can wait for a few weeks." Ten days later, the New York lockout succeeded and workers surrendered "unconditionally." Providence manufacturers' solidarity was never tested, for the union's moment had passed. Local 9's membership faded quietly away, as did the "labor question" in the regional jewelry trades. The conjunction of the New York tactical catastrophe (and its demands *for* funds from Providence workers rather than offers *of* funds to back their impending strike) with a rare unity among "Eastern" proprietors hostile to "interference" in their affairs sank the labor movement in 1903.[24]

NEMJA received a membership boost from the antiunion drive, participation passing the three hundred mark in 1904, when it incorporated as the more inclusive New England Manufacturing Jewelers' and Silversmiths' Association (NEMJSA). Yet the association continued to fumble. It was unable to assemble a group exhibit for that year's St. Louis Exposition, just as it had earlier failed to mobilize members' contributions to help match a $50,000 endowment contributed to the

Rhode Island School of Design (RISD). Though complaints continued about jobbers' abuse of discounts and the increase of cancellations (leaving firms with made-up stock that lacked buyers), though over-competition and price-cutting remained endemic and all admitted the need for thorough costing, though ideas for invigorating technical education and trade schools surfaced periodically, NEMJSA took none of these issues to heart, instead continuing its round of banquets and summer excursions. In these prosperous years it managed only to memorialize Congress on behalf of a weak National Stamping Act (prohibiting marking that overstated the karat content of gold goods) and to secure $350 from members to fund free places in twice-weekly, evening jewelry classes held in a RISD basement room. Absent a crisis, it lapsed into inactivity.[25]

Still, the Providence district remained an ideal place to manufacture jewelry. When a Taunton, Massachusetts, editorialist complained that, despite "better facilities," his town could not attract jewelry firms, a Rhode Islander responded, "Experience has shown that it is easier, more convenient and more profitable to conduct a manufacturing business in places where similar manufactures are largely conducted." Of course, this clustering in part reflected ready access to a pool of workers already "skilled in that particular branch," as at regional textile or shoe centers; but in jewelry there was something more. Nearly all attempts to establish jewelry plants in the West had failed, despite ample capital and worker-training schemes. "The chief reason for this is the difficulty of getting *supplies* promptly. No matter how well equipped a jewelry factory is, there occur every day demands for this, that, and the other line of supplies, or for outside skilled assistance, in one way or another, which is [*sic*] impossible to be obtained . . . in any towns far removed from centers of jewelry making, where such cognate pursuits are carried on." Although Taunton was perhaps near enough to anticipate some spillover from the Attleboros, it could hardly rival the flexible response of the jewelry district's networks.[26]

Three sorts of difficulties still troubled individual firms, two concerning the labor force, the last, profits. Apprenticeships had faded away in the 1880s (except at Gorham), so that the "all-round" skilled worker had become increasingly scarce, as in other metalworking trades. Meanwhile, technical change and the emergence of findings firms and other auxiliaries had inaugurated a diffuse division of labor—machine tenders overseeing "automatic" chain makers or running small die presses in findings shops, and highly but narrowly

# SEAMLESS WIRE

In All Sizes and Qualities.
Manufactured by

## EDWARD N. COOK,

Maker of

**Gold and Silver Plate, Figured Stock, Etc.**

Also Hard Enameling—Plain and Fancy.

Something New: ALUMINUM SOLDER.

TELEPHONE. PROVIDENCE, R. I. 144 PINE STREET, Jesse Metcalf Bldg.

## WM. H. LUTHER & SON,

Works, 214 Oxford St., Providence, R. I.
Salesrooms, 200 Broadway, N. Y.

MANUFACTURERS OF

**ORIGINAL DESIGNS IN**

**High-Grade Low-Priced Jewelry.**

**OUR SPRING LINE**

Of Waist Sets, Cuff Buttons, Brooches, Etc.,

WILL BE COMPLETE

**ON OR BEFORE DECEMBER 10th.**

Remember our designs are all original and our facilities are unsurpassed.

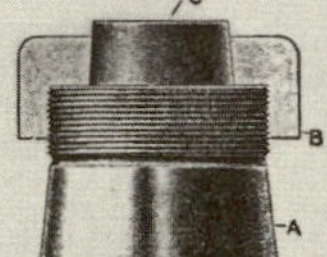

...THE...

**STEVENS' DIE HOLDER**

A Money Saver for Manufacturing Jewelers.

For particulars write to ....

**FRANK E. FARNHAM,**

MANUFACTURER OF

**Jewelers' and Silversmiths' Dies, Tools and Cutters,**

35 Potter St., Providence, R. I.

## We Will Engrave Your Portrait,

building, etc., same style as sample, up to 2½x4 inches for $2.00, larger sizes at 20c. per square inch. We have for years made a specialty of Jewelers' Cuts, Electrotypes and Printing, and give special attention to mail orders.

Telephone 1357 or Postal and will call.

**THE J. J. RYDER CO.**

47 Washington St., cor. Eddy St., Providence, R. I. *Engravers and Printers.*

## BUTTONS AND WAIST SETS

SEAMLESS WIRE RINGS.

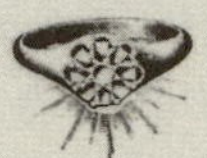

IN BELCHER AND TIFFANY STYLE

**Coral Brooches and Ear Drops.**

**2-POINT INVISIBLE SETTING**

The Only Setting for Cluster Goods on the Market.

**THE GOLCONDA GEM.**

## The R. L. Griffith & Son Co.

Factory, 144-158 PINE ST.,

N. Y., 237 Broadway. PROVIDENCE,
St Louis, Commercial Bldg. R. I.

**STARRETT'S**

## Speeded Screw Micrometer.

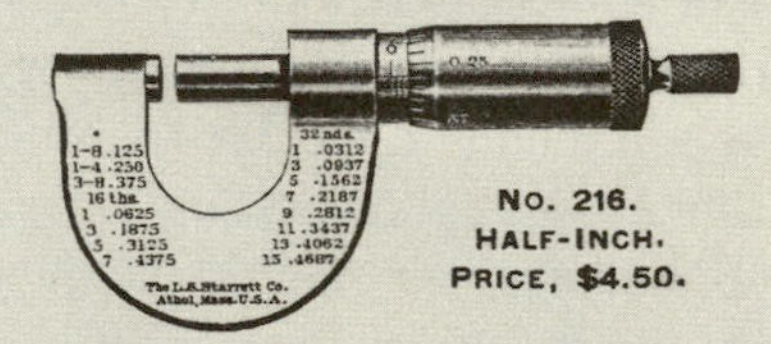

NO. 216.
HALF-INCH.
PRICE, $4.50.

WE make a large variety of Micrometers (from half-inch to six inches), and other instruments for accurate measurments. Special attention is invited to our line of fine tempered steel rules, graduated in both the English and Metric systems.

Our 112 page illustrated Catalogue is mailed free on request.

**THE L.S. STARRETT CO., Box 88, ATHOL MASS. U.S.A.**

18. A typical display advertising page from *Manufacturing Jeweler*, 1906, showing the variety of specialists and auxiliaries in the trade. Courtesy of Hagley Museum and Library.

skilled men cutting dies or coaxing quality results from plating baths in auxiliaries. However, able "bench hands" still remained crucial for producing hundreds of seasonal samples (often initiating the designs themselves), for the finishing stages in production, and for handling special orders, repairs, and rework on "seconds" too valuable to scrap. A competent bench worker could shift readily from engraving gold rings to ornamenting silver brooches, moving quickly with the vagaries of incoming orders' sizes and specifications, and could stand both the pressure of the rush season and the stress of short hours or layoffs in each year's slack months.[27] In a classic free-rider stalemate, as the core journeymen aged, no manufacturing jeweler proved willing to take the risks and incur the expenses of reinstalling apprenticeships, nor did collective initiatives appear, as at Cincinnati. Few immigrant craftsmen ventured north to Providence, either, for they found ample opportunities in the high-end New York/Newark complex, which was not incidentally more congenial to the rising Jewish segment of the incoming stream. Locally, young men with a metal trades interest seemed to favor positions at Brown and Sharpe or Gorham (together employing six thousand in 1907) over the irregularities of the jewelry industry and what trade commentators at times referred to as its lack of "manly" work. Thus the first feature of the labor impasse, a shrinking pool of skilled workers, seemed intractable.[28]

Outwork, informed by equally gendered views of the labor market, solved the second labor problem—firms' inability to host enough factory workers in rush seasons. Huge fall demand in 1905 and 1906 overwhelmed the shop capacities of Providence and Attleboro companies, particularly those in the low-end brass and chain sections, inducing them to send work out to women "who have long since retired from the jewelry industry." Though the local trade journal worried about the market's enthusiasm for cheap jewelry and makers' declining commitments to quality, proprietors with stuffed order books simply sought means to get the goods out the door before the fall surge ebbed. By 1906, this entailed sending unspecified "machines" (perhaps foot-powered die presses) to the homes of married women, along with routine piecework—assembling pendants, watch fobs, and ten-cent earrings or attaching finished pieces to cards for retail displays. Shifting these jobs "outside" cleared factory space for other uses and confirmed the value of outwork as a competitive strategy. Firm owners might bemoan the decreased supply of young men willing to regard the jewelry trade as a vocation, but they earnestly pursued "retired" young mothers to fill busy seasons' labor requirements.[29]

In the absence of usable company records, the profits question is tricky, for owners often whined about thin margins. Yet finding such complaints amid historically vigorous jewelry markets makes it worth taking them seriously. Late in 1906, "the best year for ten or fifteen years past," manufacturers reported that "many goods are really being made at a loss," for in the months since prices had been set on samples, the cost of silver, stones, and supplies had jumped. Overall, this led to a "margin of profit . . . less than usual." S. O. Bigney added that labor expenses had also risen; thus "in order to show an equal [dollar] amount of profit over some former years, a larger business had been necessary." In March 1907, prominent Newark firms announced a 10 percent price increase. *Manufacturing Jeweler* urged New England makers to follow suit and correct the problem of "large sales and small profits." The leading firms took no action; soon the fall panic threw the trade into a temporary crisis, shrinking hopes of any profits, much less enlarged ones.[30]

## Troubled Prospects, 1908–1912

Late in September 1907, Attleboro windbag S. O. Bigney assured delegates to the National Retail Jewelers' Association convention in Chicago that huge autumn harvests meant that Rockefeller and the denizens of "Wall Street and the other gambling hells of the country" could not ruin prosperity or plunge the nation into panic and depression. These brave words revealed both Bigney's proprietary populism and his ignorance of economics. Six weeks later, at the height of the panic, Providence's Union Trust Company, with $28 million in deposits for 25,000 accounts, closed its doors and entered receivership. The Jewelers' National Bank in North Attleboro failed in December, its cashier a suicide. Thus opened an unsettling period in the New England jewelry centers, years in which a gradual accumulation of sour news eroded Bigney's confidence and that of many among his colleagues.[31]

The regional banking crisis took six months to resolve. Union was the preferred financial institution for jewelry proprietors, and its blocked accounts caused them immediate trouble in amassing payrolls and covering accounts payable. There was a further difficulty. Firms had, as usual, borrowed substantially from Union Trust to handle materials and supplies expenses during the fall rush, giving in promissory notes normally redeemed as revenues trickled in, often requiring a series of renewals for shrinking balances due the bank. Proprietors feared that the bank's receivers would decline renewals as reorganization proceeded, expecting that spotty winter collections would impede timely redemptions. However, manufacturers reached an accommodation with the bank overseers (by unrecorded means) that averted multiple defaults, and Union reopened under new management in May 1908. The ten Attleboro men who had sponsored the collapsed Jewelers' National fared worse, for the comptroller of the currency assessed the bank's shareholders "100 per cent on the par value of their stock" in order to cover four-fifths of an estimated $100,000 shortage. By May, a 60 percent dividend on depositors' accounts was authorized and a newly chartered bank cheerfully received some $250,000 issued to the claimants. The panic froze funds and slowed trade, causing considerable anxiety and some real losses; but the jewelry district's larger troubles arose from other quarters—market conflicts, "garret" new starts that created "Ruinous Competition" in cheap jewelry, and a rapid expansion of outwork, none of which trade leaders could arrest.[32]

Controversies between producers and jobbers were standard fare, but between 1909 and 1912 an old abuse resurfaced. Jobbers began refusing to order seasonal styles for their own stockrooms, purchasing only sets of samples for their salesmen to show retailers. Now, instead of initial invoices that identified the subset of winners among each firm's new styles, followed by further calls for the most successful designs, manufacturers encountered pressure to make and hold inventories of their entire lines, ready for a stream of rush orders demanding instant shipment (as in the 1880s). Equally unnerving, jobbers who had gathered samples at trivial expense were perfectly positioned to have designs made up at cut prices by companies competing with their originators, especially if the latter declined to make advance stock or if a style proved a market hit. Moreover, in order to get the earliest possible look at samples, jobbers started circulating among the factories weeks before the informal seasonal opening dates, May 1 and December 1. A sufficient number of manufacturers unveiled their new designs that seasons began creeping backward to early April and November. In 1911, leaders of NEMJSA and the jobbers' association agreed that new lines would be opened only on the firsts of May and December, but both members and outsiders ignored them. Increas-

16

# Fine Gold Filled Brooches and Chatelaine Pins

Quality Guaranteed. All Stones Are Imitation Unless Otherwise Stated. Each Chatelaine Pin Is Fitted with a Safety Catch. Illustrations Show Actual Size. Prices Each.

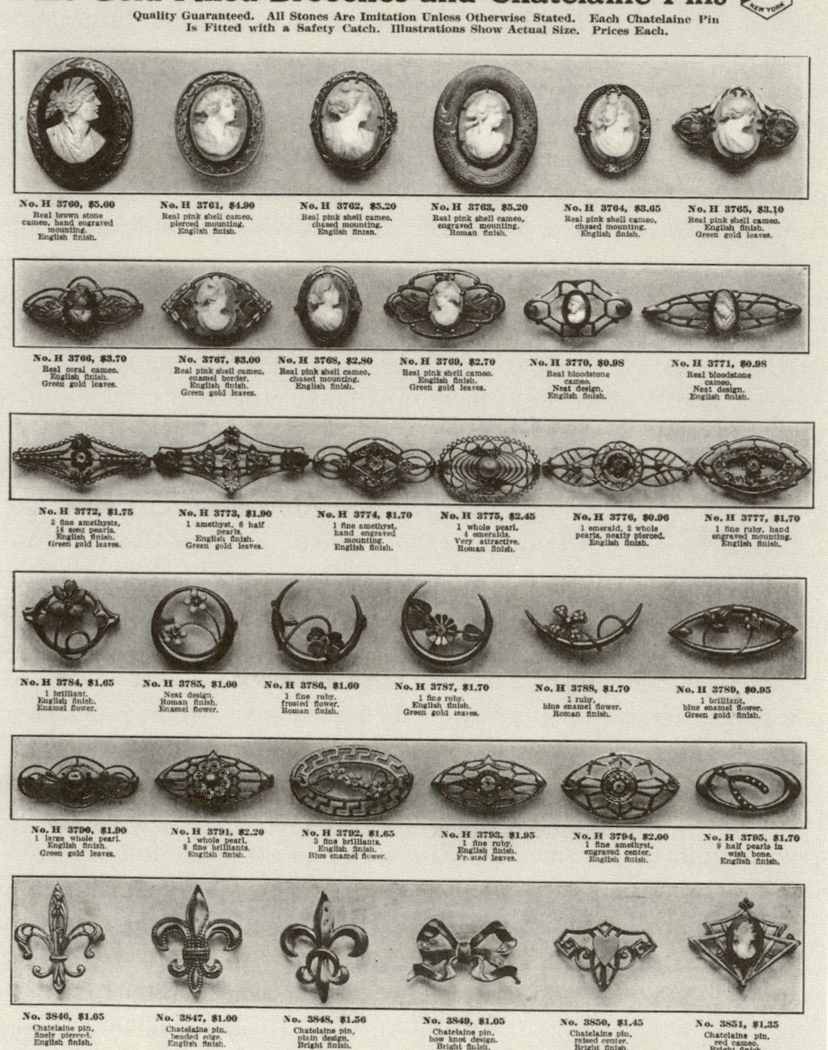


19. Style diversity in jewelry. One plate from a New York jobber's 150-page catalog, ca. 1910–20. Courtesy of Hagley Museum and Library.

ingly chaotic competition resulted, in which (1) final goods firms responded to design piracy, inventory demands, and price shaving in part by passing these viruses to the findings makers who furnished components; (2) jobbers claimed that they too faced a crisis of rising expenses and slipping margins; (3) retailers protested manufacturers' inability to organize to solve "the question of standard quality [and] equitable selling prices"; and (4) outwork spread steadily as the mesh bag craze ballooned.[33]

Mesh bags, copied from European novelties, consisted of a fabric made from interlocked metal rings fashioned into purses of varied sizes and topped with ornamental bar clasps attached to the uppermost row of rings. Though the two halves of the clasp could be formed in die presses, the ring mesh had to be fabricated one link at a time, by hand—a classic low-wage labor-intensive process. Facing huge demand, jewelry manufacturers engaged subcontractors to secure a homeworking labor force. By 1910 thousands of the district's working-class women "engaged in the production of the mesh bag," with many "hustling concern[s]" employing "three or four hundred persons who devote[d] most of their spare time" to it. In one day's Providence newspapers that year, four contractors advertised for 1,600 outside hands. Several years later, Massachusetts authorities counted over 9,000 outwork mesh makers in the Attleboro area alone, observing that "[a]ll but two of the [100] contractors found in this industry were women, nearly all married women," some of whom realized "an annual profit of $4,000 or $5,000." By contrast, "nine-tenths" of home workers "earned less than $150" yearly.[34] Though at least four-fifths of the area's estimated 20,000 outworkers in 1912 were purse makers, moving other tasks outside the factory also had attractions. Linking varieties of chain that were not machine producible, attaching "bars, drops, swivels, barrels, catches, or ornaments," painting designs on enameled brooches and pins, and low-end stone setting, beading, burnishing, and wirework—all these tasks occupied women and their children in piecework home production. Proprietors cloaked these tactics in charity garb ("of especial value to the unemployed or families where sickness has left them almost destitute for the necessities of life"). Yet their importance to the industry became clear at tariff hearings in 1909, where New England delegates fought (successfully) against any lowering of the barriers to German imports, lest cheap Pforzheim mesh bags derail their sudden market success.[35]

Outwork's magnetic appeal and cost-effectiveness further deranged market and pricing practices in the district. It encouraged a spate of new starts by craftsmen hoping to pick up on the mesh bag's persistence—tiny, almost phantom enterprises that fed jobbers' eager demands for ever-cheaper bags by copying designs, playing the outwork game, and engaging finding firms and auxiliary die-cutters and platers, thus requiring a minimal initial investment and an in-house staff only for final assembly.[36] By 1912, small shops offering low-end, yet stylish, fashions threatened the viability of older companies that were wary of making advance stocks and were having trouble keeping their regular employees occupied filling jobbers' erratic orders. The copying tactic became so prevalent that ever-boisterous S. O. Bigney underwrote a series of trade journal ads decrying the devaluation of styles, falling quality, and the widespread defiance of the National Stamping Act. NEMJSA responded that this was merely Bigney's self-promotion, for such issues were properly its concern. His ads ceased, but the association—typically—did nothing.[37]

The disaggregated production network's decay into a hyperflexibility that undercut the capacities of veteran producers, the multiplying "garret" new starts and subcontractors, and the market power wielded by jobbers had implications that were soon apparent. Demand for mesh bags from Providence alone surpassed a half-million units in 1911, but the rest of the trade began to flounder in 1912. A journal writer again worried about the collapse of season-opening orders for wholesale stock: "The custom that has prevailed for the past two or three years on the part of the jobbers to send in their orders to the manufacturers for exactly the number of articles needed . . . and thus make the manufacturer carry the entire stock, has grown to a larger extent than ever this year. This has resulted in the manufacturer having . . . to fill orders at a moment's notice and then carry the book account for an indefinite period. In [addition] there has been a constant string of failures, bankruptcy proceedings, extensions, and other financial difficulties that in the aggregate have amounted to a considerable total and, generally speaking, it is the manufacturer who gets the short end."[38] Exactly so.

At NEMJSA's April banquet, proprietors muttered about the "unusually large number of failures [among] their customers." That spring most jobbers unnervingly delayed buying even samples of new lines until late May, perhaps in reaction to massive design theft, effectively "curtailing [makers'] activities." Orders recovered moderately in the fall, but 1913 was a disaster. A general economic recession mashed jewelry demand as fully as it undercut machine tool sales, but jobbers' market power punished jewelry manufacturers more severely. Thirteen Providence firms sought bankruptcy protection in 1914; others scurried to make arrangements with their creditors, for the trade contraction now fell wholly upon producers holding quantities of styles not wanted and a dead loss. The 1913–14 crash proved to be the trade's worst reversal in twenty years, and the worst ever for makers. "In 1893, while the manufacturers suffered severe losses through failures [among clients], it was the jobbers who were in straitened circumstances, while it is now the manufacturers that are being driven to the wall." Holding seasonal stock and extending long credits in a fashion trade constituted a recipe for disaster.[39]

By 1912, the Providence and Attleboro jewelry industry had taken a critical misstep. With mutual distrust and subsectoral specialization obstructing any coordination to moderate either the effects of outwork or the pressures from wholesalers, leading enterprises lost novelty's price advantages to rivals seeking short-term gains through copying and outwork. Particular new starts might temporarily reap quick returns, but "ruinous" competition's effects on the New England jewelry industry would be rued for the next half-century. A capacity for swift adjustment to fashions, forged in the years after 1880, devolved into a hyperflexible network that advantaged eager copyists, cynical jobbers, and large retailers.[40] Production did not collapse in the region, for there was no other American center that possessed all the requirements of a disaggregated jewelry manufacturing district and little incentive for any locale to attempt to amass them. Nor did all firms slide into the slough; indeed, a few innovated—for instance, framing exclusive contracts to provide jewelry to schools, fraternities, and colleges, which carved a durable niche for the Balfour Company. Though there would be occasional busy years ahead, weak margins for most companies marked them as seasons of "profitless prosperity." Ultimately, the Rhode Island and southern Massachusetts jewelry districts never regained the initiative and momentum established in the mid-1880s and revived briefly in the early twentieth century.

## *Chapter 11*

# WORKSHOP OF THE WORLD: PHILADELPHIA

NO AMERICAN industrial center so fully exemplified the dynamics of specialty manufacturing as did Philadelphia in the late nineteenth and early twentieth centuries.[1] The Quaker City's textile trades defined fashionable fabric production nationally. Its diverse metalworking capacities matched those of any American center, and its manufacturers of locomotives and ships plied international markets. Though much smaller, Philadelphia's apparel trade resembled New York's with outwork shops making both custom clothing and ready-to-wear styles. Compressing two decades of this city's elaborate industrial history into one chapter borders on the absurd but is here necessary. Key developments in textiles and metalworking will first be outlined, followed by a discussion of local institutions that undertook political action, marketing initiatives, and, less often, governance roles. A review of technical education in the city's two leading industries will precede a closing analysis of trades anchored by giant specialists (ship and locomotive building) or chiefly by networked small businesses (apparel).[2]

## Textiles and Metalworking

In December 1900, a federal Industrial Commission examining "the relations and conditions of capital and labor" held hearings at Philadelphia's Manufacturers' Club. The heads of Baldwin Locomotive, the Cramp shipyards, and Midvale Steel testified, as did factory inspectors, proprietors of specialty firms making textiles, lighting fixtures, and scientific instruments, along with owners and workers from the city's ever-contentious shoemaking sector.[3] Frank Leake, a pile fabric company president, represented area woven goods interests and characterized Philadelphia as "a place for the manufacture of high-grade novelties and specialties in the textile line." To achieve the style effects crucial to success, local mills employed staff designers ("very frequently paid more money than the [factory] superintendent") and subscribed to European agencies that supplied reports on and swatches of each season's Continental novelties, which mills "adapted" to forms "desirable in this market." Though some firms did copy others' best-sellers ("making another fellow's goods cheaper"), this was a minor bother. The unavoidable full season lag in weaving duplicates put followers "behind the market on style," causing area leaders as little trouble as Philadelphia companies caused Paris and Lyon houses. Leake quietly underscored textile specialists' versatility by noting that his mill used cotton, linen, silk, wool, and worsted yarns to weave Turkish towels and plush fabrics. He added that few companies spun their own yarns "because of their changing output, which requires one year one class of yarn and another year another class."[4]

The long depression initially had hit textiles harder than other sectors; one trade journal correspondent estimated in 1894 that 90 percent of Philadelphia's capacity was "absolutely at a standstill." Leake observed: "Many novelties ceased almost altogether. People in depressed times are not buying luxuries; many high-priced goods were dropped out of the lines for this reason." Though flexible textile firms moved down-market (reversing Providence jewelers' upgrade to silver novelties), the district's chief low-end specialty, wool ingrain carpets, nearly collapsed. Cash-poor, its farm and factory clientele bought rag rugs and Chinese straw mattings at half the price of ingrains. Until 1898, local textile workers, used to in-season overtime and short hours in slack months, contended with long stretches of three- and four-day weeks, punctuated by extended layoffs. Piece-rate reductions (ca. 10 percent) further shrunk their earnings. Several respected mills failed and liquidated, but most enterprises devised tactics to maneuver through the slump.[5]

The simplest step was to shun wholesalers, who grew reluctant to order and hold seasonal stocks, and undertake direct sales to retailers and the apparel industry. Department stores had become major outlets for carpets, hosiery, knitted outerwear, and yard goods; whereas in men's and women's ready-to-wear, garment firms in New York, Philadelphia, Rochester, Cleveland, and Chicago could be approached economically. Makers offered samples priced between their old rates to jobbers and the latter's quotes to retailers or clothiers; and by 1900, direct selling was "the prevailing tendency" in marketing styled fabrics. Less honorable, but no less effective, was "cheapening the goods." If deflation and soft demand meant lower prices, firms could reap profits by redesigning styles to use less yarn per square yard, by working wool or silk filling over cotton warps, or by using reprocessed wool (called shoddy) as part of the spinning mix. Though one manufacturer growled that hard times taught Philadelphians "how to make a most miserable piece of goods," a Textile School graduate appreciated that they "had to do it to survive."[6]

Most assertive were those firms that plowed accumulated reserves into new textile lines during the doldrums. Much like machine tool makers experimenting with improved lathes in dead markets or electrical giants pursuing new motor and turbine designs, Philadelphia textile mills found fresh technological terrain to explore in the 1890s. Leake suggested that the main line of development lay "in jacquard loom products, such as upholstery goods, lace curtains, [and] carpets," but he overlooked technical advances in knitting women's hosiery, spinning fancy yarns, and fashioning braided trims. Having started in carpets, the Bromley family reinvested a portion of its surplus in new lace-making factories and rapidly became this subsector's national leaders. Other firms moved out of ingrains to higher-grade carpets or added upholstery Jacquards. By 1900, Leake averred that "fully 95 percent of the upholstery manufactured in the United States is made in Philadelphia," while diversity "in fineness of texture" ruled the booming lace trade.

Philadelphia textile specialists' resilience was plain to all. Indeed, once news of the American Woolen merger percolated through the industry amid market revivals, promoters descended on the city in 1899 bearing consolidation schemes for upholstery, carpets, spinning, and knit goods sectors. None materialized, for local proprietors placed extravagant valuations on their plants and demanded cash, not securities, for their sale. Many firms endured years of low or no profits in the 1890s, but workers suffered more by far. Although rate reductions had been restored by 1900, generally passing 1893 levels, prosperity brought the return of wearying sixty-hour weeks. Trade unions, toward which manufacturers professed a benign, if not dismissive, attitude, soon forced the issue of shorter hours.[7]

The mobilization followed a burst of labor activism in 1898–1900 among the city's twenty-six textile craft unions, a flurry that had helped elevate piece rates. Loosely united in the Central Textile Workers Union, operatives finally called for a fifty-five-hour week in 1903; but despite a general strike that idled nearly 50,000 men and women that summer and sparked Mother Jones' famed children's crusade, the effort disintegrated. Male activists amassed a pitifully small strike fund ($25,000) and, consistent with long-standing gender attitudes,

20. Jacquard carpet weaving at William Scholes and Company, Philadelphia, 1912. Devoted to novelty, Scholes devised a means to weave fancy carpets from paper, for use on covered porches or in sunrooms. Courtesy Pennsylvania State Archives (MG 219, Philadelphia Commercial Museum, Item #3720).

completely failed to organize working women (half the labor force). The short hours demand also ignored women's interests, for all but a few women worked at hourly rates, would lose earnings were the week shortened, and thus had no economic incentive to support the strike. Mother Jones notwithstanding, unpersuaded women returned to their jobs in droves. The Textile Manufacturers' Association, assembled by mill-owning members of the Manufacturers' Club, gathered forfeit money, refused to negotiate with the skilled men's unions, and held solid. Other than a sympathy walkout during the 1910 Philadelphia transit strike, textile labor conflicts were few and scattered for the next decade.[8]

By 1905, over seven hundred firms and 60,000 workers crowded the city textile industry's dozen divisions; another hundred mills and 15,000 operatives dotted towns in adjacent regional counties, together reporting $100 million in invested capital. Although five mills employed over 1,000 individuals (the Dobson brothers' carpet and worsted mills nearly 3,000), mean employment remained under 100 per firm. The classic network pattern of separate specialist operations continued, soon augmented by the introduction of machinery for knitting women's full-fashioned hosiery—the tapered, seamed silk stockings that propelled the district's postwar surge. Profits continued hearty, averaging about 12 percent of sales; and the 1907 panic was a hiccup—failures here and there, then renewed expansion within a year. Lace and upholstery leaders commenced national magazine advertising to establish brand names, built new mills, and ordered more machinery. Dyehouses secured innovative "package" vats that colored five hundred pounds of yarn at a time, faster and more reliably than earlier "dip" methods. In knitting, novel technologies and the vogue of sports, nature walks, and beaches interacted to widen markets for machine-made jerseys (sweaters) and bathing suits. Meanwhile, specialist auxiliaries multiplied in card cutting, hosiery printing, and fancy spinning. Output ballooned as the workforce stabilized through 1912, suggesting rising productivity. Still, there were complications.[9]

First, the ingrain carpet sector was dying by degrees. Manufacturers who rejected shifting to higher grades either enjoyed rising shares of a fading market or simply went broke. Ingrain makers convened in 1905 to coordinate price levels and synchronize changing them, but by 1912 few of them remained in business. Skilled workers sensibly migrated to other lines; when a brief ingrain revival occurred in 1907, millmen searched in vain for veterans to activate hundreds of long-idle looms. The most alert companies phased in production of other carpet constructions, particularly cut-pile Axminsters that were sweeping the market, and gradually scrapped their ingrain looms. A similar dynamic would later undermine seamless hosiery (as fitted, full-fashioned stockings gained favor) and silk fabrics (as rayon advanced), pressing firms to engage technological novelty, not just track machine improvements and style shifts.

Second, the depression legacy of "cheapening the goods" combined with new retail pricing patterns to create problems for wool and worsted fabric firms. As retailers simplified their marketing and record keeping by laddering men's suits at $10, $15, and $20, price rigidities pushed back through the production system. Apparel firms delivering suits at fixed wholesale prices soon realized that they could pay no more than a dollar a yard for a $15 suit's fabrics. Manufacturers "making for the price" dealt with rising materials costs by "manipulation," calling on spinners to mix shoddy or cotton with new wool, thus degrading quality and infuriating piece-rate weavers, for frequent broken yarn stoppages during the running of "rotten work" threatened

21. The carpet designers' corridor at Hardwick and Magee, Philadelphia, ca. 1910. For decades, Hardwick was one of the city's largest weavers of fine figured woolen floor coverings. Courtesy Pennsylvania State Archives (MG 219, Philadelphia Commercial Museum, Item #2476).

their earnings. When fixed selling prices forced makers to absorb cost increases, productive flexibility had perverse implications, if profits could only be achieved through the sacrifice of both style and quality. Other Philadelphia textile sectors roared ahead into the twentieth century's second decade, but ingrain carpet and wool fabric firms were not among them.[10]

Similar unevenness could be found in the city's metalworking industry, which, if anything, was more complex than textiles. Whatever their specialties, weavers and knitters worked for a single, if vast, market of American household consumers reached through apparel firms, wholesalers, and stores. The city's metal trades, by contrast, supplied heavy tools and machinery to industry and railroads, thousands of components for manufacturing and construction, motive power to railways and transit systems, conduit, pumps, and meters to utilities and municipalities, and hand tools, hardware, and accessories to skilled workers and households. Still there were profound resonances. Though metalwork largely excluded women, production relied crucially on versatile, skilled craft workers—notably foundry molders and shop machinists in their thousands. Proprietary ownership and partnerships were standard; with few exceptions incorporations confirmed insiders' control. Intricate contracting networks among foundries, machine shops, assemblers, and auxiliaries (e.g., for heat treating or pattern making) facilitated much of the sector's output. Moreover, as a recent analyst explained, concerning local metal trades employers: "Some had trained in others' firms; many had received their engineering education at the University of Pennsylvania; in maturity they joined the Engineers', Foundrymen's or Manufacturers' Clubs; and many were active members of the same religious denominations. . . . The Philadelphia metalworking industry was a metropolis-wide complex of specialized, interrelated firms. Its factories were located in the same neighborhoods of a sprawling city. . . . But the thing [owners] had most in common was their dependence on the same labor supply." In 1900, their sector included 660 firms with 35,000 employees and capital above $70 million.[11]

Given this diversity, the 1890s depression had had a highly differentiated effect on metalworking sectors. Heavy tool companies, railway suppliers, and firms building infrastructure for utilities, transit, and construction battled over sparse contract offerings for five years, dismissing half or more of their workers, who in turn scrambled for jobs. Companies making small tools, hardware, and housewares more rapidly found niches of opportunity and sooner returned to full employment. Small machine shops, knitting machine makers, and firms oriented toward high-end lighting equipment and scientific instruments perhaps managed best, but they were a tiny minority. As in textiles, there were few shocking failures, whereas many firms accepted barely profitable orders to hold their reduced core workforces.

Diversified machinery firms, like Pusey and Jones at nearby Wilmington, Delaware, pursued multiple strategies to secure work. While chasing regional jobbing orders (completing 36,000 of them, 1893–1900, or 90 per week), its owners also bid repeatedly on scattered proposals for coastal and lighthouse ships (winning only two). Most important, P&J's Thomas Savery, a designer of patented paper-making machines and pulp grinders, entered upon "A Personal Experiment" in 1894, colluding with two other U.S. paper machinery companies to rig bids on paper mill equipment contracts. Sharing the available trade kept all three active and assured profits on each machine delivered. Thus P&J constructed fourteen Fourdriniers in the first half of 1897, logging $41,000 in profits on contract prices of $458,000. However, with returning prosperity, the arrangement broke down in 1901–2 when Worcester's Rice, Barton, and Fales commenced telegraphing lowball bids to potential clients after receiving P&J's quotes. Worse, the pricing scheme collapsed just as materials expenses escalated. Thus in 1902 it cost $602,000 to complete fourteen contracts Savery secured with bids totaling $544,000, netting a loss of $58,000. As Savery celebrated his sixty-fifth birthday at Wanamaker's grand Crystal Tea Room that May, he had ample reason to regret the restoration of competition.[12]

Of course, great enthusiasm accompanied the economy's rebound, but at the Industrial Commission hearings, Midvale Steel's Charles Harrah grumped about metalworkers' loyalty to other companies that had laid them off during the slump. Asked about his employees' permanency, Harrah observed that Baldwin Locomotive was better placed than Midvale, "because when work is slack with them we take on a lot of their men; but the moment work picks up with them again the men leave us and go back to Baldwin's, no matter whether they are getting big wages—no matter what happens they go back to Baldwin's." Though the locomotive works was a "slaughterhouse" that lacked medical facilities appropriate to its hazards, a firm innocent of welfare capitalist programs, and one run through a tough "drive" system and prone to layoffs, it remained in myth or custom the source of the most challenging jobs in Philadelphia metalworking—a place that prepared craft workers for supervisory positions or the step into entrepreneurship. Providing the cautious with an assured callback and the ambitious with access to a rich learning environment, Baldwin commanded its workers' loyalties.[13]

However, only a minority worked for the few giant metalworking anchors. Given varied seasonal and cyclical demand in different trade sections, firms irregularly dismissed and added men throughout the year. When the labor market was tight, companies bid up day-wage rates to secure a force adequate to meet orders. Sensibly, workers strove to prevent surplus conditions that yielded lower offers. Philadelphia foundry craftsmen organized and sustained International

Molders Union locals from the 1870s, in considerable part to control the flow of apprentices who eventually would compete with them for positions.[14] Local manufacturers organized a founders' association in 1891 to contest craft regulation, set poundage prices on castings, and defend their entrepreneurial prerogatives. This group "went national" two years later and in 1896 sponsored the first convention of the American Foundrymen's Association, whose 1898 offshoot, the National Founders' Association, focused centrally on labor issues. Members paid "regular dues and special assessments" in proportion to their workforce sizes, and a sizable entry fee to the group's "reserve [fighting] fund." Firms breaking association rules or resigning sacrificed that "deposit." Following the New York Agreement in 1899, a national NFA-IMU pact to negotiate conflicts without strikes or lockouts,[15] the Philadelphians concluded a regional treaty that suspended certain union rules, established a minimum day rate, overtime supplements, and a grievance procedure, and laid the base for four years of foundry labor peace. Owners respected the IMU's limitation of apprentices, a step that produced "a nice, tight labor market, permitting continuing advances along the roads toward shorter hours [and] higher wages."[16]

Meanwhile, metal fabricators had created the National Metal Trades Association to prosecute shop control struggles against the International Association of Machinists. Machinists were poorly unionized in Philadelphia; thus the NMTA initially had few adherents there. However, its increasingly militant "open shop" campaigns, complete with extensive, effective strikebreaking actions, offered area foundrymen dramatic examples of a confrontational alternative to conferences and conciliation. In December 1903, a cluster of city foundry proprietors gathered with owners of midsize fabricating companies to form the Metal Manufacturers' Association—not a local branch of the NMTA, but an independent body that would challenge unions in the labor market without attempting to eradicate them. When members fought the IMU in a series of 1904–6 foundry strikes, the MMA's Labor Bureau provided replacement workers—the key to victory. The 1907 crash yielded layoffs for half the city's metalworkers; MMA members experienced few shop-floor control fights until the war years.[17]

Like textiles, the Philadelphia metal trades added new specialties: most notably, scientific instrumentation at Leeds and Northrup and Brown Instrument, along with Link Belt's customized conveyor systems and equipment. Each of these innovation tracks buttressed the extension of bulk and mass production. Leeds' pyrometers, which measured internal temperatures in open hearths and retorts with gradually increasing accuracy, aided steel mill and chemical plant engineers seeking to control basic processes and to displace craftsmen's rule-of-thumb empiricism. Link Belt's conveyors and chain-drive bucket systems speeded materials handling in flow production, whether for soup cans and auto parts or coal and ores. Yet like machine tool firms, these companies constructed their specialty goods in batches or as custom products while standardizing components, refining accuracy, and improving reliability. Thus their needs for skilled workers, their placement in highly volatile markets, and their concern for quality and competition based on factors other than price linked them structurally to Philadelphia's specialty machinery makers.[18]

The city's textile and metal trades achieved record output and employment early in the new century, together mobilizing a workforce that passed 100,000, and shipping goods worth a quarter of a billion dollars annually. They operated within an unmatched network of partial process firms and auxiliaries, profiting from regional externalities and responding adroitly to diverse currents of demand for stylish consumer items and sophisticated industrial products. In addition, individuals and firms from these sectors initiated and sustained a range of institutions that sought to manage the uncertainties this extensive system of production entailed, to improve marketing and training, and to organize common fronts against labor, governments, and foreign competition.[19]

## Collective Institutions and Technical Education

The organizational surge of the 1880s and early 1890s transformed an employers' emergency antiunion coalition into the Manufacturers' Club, promoted construction of the Bourse as a central marketing emporium, and built the Textile School into a viable educational institution. Over the next twenty years, each of these key institutions solidified its role on the regional industrial scene. The club's membership reached one thousand by 1895, fewer than a quarter from its originating textile trades. Clubmen gathered at its elegant Pine Street rooms to realize their "earnest desire . . . to purchase of each other," to "mold the conditions of trade . . . in their favor," and to exert influence on governments for policies favorable to manufacturing. The club's private meeting rooms for dining and discussion and its published roster of members by "business classification" facilitated deal making. Forming "sections" (through which allied manufacturers could treat with shippers, suppliers, or wholesalers) and the Bourse, "born in the Manufacturers' Club," helped shape trade conditions. Moreover, city hall, Harrisburg, or Washington rarely ignored members' sectional or general resolutions conveyed by "special committees." The club's political activism blocked local ordinances that would have increased production costs, supported infrastructure projects (elevated rail transit) that widened labor sheds, and forced revisions of state laws on factory inspection and workers' compensation. Though most members were reliably Republican and protectionist, they differed on bimetallism, unions, and tactics for securing railway rate revisions, topics debated at "Free Speech Club" meetings.[20]

A decade after hosting the Industrial Commission's hearings, club members financed construction of a ten-story office building a few blocks south of city hall, then leased the lower floors and relocated the club to its top levels. There, during the reform administration of mayor and member Rudolf Blankenburg, the federal Commission on Industrial Relations held its 1913 Philadelphia sessions. Yet for all its prominence and visibility, the Manufacturers' Club was an elegant convenience and a platform for subgroup activities and political initiatives, rather than an agent for industrial governance. It set no standards for business dealings, took no position on labor relations, and expelled members only for outrageous conduct or failure to pay their bills.[21]

Formally opened late in 1895, the Bourse housed several hundred firms' marketing arms. Its nine floors of offices opened onto a spectacular central atrium above ground-level display spaces and a high-ceilinged basement filled with machinery samples rigged for powered operation. Textile and final-goods metalworking companies filled its rooms, either offering seasonal novelties to fabric, fixtures, and hardware dealers relieved of tramping through the factory districts or demonstrating working tools for industrial buyers' inspection. The Board of Trade located its offices there, later joined by the Commercial and Maritime Exchanges, making the Bourse Philadelphia's pivotal site for specialty merchandising by 1910. Meanwhile, in 1893 the Textile School joined the local School for Industrial Art in spacious quarters near the Manufacturers' Club and developed a dual curriculum—three years of day classes in materials, design, and technology for full-time students, chiefly manufacturers' sons and/or college graduates, and a set of evening technical courses for workers, some of whose employers paid their tuitions. As Theodore Search stressed to the Industrial Commission, by 1899 the Philadelphia school had become a model for state-sponsored textile institutes in Massachusetts and the South, but remained private, offering scholarships to Pennsylvanians in return for annual city and commonwealth appropriations. By 1910, despite additions, the school building was overloaded with machinery, labs, and students. At its annual Manufacturers' Club banquet, the 350-member Alumni Association heard plans described for relocation and further expansion on a North Philadelphia site, midway between the eastern Kensington/Frankford textile districts and the western Germantown/Manayunk concentrations. Each in its own sphere, the club, the Bourse, and the Textile School had become regional industrial assets.[22]

Dozens of other organizations and associations supplemented these three successful institutions after 1893, even as earlier ones grew apace. The revitalized Franklin Institute, for example, conducted evening technical classes at branches near the works of sponsoring manufacturers. The Engineers' Club (founded 1877) neared five hundred members in 1901. Attendance at twice-monthly professional meetings averaged seventy while its quarterly *Proceedings* circulated nationally. Among the newer groups were Kensington's Cosmopolitan Club, inaugurated by textile factory masters, and a host of subsectoral trade bodies bringing together, usually at the Manufacturers' Club, aggregations of carpet yarn spinners, upholsterers, shoe last makers, stove builders, shirt manufacturers, or master dyers. The Commercial Museum opened in 1896 as a permanent venue for import-export information exchange. It exhibited both Philadelphia products that might interest foreign buyers and samples of raw materials from overseas sources, and forwarded member firms information on importing agencies, credit matters, trade regulations, and market customs in Europe, South America, and Asia. Thus the museum substituted staff expertise in collating useful business data for the State Department's weak consular service reports. It also boosted Philadelphia industry in its monthly journal, *Commercial America*, and drew its funding from business subscriptions, exhibition fees, and local and state government subsidies.[23]

Owners and managers circulated through this welter of institutions, within and across sectoral boundaries, gathering information, seeking business, and establishing shared stances toward a variety of political and industrial questions. Consistent with the city's proprietary capitalist milieu, except in rare strike emergencies, owners shunned proposals that would cede meaningful authority to any central body, rejecting mergers and strong trade associations with governance capacity. Nor did they identify and follow sectoral leadership; Philadelphia textiles and metalworking spawned no Lodges or Geiers, as in Cincinnati, and no informal core group, as at Grand Rapids. Individuals who tried to be trade spokesmen earned the same chilly reception that bombastic S. O. Bigney sparked in Providence, for the complement to refusing labor's "dictation" was rejecting "instruction" by other mill operators. Collective action thus was either situational and short-term or located in carefully circumscribed institutions that provided specific services at modest costs.

Three examples should suffice. The Philadelphia Textile Association, first convened to battle the Knights of Labor, reinvented itself as a "permanent" trade body following its revival and success in the 1903 general strike. It faded rapidly away, however, having failed to establish any broad base of common interests among firms in the industry's many subsectors, for whom even labor markets were dramatically segmented (recall the twenty-six textile craft unions). Characteristically, it was reconvened as the Philadelphia Textile Manufacturers' Association during wartime labor struggles. The Metal Manufacturers' Association was better placed on this count, given the predominance of only two skill groups, molders and machinists, in sectoral workforces. Its Labor Bureau glued members to the association; for it inexpensively referred thousands of skilled men each year to participating companies and certified their experience and qualifications. The bureau placed few requirements on firms or workers. Employers had no obligation to hire men referred; unionists were not barred, nor were they asked to take pledges of nonaffiliation. No placement fees were charged either party. For middling and smaller companies, the bureau provided a crucial service that they could not furnish so well or so cheaply on their own. Yet beyond managing labor supply and providing strike support, the MMA little engaged owners' or managers' attention. Later efforts to encourage contracting among member firms or "common cost accounting techniques" went nowhere. As its historian summarized, "proprietary 'niche producers' were scarcely interested in, or capable of, cooperating on any project which did not have the immediate object of sustaining their dominant position in the labor market and maintaining the short-term viability of their labor process."[24]

The Pennsylvania Manufacturers' Association illustrates both leadership and service issues. Created in 1909 to pressure the legislature, the PMA was the personal vehicle of Joseph Grundy, a willful wool manufacturer from outlying Bristol, Pennsylvania. Its early meetings, the second at the Manufacturers' Club, drew none of Philadelphia industry's central figures. Though several associations sent delegates, most soon dropped away. Grundy, who dominated the PMA for forty years, explained its purpose: to "*prevent*, as far as possible, vicious, unfair, and unwarranted legislation . . . affecting the employment of labor." An abrasive Republican machine spokesman, Grundy initially offered nothing to Philadelphia manufacturers that could not be achieved through existing channels. However, when a workmen's compensation statute moved toward enactment, the MMA and many textile firms joined the PMA, first to weaken the bill, then to secure liability coverage "at a group discount rate" through its Casualty Insurance Company. Specific services, political and financial, overcame resistance to Grundy's bluster and his obsession with control. Philadelphia manufacturers paid their dues and insurance bills, but otherwise ignored the wind from Bristol.[25]

In technical education, the Textile School served specialty manufacturing competently. By 1910, 175 graduates, half its day-course alumni association, owned or managed Philadelphia-area mills, with others clustered at New Jersey and New England styled-textile centers. Over three thousand textile workers had attended its evening division, though an unknown proportion completed its curriculum.[26] The University of Pennsylvania's Wharton and Engineering Schools and, more modestly, the new Drexel Institute well served the metal trades' needs for commercially and technically trained managers and proprietary successors.[27] Developing skilled metal workers remained an unsolved problem, however, despite leading firms' friendly takeover of the Spring Garden Institute by the 1880s, when it was reoriented toward practical mechanical training.

A 1908 Engineers' Club survey outlined the situation. A poll of 44 metalworking companies, including most of "the larger establishments," revealed that among their 12,700 "mechanics," a stunning 8,500 were "all-around" machinists, foundrymen, or boilermakers (67 percent), and a third "single-tool machinists" or "foundry specialists."

The Labor Bureau helped hold this critical skilled force in the city by placing laid-off men in vacant positions, but it had no role in training youths to replace them as age, injuries, and departures accumulated. Survey returns documented apprenticeship's continuation; but with only 739 apprentices reported (one for every eighteen mechanics), at best 200 new journeymen might complete their indentures each year. In contrast to Cincinnati, only three in ten working "boys" had been signed to apprenticeships. Club president Henry Spangler, using railway shop figures for comparison, estimated that the Philadelphia metal trades needed to triple the present number of apprentices to replenish its workforce over the long term. However, he little expected that dramatic increases would occur. Ignoring union restrictions, Spangler quoted an industry veteran: "Employers do not want apprentices because it does not pay." Shops facing demand for rapid and precise production were "the most expensive school[s] in which to teach a boy a trade." Individual manufacturers simply had no practical incentive to expand their limited shop-training commitments.[28]

What of technical and public education? Evening classes through the Franklin Institute or Drexel focused on mechanical drawing and advanced subjects, thus could "only be classed as valuable aids to those already in the trades." Sadly, the public schools' day programs in "Manual Training" were "a disgrace to the city," "a farce" in which "pattern making [was attempted] in a plant without a single pattern-makers' lathe . . . [and] drawing taught in a room packed so full that it reminds one of a sweat-shop." Spangler urged that they "be abandoned forthwith" and the boys returned to regular instruction. Public evening classes drew "a keen, alert crowd of young men and boys," but facilities were miserable and the teachers ("mostly mechanics and good ones"), though earnest, "were absolutely untrained in the art of teaching their trades." "[T]here is all the difference in the world between knowing how to do a thing, and knowing how to teach others to do it," he added. Surveying this bleak situation, Spangler urged local manufacturers, "preferably [those who] have come through the ranks themselves" to take "a large hand" in designing publicly funded, two-year schools where "trades are taught . . . and nothing else." His model was the local, well-regarded Williamson Free School of Mechanical Trades. Williamson's three-year course produced small numbers of "first-class" journeymen knowledgeable in tool and die work as well as "jobbing," taking a task "directly from the beginning to its completion, on all the machines." Nothing resulted from Spangler's appeal. Public education, however ineffectual, was a politically intricate and contentious issue in Philadelphia. Metal manufacturers rejected expending the effort and political capital needed to reshape trade instruction, nor did they strike out on their own as textile millmen had a quarter-century earlier. When a severe skilled labor shortage at last emerged in the mid-1920s from a combination of workforce aging and thinned immigration, the MMA finally initiated "cooperative training programs for foremen and machinists' apprentices," largely paid for with tax revenues—a generation after the moment for timely intervention and innovation had passed.[29]

Too often, technical education and the reproduction of skilled labor stood out as specialty manufacturing's weak points. Only Westinghouse and General Electric had the resources to train engineers, technicians, and skilled workers within the enterprise, though a group of anchor firms sustained apprenticeship programs into the war era, as did many networked machine tool builders. In sizable urban centers, like Cincinnati and Philadelphia, engineering and business education took root in universities, but far less often at sectorally focused institutions like Philadelphia's textile and industrial art schools or Grand Rapids' later furniture design institute. Perhaps the distance between engineering principles and practical realities in custom and batch production was also one reason why firms preferred family members and able workers for managerial and technical posts. Moreover, at the time, specialty firms relied heavily on immigration and in-migration for labor force expansion and worker replacement, a habit that relieved them from acting on repeated rhetorical invocations of the need for long-term planning regarding skilled labor shortages.

In textiles, jewelry, and furniture, apprenticeship had decayed sharply by 1900. Workers built skills by moving among enterprises or developed a narrower specialization than had been customary in earlier operations; in apparel, they were expected to possess on entry basic hand or machine sewing experience adaptable to the firm's needs. Broadly, public education for skilled vocations was a nullity; the few effective private trade schools were easily counted and commended, but not widely replicated. Collective initiatives among American industrial specialists were numerous but most often, as in Philadelphia, targeted immediate problems, not issues critical to sectors' long-term vitality.[30]

## Polar Extremes: Locomotives and Ships versus the Clothing Trades

Two immense firms crowned Philadelphia manufacturing in these decades—Baldwin Locomotive and the Cramp shipyards. Hundreds of apparel factories and contractors occupied its lower echelons. Baldwin and Cramp sold complex and hugely expensive goods, made by an overwhelmingly native-born workforce, in global product markets; whereas the "rag trade" depended on transatlantic migrations to supply laborers to fashion clothes for a wholly American clientele. Yet Baldwin, Cramp, and the clothing makers were all manufacturing specialists who rode through market, technical, and political changes, confronting erratic demand, craft unions, and complex vectors of product competition. Though they represented the preeminent anchors and the bottom strata of networked specialists in Philadelphia's flexible industrial complex, the resonances between their experiences are perhaps as significant as the contrasts.

Baldwin Locomotive, which remained a partnership until 1909, built nearly 40 percent of American railway motive power constructed between 1890 and 1910. Its workforce fluctuated with engine demand—from 2,000 in 1894 to 18,500 in 1907. During the depression, like Pusey and Jones, Baldwin entered a price-management compact with its smaller competitors, respecting elaborate price lists issued for dozens of basic forms and, "[w]ithin each type, variations in size, cylinders, track gauge, and extra equipment," at least through 1897. In 1899, Baldwin delivered 901 engines in the several hundred different designs required by railroad master mechanics, who crafted locomotive specifications fitting their roads' terrain, freight and passenger mixes, and so forth. Record production in 1900 (1,147 engines worth $20.5 million) opened the company's greatest boom period; in both 1906 and 1907, Baldwin completed over 2,600 locomotives. Its design team expanded from 125 to over 400 engineers and draftsmen who elaborated as many as 125 new engine variants in a single year. Baldwin's 1908 catalog specified 379 types of "standard" steam locomotives, not counting its compressed air and electric lines. Its 1915 catalog listed 492 basic types that could be altered to meet clients' requirements. Perhaps only GE and Westinghouse practiced flexible specialization on a comparable scale.

Technical and work process innovation was routine at Baldwin. Superintendent Samuel Vauclain patented a steam-saving "double compound" locomotive in the 1890s (which drew a host of orders) and encouraged a raw materials transition from wrought iron to more durable cast steel for many components. Despite slack markets in the 1890s, Baldwin partners deepened their early investment in electric drive, installing motors totaling 3,500 h.p. by 1900, and spent several hundred thousand dollars on electric cranes, hoists, and elevators that eased materials handling. Eliminating overhead shaft and belt systems also added to the flexibility of machine placement and augmented efficient power use in the company's increasingly cramped Spring Garden district quarters. Both the post-1897 demand surge and the rising size of locomotives added to the spatial pressure on Baldwin's facilities. The average engine produced in 1890 weighed 46 tons; by 1900, 65 tons; and in 1905, 78 tons. More challenging, the largest locomotives grew immense: 144 tons in 1905, 177 tons the next year, and 250 tons by 1912. Constructing such monsters involved more than widening and elevating doorways; it also demanded elaborate work-flow routing complicated by intense time pressures. In 1906, Baldwin determined that it could no longer reconfigure its existing plant and began constructing a much larger works at Eddystone, an open-field site beyond the city's southwestern boundary.

Technically progressive and inventive in design, Baldwin was managed through a sophisticated version of proprietary capitalism. By the 1890s, firm partners installed scientific materials and product-testing facilities, extensive component standardization across types, and detailed reference systems for tracking specifications of an engine's 6,000 or more parts. However, production remained governed through direct agreements between partners and inside contractors, who bid for portions of each order, then oversaw their rapid completion by teams ranging from a few dozen to over 200 workers. Middle management was almost nonexistent; instead, partners tracked contract costs religiously, for their profit shares were at stake. Moreover, constantly changing designs, expectations of precision, and contract deadlines, on one hand, and regional metalworking's highest wages, earned without managerial piece-rate chiseling, on the other, meant that Baldwin *needed* and *got* the pick of Philadelphia's all-around metal craftsmen, decade after decade (some 4,000 machinists in 1911). Throughout this era, Baldwin represented specialty manufacturing's maximum case for

# Six Coupled Locomotives

## Gauge 4 Feet 8½ Inches

With Separate Tenders

Class 6-D | Type 0-6-0

| CODE WORD | Class | Cylinders Diam. Stroke Inches | Diameter Driving Wheels Inches | Boiler Pressure Pounds per Square Inch | Cylinder Tractive Power Pounds | Weight in Working Order Pounds | Wheel Base | Capacity Tender for Water 8⅓-lb. gallons | LOAD IN TONS (2000 POUNDS) OF CARS AND LADING | | | | | | | |
|---|---|---|---|---|---|---|---|---|---|---|---|---|---|---|---|---|
| | | | | | | | | | On a Level | On a Grade per Mile of | | | | | | |
| | | | | | | | | | | 26.4 ft. or ½ % | 52.8 ft. or 1 % | 79.2 ft. or 1½ % | 105.6 ft. or 2 % | 158.4 ft. or 3 % | 211.2 ft. or 4 % | 264.0 ft. or 5 % |
| Martigena.......... | 6-12 D | 9 x 16 | 33 | 160 | 5,340 | 30,000 | 6′ 9″ | 800 | 575 | 270 | 165 | 115 | 85 | 55 | 35 | 25 |
| Martignone......... | 6-14 D | 10 x 16 | 33 | 160 | 6,590 | 35,000 | 7′ 7″ | 1000 | 710 | 335 | 205 | 145 | 105 | 65 | 45 | 30 |
| Martillada......... | 6-16 D | 11 x 16 | 33 | 160 | 7,970 | 40,000 | 8′ 0″ | 1200 | 860 | 410 | 250 | 175 | 130 | 85 | 55 | 40 |
| Martillazo......... | 6-18 D | 12 x 18 | 37 | 160 | 9,520 | 45,000 | 8′ 1″ | 1500 | 1030 | 490 | 300 | 210 | 160 | 100 | 65 | 50 |
| Martillen.......... | 6-20 D | 13 x 20 | 42 | 160 | 10,930 | 50,000 | 8′ 6″ | 1800 | 1190 | 565 | 345 | 240 | 185 | 115 | 80 | 55 |
| Martinella......... | 6-20 D | 13 x 22 | 44 | 160 | 11,480 | 54,000 | 8′ 9″ | 2000 | 1240 | 590 | 360 | 255 | 190 | 120 | 85 | 60 |
| Martingal.......... | 6-22 D | 14 x 24 | 44 | 160 | 14,530 | 64,000 | 9′ 6″ | 2200 | 1580 | 750 | 475 | 325 | 245 | 155 | 110 | 80 |
| Martinho........... | 6-24 D | 15 x 24 | 44 | 160 | 16,690 | 72,000 | 9′ 9″ | 2400 | 1810 | 865 | 535 | 375 | 285 | 185 | 125 | 90 |
| Martiniano......... | 6-26 D | 16 x 24 | 44 | 160 | 18,980 | 82,000 | 9′ 9″ | 2600 | 2070 | 985 | 610 | 430 | 325 | 210 | 145 | 105 |
| Martiniega......... | 6-28 D | 17 x 24 | 50 | 170 | 20,040 | 90,000 | 10′ 6″ | 2800 | 2180 | 1035 | 640 | 450 | 340 | 220 | 150 | 110 |
| Martinisme......... | 6-30 D | 18 x 24 | 50 | 170 | 22,460 | 100,000 | 10′ 6″ | 3000 | 2450 | 1165 | 720 | 510 | 385 | 250 | 175 | 125 |
| Martinstag......... | 6-32 D | 19 x 24 | 50 | 170 | 25,040 | 110,000 | 10′ 6″ | 3500 | 2720 | 1300 | 800 | 565 | 430 | 275 | 190 | 140 |
| Martinus........... | 6-34 D | 20 x 24 | 50 | 180 | 29,380 | 122,000 | 11′ 2″ | 4000 | 3060 | 1490 | 925 | 650 | 495 | 320 | 225 | 160 |
| Martios............ | 6-34 D | 20 x 26 | 50 | 180 | 31,810 | 132,000 | 10′ 10″ | 4000 | 3390 | 1615 | 1000 | 710 | 540 | 350 | 245 | 180 |
| Martirizar......... | 6-36 D | 21 x 26 | 50 | 180 | 35,080 | 141,000 | 11′ 0″ | 4000 | 3620 | 1730 | 1070 | 760 | 580 | 375 | 265 | 195 |
| Martisie........... | 6-38 D | 22 x 26 | 50 | 180 | 38,500 | 154,000 | 11′ 0″ | 4000 | 3960 | 1895 | 1175 | 835 | 635 | 415 | 290 | 215 |

.23 and 8 T x 20

22. A page from Baldwin Locomotive's 1908 catalog, documenting the sixteen variations on one of the firm's basic engines that the works was prepared to fashion, ranging in weight from fifteen to seventy-six tons. The "code word" column was for purchasers' use in telegraphing or cabling orders. Courtesy of Hagley Museum and Library.

profitably interweaving skilled labor, the custom and batch strategy, technical virtuosity, and product diversity.

A crucial merger, the Eddystone decision, and secular trends in railway purchasing soon combined to undermine Baldwin's preeminence and derail its momentum. In 1901, ten of its partners in the 1890s pricing pool amalgamated to form the American Locomotive Company, which instantly became America's largest producer, holding about 45 percent of engine sales. Baldwin rejected their merger invitation, then faced sharp bidding competition that depressed profits and helped impel the company toward building the vast Eddystone plant. Baldwin's historian has astutely observed that ALCO's need for orders to fill its multiple facilities and generate interest and dividend payouts instigated unwelcome price competition that challenged Baldwin's reputation for quality and timely deliveries. Theoretically, in this situation, Eddystone could speed materials flow and lower unit costs. However, at this critical juncture, the company's leaders erred strategically and seriously.

Baldwin's partners incorporated in 1909 and successfully floated a $10 million bond issue to fund new plant and tools, raising fixed expenses in anticipation of capturing huge railway orders for cost-effective motive power, then sold "7 percent preferred" and common stock in 1911 to increase their capital to $40 million. To their dismay, it took two decades to complete the Eddystone facilities, and orders adequate to fill Baldwin's new 3,000 engines/year capacity never developed. Railways had generated peak 1906–7 demand by massive replacements of smaller locomotives with new engines having far greater power. Soon, the average age of the nation's locomotives was somewhat over ten years, but their use expectancy was thirty years. Unless extraordinary technical or shipping market changes materialized, demand would surely drop. That outcome was assured by Interstate Commerce Commission regulators' resistance to increased railway charges, following the Hepburn Act (1906), a resistance reinforced by Louis Brandeis's pointed critique of railroad inefficiencies (1911). Railways, facing static rates, chose to work with the relatively new motive power they already owned. Hence orders slid until the war emergency, then fell again thereafter. Baldwin used slack demand to defeat labor organizing during the city's 1910–11 strike wave, but that was beside the point. To make locomotives, one has to have clients for locomotives, and clients became increasingly hard to snare after 1907. Though its new stockholders little realized it, Baldwin had peaked.[31]

Like Baldwin, the Cramp shipyard was a generalist in a specialty industry, building all varieties of vessels from tugs to warships. There too skill-centered employment moved with contracts, soaring to over 6,000 or plunging to fewer than 1,000 in these decades. Like the locomotive works, Cramp combined partial integration (its I. P. Morris works built engines and boilers) with contracting for many components, most crucially ship hull plates formed and bent to specified shapes at regional specialty steel mills (e.g., Lukens, Phoenix, Midvale). Yet if Baldwin had only a few hundred clients, Cramp worked with only a few dozen, chiefly shipping lines and governments, and faced superior British competition for commercial and foreign military contracts. Though seven substantial shipyards made the Delaware Valley "the American Clyde," referencing the U.K. concentration near Glasgow, U.S. shipbuilding only faintly echoed its transatlantic rivals' scale, importance to the national economy, and level of government support. Clients' diverse vessel needs, technical and design changes, and the inconstancy of demand meant that every ship was a custom product composed of purpose-built elements. With rising international trade and competition among navies, each ship launched was also larger than predecessors used for similar purposes.

The Spanish-American War pulled Cramp and other yards out of their depression lethargy. Steel ship output multiplied fivefold, 1897–1902; and the navy ordered 91 warships, 1898–1906, crowding the ways at American builders for the first time in a decade and the last until the next war.[32] The commercial surge largely derived from shippers' expectations that federal subsidies for American registry vessels, included in the 1900 Republican platform, would be legislated during McKinley's second administration. Cramp and Camden's upstart, but well-financed, New York Ship Corporation soon gathered major contracts for transoceanic passenger liners and cargo steamers, as well as for the "new" navy's ships. The commercial boom evaporated first, once Congress killed the subsidy bills. Naval contracts ebbed by middecade, and a bustling trade in coastal vessels died in the 1907 panic. This boom-and-bust sequence had serious implications.

Clustered calls for larger ships at the turn of the century led Cramp to invest $2 million in new tools, a huge machine shop, massive electric cranes, and longer berths for erection—all aimed at more rapid completion of bigger vessels. By 1903, the yard had $24 million in contracts on its books. However, financing this construction was a major problem for a firm capitalized at only $4.5 million. Private clients made partial payments only after sections of the work were finished, and the navy, a notorious slow payer, demanded numerous specification changes, whose costs could take years to recover. Thus in 1899, company president Charles Cramp attempted to organize a merger between his yard and Vickers, a major U.K. steel and military ordnance producer, in order to purchase an American ship plate plant and thus speed component deliveries while deepening his capital base. This scheme collapsed when Charles Schwab gained control of the target company, Bethlehem Steel, in 1901 and negotiated a forward union with the troubled U.S. Shipbuilding Corporation, a product of the first merger wave. Balked, Cramp ran head-on into a working capital crisis, passed its regular stock dividend (which halved share prices, 1902–3), and, to avert insolvency, worked out a $5 million rescue deal with Philadelphia and New York banks whose representatives effectively took over management of the firm. Their financial expertise proved useless. Ship demand faded, delays haunted completion of navy vessels, and anticipated cost reductions failed to materialize. Between 1904 and 1912, Cramp's total surplus on all construction amounted to roughly $1.6 million (as contrasted with nearly $1 million, 1901–3), most of which went to pay bond interest. Aggressively low bids secured a series of naval contracts (two-thirds of all Cramp ships, 1907–12) but assured marginal profits. Despite its technical improvements, the yard could only limp along. Shareholders' dividends rested at zero until 1915; Cramp kept going largely on its Morris unit's profits from turbine and pump manufacturing for public utilities.

Camden's New York Ship, by contrast, looked to be the model of a modern shipbuilding establishment. Its founders, appalled by New York–area land costs, transferred their affections to an open site twice Cramp's acreage on the Delaware opposite Philadelphia. There they constructed a state-of-the-art facility including covered ways (to defeat harsh weather), multiple cranes and tracks (for swift materials movement), and a central engineering building (for design and production control). NYS's visionary president, Henry Morse, recognized the Philadelphia district's valuable labor market and contracting possibilities, and planned to mobilize them to build "fabricated" ships—that is, standard vessels constructed from interchangeable parts made on the "American system" of duplicate production. However, it proved overwhelmingly costly to create templates, jigs, and fixtures for ship components that might be used twice in a decade and were readily subject to technical supersession, especially given that buyers' ship specifications varied wildly. Morse quickly backtracked to metal specialists' well-trodden path of fashioning production systems and a number of staple components for varied final goods. As an observer explained, Morse "proved that a greater proportion of the ship could be fabricated than anyone had believed possible," but "tried to do the whole thing and found that he could not." His notion of a standard ship would at last be realized at Philadelphia's Hog Island works during World War I but even then would be a source of production miseries and industry controversy. In the interim, the firm he left to his successors on his death in 1903 moved to batch production of tankers for Gulf Oil and uniform lighthouse vessels, plus customized naval warships for the United States and Argentina.[33]

In shipbuilding, only war emergencies, and in locomotives, only the market power of General Motors (which stepped sideways into the trade a generation later),[34] could force standardization on users of these highly specialized capital goods. In this era, makers ably met clients' specifications and frequently initiated new designs or guided required particulars into familiar and cost-effective channels. Interplay between makers and users was fundamental to heavy transport equipment sectors' vitality, though it did not always yield profits. If bidding for the immediate sale shaved revenues necessary for sustaining and technologically updating producing firms, the trades' foundations were undercut. As ever, price competition was fundamentally antagonistic to profitability in specialty manufacturing, a difficulty that management errors in investment and finance and lowball bidding practices compounded.

Placing shipbuilding and locomotives alongside electrical manufacturing also makes it clear that specialty production was not simply a "small business" phenomenon. Very large enterprises (by then-current measures) practiced specialty manufacturing, expecting that clients would pay a premium for precision, timeliness, or product capacity. By extension, specialty metal production and fabrication were distinct structurally and technically, as well as in market orientation, from bulk or tonnage metal manufacturing. Steel firms near Philadelphia's markets for the building of machine tools, machinery, locomotives, and ships fashioned links with regional specialists that shaped their own tactics for technological investments and labor recruitment. This network effect defined the region as a center for skilled metalworking employment and, at the same time, as a district of erratic prospects for firms and individuals.

Making clothing was distant from ship and locomotive building in scale, technology, and competitive dynamics, but the apparel trade represented a newly developing area of specialty manufacturing in Philadelphia. As elsewhere, factory and outwork manufacture of men's and women's garments challenged tailors' and dressmakers' craft enterprises. For fashionable shop owners, male or female "ready-to-wear" goods represented both ill-sized clothes and grinding price competition—the first an aesthetic, the second a practical, misery. Still, despite handicraft employers' disdain, ready-made apparel climbed the ladder from antebellum slop-goods for sailors and laborers to respectable plain suits and work clothes for men and, by the late 1890s, stylish shirtwaists, skirts, and coats for women. National output of men's ready-made apparel reached $250 million in 1890; women's clothing sewn in factories was valued at $68 million that year but jumped to $160 million by 1900, after a decade of stable to falling prices.[35]

As New York shops made half the nation's ready-to-wear, Philadelphia was clearly a secondary center, selling $31 million in men's goods and $5 million in women's apparel in 1890, three-quarters of which flowed from factories. During the 1880s, downtown manufacturing wholesalers acquired new sewing, fabric-cutting, and steam-pressing machines and installed the "progressive" or sectional assembly system, in which bundled garment components were passed sequentially between work stations. By 1890, menswear entrepreneurs' invested capital doubled 1880's figures, while their workforces dropped by almost half to 9,700 of the city's 14,000 clothing employees. Ninety factories with 37 percent of men's apparel operatives accounted for 74 percent of sectoral revenues; 900 small outwork shops and custom tailors (averaging seven employees each) produced the balance, suggesting the productivity of "inside" shops' redesigned labor process. Though outwork profits are unclear, custom clothiers earned mean gross profits of $2,500 on average sales of about $10,000 in 1890, whereas factory operations reaped on average $40,000 from output worth $260,000. At the sector's top end, in 1891 menswear maker Nathan Snellenberg "went down in his clothes, . . . pulled out the modest sum of $178,000, and paid it over for two stores on Market Street" in the central shopping district, plowing his returns from manufacturing and wholesaling into retailing.[36]

The next year, Abraham Kirschbaum invested over $300,000 in a fireproof, nine-story, double-elevator sales and factory building a few blocks away. After experimenting with cheap staple lines and subcontracting, Kirschbaum resolved to move toward quality and variety, striving "to do for the medium grades . . . what is done for the highest grades." By bringing all operations inside, this family firm expected to achieve economy and speed, as well as "excellence," for outwork "unavoidably lacked that element of constant personal supervision necessary [for] absolute perfection." Thus Kirschbaum anticipated New York manufacturers' return to the large inside shop, which commenced in the late 1890s in reaction to the shabby character of "tenement" goods and to reformers' attacks on outwork.[37]

The depression prevented others from copying this strategy; indeed, by 1894 Kirschbaum's inside employment had dropped from 300 to 90 and the firm returned to outwork relations with 14 small contract shops. Throughout the district, displaced apparel workers sought orders from manufacturing wholesalers or department stores, assembling garments at home or in rented South Philadelphia shops. John Wanamaker alone dealt with 39 subcontractors in 1894; together a dozen of the largest houses engaged 277 of them, roughly half the 600 outwork shops factory inspectors documented that year. Pennsylvania legislators added "sweat-shops" to inspectors' rounds in 1895 and required contractors to secure a state permit. Through thousands of visits, agents forced slow improvements in working and health conditions; but garment contracting continued to be a trade in which "almost anybody can start without any money," as an inspector told the Industrial Commission in 1899. Yet the depression rush to contracting did not signal a decaying trade with multiplying sweatshops, nor did the return of active markets wipe out this expedient tactic. Rather, outwork became a continuing presence in low-end staple menswear. It declined gradually over the next decade while inside shops rose to dominate growing styled-clothing production, particularly for women.[38]

In 1900, women's clothing represented almost 30 percent of Philadelphia apparel's $40 million sales, double its 1890 share—one consequence of the shirtwaist revolution that brought wide acceptance of the broad-shouldered, fancy cotton blouse and the long, pleated skirt as women's basic work and shopping wear. Factory inspectors in 1902 found only 6 percent of the women's apparel workforce in contract shops (as contrasted with 44 percent for men's garments); factories could complete seasonal orders for women's fashions faster and with better quality. The Kirschbaums soon built a six-story menswear fac-

23. Portrait photograph of the workforce at one of Philadelphia's many small, loft apparel shops, ca. 1910. If trade customs are any guide, the proprietor stands at center, third row, the only man wearing his suit jacket. Courtesy of the National Museum of American Jewish History, Philadelphia.

tory in South Philadelphia and restructured their outwork teams to create an inside contracting system analogous to Baldwin's. Men's apparel manufacturer and wholesaler Arnold Louchheim began publishing an annual style book "with leaves of grey photograph board on which the fashion plates are mounted," while Fleisher Brothers boosted their "high grade garments [in] cloths of the latest styles" and the Blumenthals announced their "original creation of novelties [in] suits for young men." Promoting fashion and variety was the coming strategy in men's clothing, for it moved a firm's goods toward the upper levels of emerging retail price lines and caught buyers' attention by employing stylish fabrics and designs.[39]

Outwork was of less value in this arena than in staple garments, for fashion manufacturing placed a premium on rapid response to orders, quick production shifts among styles, detailed supervision, careful inspection, and finished quality consistent with higher prices, all of which could best be accomplished at inside shops. As Philadelphia menswear makers battled New York houses offering up to "one thousand different styles in new lines" yearly, they agreed that "the chief requirements for success are special goods, special patterns, a thorough knowledge of every detail . . . , and the ability to keep expenses down." Outwork certainly cheapened costs in price-competitive staples; but in seasonal fashions it often increased expenses from cancellations due to delays or returns of goods "not up to samples" and allowed rivals to pirate designs being made up outside one's own shop. The more Philadelphia men's clothing firms duplicated the fashion commitments of women's wear manufacturers, the less outwork figured in their production plans. Contractors who employed 30 percent of the city's 24,000 clothing workers in 1900 declined in tandem with staple clothing's significance, even as the apparel workforce expanded to 28,000 by 1915.[40]

The 1907 panic's effects confirmed the shift to style and higher-grade products in menswear. One manufacturer later commented: "There is no way of estimating the increase in the production of suits retailing at $20 or more in the ten years since the movement began, but it was probably from ten to twenty times as great in 1905–07 as it was in 1897. . . . One significant aspect of the business contraction . . . was that the high-grade end of the business suffered least of all. This was contrary to all precedent [and] created great surprise." As contracting and staple production diminished in Philadelphia, men's clothing employment shrank (from 14,200 in 1902 to 10,500 by 1915), but women's apparel's growth more than compensated for the job loss (rising 109 percent to over 17,000). Women's dress, coat, and suit firms, many led by Eastern European Jewish entrepreneurs, became the new powers in the city's clothing sector. Their evident prosperity sparked workers' campaigns for shorter hours and higher wages in the wake of New York garment trade organizing after 1911. Like other Philadelphia proprietors, clothing factory owners formed a local trade association, resisted all demands, and weathered a summer 1913 strike at the height of the production season. The following year, however, they inked New York–style labor agreements that, among other things, abolished homework and established an "impartial chairman" for adjusting grievances.[41]

In the early 1890s Philadelphia's garment industry had consisted chiefly of factory and outwork shops producing staple menswear, plus tailors' and dressmakers' custom work. By 1912 this configuration had been thoroughly transformed. An upmarket fashion orientation at men's ready-to-wear firms and at a host of new women's and children's clothing enterprises emerged, overwhelmed custom sewing, undercut contracting, and pushed staple garments to the margins of the local trade. Through that process, the Philadelphia apparel industry joined the city's specialty manufacturing sectors and avoided duplicating Providence jewelers' slide into ruinous competition and sweated outwork. Why the difference? Two issues stand out. First, the principal Philadelphia apparel makers themselves either were wholesalers with direct contacts to retailers nationally or operated as retailers (Snellenberg, Wanamaker, and Strawbridge and Clothier)—some with wholesale departments like that of Chicago's famed Marshall Field. This structural position obviated the sort of jobber/manufacturer wars that were endemic to Providence, limited design piracy, and eliminated jobber-sponsored start-ups. With the trend toward styled goods, even those maker/retailers who had experimented with outwork in the 1890s brought apparel production to inside shops. Second, market demand for both quality and timeliness in middle- to high-grade men's and women's fashions placed technical and speed requirements on producers that Philadelphia's outwork connections could not master. Cheap staples remained a contractor's haven, and custom work for the "best people" continued in a modest way, but outwork proved too slow and unreliable to enable its practitioners to undercut the accumulating dominance of factory-based, ready-to-wear manufacturing or to allow the copying of styles to make any significant dent in ready-to-wear sales.[42]

## A Philadelphia Conclusion

The turn-of-the-century decades showed Philadelphia's mature capabilities for specialty production in their best light. Regional firms kept pace with technological advances (indeed initiating many), conserved their crucial skilled labor forces, sustained extensive interfirm production networks, and built institutions valuable to individual trades and the manufacturing interest as a whole. Proprietors resolutely defended high tariffs, which they believed were tightly coupled to high wages, expanding consumption, and business prosperity; yet they fought any incursion on their prerogatives by governments, unions, or colleagues who might limit their control of their enterprises. The Manufacturers' Club operated as an effective forum for intra- and cross-sectoral contacts and alliances, the Bourse gathered clients, and area clubs and schools fostered technical exchanges, though far less technical training than was optimal. Fashion apparel, instrumentation, conveyor systems, lace, upholstery, and fine hosiery and knits emerged as the new growth poles for flexible production. All specialty trades weathered the long depression, the nasty panic, and a series of labor upheavals with minimal damage.

With hindsight, cracks in Philadelphia interactive workshop networks are evident, however. The city's two leading metalworking companies had, with reason, solved sectoral production problems by retooling and expanding their facilities, thereby creating debt obligations to bankers and bond/shareholders and incurring higher fixed costs. Each misestimated its sector's future prospects. Middle-sized metal manufacturers failed to devise any effective scheme to train successors for their skilled laborers. Textile mill owners, who relied on their trade school's evening programs for that purpose, little considered the profit squeeze that retailers' fixed, laddered pricing policies could create. So long as product distinctiveness, timely delivery, and quality governed markets for consumer specialties, this was a nonissue; but should price rule sales, there could be ugly consequences. Once premiums for attention to detail, precision, and style evaporated, more than a few firms would face disaster, as would their workers. Before World War I, Philadelphians hardly took such prospects seriously. No Cassandra appeared to challenge area firms' settled and usually profitable practices; eventually, other forces would do so, with devastating results.

## Specialty Production, 1893–1912

The custom and batch manufacture of specialty goods was ubiquitous and significant in industrializing America through the decades during which the "modern business enterprise" developed in oil, tobacco, primary metals, foodstuffs, and other sectors. Across the nation's manufacturing belt and at scattered outlying sites, specialists crafted both minuscule and massive products that met users' diverse needs. Thereby they created hundreds of thousands of jobs, generated millions in profits and investments, refined technologies, and sustained dozens of focused, flexible industrial districts and their related institutions. Yet, as this and the preceding sections have illustrated, specialty sectors and firms were strikingly diverse, as were their urban constellations and their trajectories.

Consider first the three classes of specialists. Integrated anchor companies, making transportation and infrastructure goods, worked their way ably through the depression, updated their products, installed more effective technologies in expanded facilities, effected managerial and financial reorganizations, and in all cases overmatched labor unions. However, Baldwin and Cramp, unlike Pullman, GE, and Westinghouse, began struggling by 1912, owing to both external shifts they could not control and internal decisions that, seemingly prudent, later proved to be serious errors in judgment. The electrical giants fed a national power revolution that gathered strength as ship- and locomotive building stagnated, then declined. GE and Westinghouse also initiated timely mergers and acquisitions that reinforced their market dominance, as well as their enduring rivalry. Neither Cramp nor Baldwin did so. Schwab's Bethlehem outmaneuvered Cramp, and Baldwin, stumped by the ALCO consolidation, pushed into what became its Eddystone swamp.

For these anchors, proprietary values and commitments seemed to be shifting their valences from positive to negative as the Progressive Era closed. By contrast, Milwaukee's Allis machinery works' merger with Chicago's Chalmers heavy equipment company and its later acquisition of Cincinnati's Bullock Electric, or the three-way heavy tool merger of Niles in Ohio, Philadelphia's Bement firm, and Pond in New Jersey created multiplant specialty corporations that held national significance for generations. The increasing fragility of the freestanding Philadelphia duo suggests the importance of prewar mergers for classic integrated anchors. The financial adjustments Cramp and Baldwin undertook proved insufficient to solve their sectoral dilemmas, which worsened as demand faded.

Networked specialist operations in machine tools, furniture, textiles, and metalworking had different concerns. They too handled depressions and union challenges, added technical capacities, and developed relations with suppliers and auxiliaries, but they faced more difficult problems in marketing and in defining sectoral leadership. Here associational activity proved important at regional (and, occasionally, national) levels. Trade organizations, chiefly of midsize enterprises, generated novel collective selling tactics, presented a common front to railroads, labor, and governments, and sponsored innovative projects in technical education. Sectorally, however, these institutions varied dramatically in their capacities to function effectively outside crisis situations or beyond familiar antiunion or pro-tariff stances. In large part this differentiation derived from trade-specific characteristics that either limited or created opportunities for associations to develop trust-based interfirm relations and deliver services which members regarded as sufficiently valuable to render opportunistic actions unappealing. Thus in fine furniture, Grand Rapids' associations constructed marketing and transportation-related services for a tightly clustered set of firms, which made restrictions on poaching workers or shading terms palatable. By contrast, the diversity of Philadelphia's metal trade markets and products defined far narrower lines of common interest and affiliation, so that the regional association could rarely move beyond simply operating its Labor Bureau to meet members' fluctuating workforce needs.

Moreover, networked enterprises in industrial districts contended with complex competition in national markets from other sectoral agglomerations. This too was uneven. At one extreme, Grand Rapids had to battle for sales with sizable, organized concentrations in New York, Chicago, and Jamestown, along with smaller groups in Boston, Philadelphia, Evansville, and High Point, North Carolina. This situation may have contributed to their special solidarity, the duplication of their associational strategies by others, and the industry's repeated failure to establish an effective national trade body. At the other pole, the breadth and scale of Philadelphia's styled-textile complex far outpaced Providence's and Paterson's modest specialty clusters, which may help account for the Quaker City's proliferation of subsectoral groups and the relative weakness of its regional "peak" association. On a tangent, Cincinnati's machine tool district faced no comparably extensive rival complex (Cleveland's tool builders focused on metal-forming machines, a distinct specialty). With competitors scattered individually or in small groups from Vermont to Wisconsin, Cincinnatians quickly assembled a national trade body in which they played a leading, but not a ruling, role. By contrast, general metalworking was so diffuse that national associations (the National Metal Trades Association, the National Founders' Association) limited themselves to anti-unionism and the circulation of information on advances in managerial and technical practice.[43]

If associations offered promising venues for some networked enterprises, others in apparel and jewelry found them much less useful. As garret or sweatshop competition and subcontracting spread, no alliance in these sectors could raise barriers to entry, control design theft, organize marketing's chaos, or generate the salient services that might restrict opportunism. Neither individual nor affiliated firms could obstruct the restructuring of wholesaling and persistent "abuses" or the price pressures from direct-sales outlets (department stores, mail-order houses, retail chains). A sizable cohort of Philadelphia's apparel firms found a way out of this tangle by moving upmarket toward fashion lines and "inside," where design ideas could be protected, but eastern jewelers, facing the entrenched New York/Newark agglomeration, devised no comparable strategy. The most durably successful among them sought niche markets instead. In apparel, where trends toward standardized goods materialized (i.e., mid- to low-price men's and women's clothing), locational dispersion regularly followed as firms chased cheaper labor and facilities to meet price competition. Alternatively, urban enterprises mining this vein tried versions of "scientific management" to speed their output of staple blue serge suits and cotton dresses. Few of these clusters could manage competition, until labor organizations or state regulation laid down and enforced collective rules, which yielded their own unexpected consequences.[44]

Specialist auxiliaries depended on synergies with networked industrial districts or on contracts with anchors that had declined to internalize their specific capabilities. Through 1912, they experienced few threats to their livelihoods. Cincinnati tool builders routinely sent their castings orders to area specialists rather than invest in integrated facilities.[45] For similar reasons, Grand Rapids furniture firms bought tons of fine veneers from nearby makers, patronized the specialty hardware firms that sprang up in the city, and relied on a local machine works for the latest in woodworking tools and accessories. Still, the decay of some industrial districts would test their capacity to expand their contracting scope to new markets or widen the spatial shed of their business relations to include more distant firms.

This discussion will close with a few comments on urban trajectories, building on Part II's reference to Jane Jacobs' concept of supple, flexible industrial city-regions. There, cities with a substantial presence of specialty manufacturing in 1890 were characterized as exhibiting distinctive interactive or parallel dynamics of sectoral development, as well as derivative and narrow-focus patterns. How do the cities and districts sorted along those lines measure up twenty years later?

First, among the three interactive cities specified, only Philadelphia and Worcester carried forward their potentials. Philadelphia manufacturers both initiated new specialty sectors growing out of their interactive bases and created a set of regional institutions that stretched across sectors to provide collective services and occasions for information exchange. Worcester's diversified metal trades multiplied their specialty operations as well, but a more limited confluence took shape regarding the Polytechnic Institute and the regional branch of the National Metal Trades Association. Outside metalworking, however, few area firms seem to have been active in collective initiatives.[46] At Providence, the American Woolen consolidation, the jewelry trade's trials, and the separate advances of Brown and Sharpe, Corliss, and the Builders' Iron Foundry in metalworking suggest a division of interests. By 1912, the latter two Providence sectors occupied "parallel" spheres, whereas the textile enterprises had separated into differently oriented AW branches and independent mills. Thus just as Britain had but one Birmingham, so too did the United States, in Philadelphia.

"Parallel" specialty locales continued their patterns of distanced sectoral development. By 1912, little effective cross-fertilization among specialists appeared in New York, Newark, Paterson, or Wilmington.[47] "Derivative" districts on occasion transcended their earlier status. Cincinnati's machine tool complex, commenced in relation to engine building and woodworking, matured into a central engine for national machinery markets. Similarly, the Chicago furniture trade, spawned in order to use readily available Great Lakes lumber and packinghouse residues, now featured a national production and distribution system, rivaling Grand Rapids, and concentrated at the American Furniture Mart. Elsewhere a new derivative district took shape in Detroit's tool and die trades, linked to the rising automobile industry.[48] "Narrow focus" locales are the small gems in this collection of districts. These second-echelon cities with sizable commitments to particular segments of specialty production moved solidly forward in the generation after 1890, increasing output, workforces, sales, and, outside heavy capital goods, the number of firms in batch sectors. Grand Rapids and the other furniture cities, Reading, Pennsylvania, in knits and knitting machinery, the "Brass Valley" cities in Connecticut, and New England's coastal shipbuilding towns (Bath, Maine; New London, Connecticut; Quincy, Massachusetts)—all these showed the economically generative capabilities of custom and batch production in small-city settings.[49]

For many such sectors and districts, substantive change arrived during the decade after 1912, first as a consequence of the Wilson administration's public policy revisions and the onset of World War I. Inflation and conflicts that carried over to the postwar years and the punishing depression of 1920–21 reinforced the shifting environment for specialty producers. This extended tour will thus conclude with final visits to Cincinnati, Providence, and Grand Rapids to assess the course of machine tool, jewelry, and furniture manufacturing.

# *PART IV*

## DIVERGING PATHWAYS, 1913–1925

*Chapter 12*

# WAR, DEPRESSION, AND SPECIALTY PRODUCTION INTO THE 1920s

WOODROW WILSON recited the presidential oath of office in 1913, the first Democrat to do so since Grover Cleveland's ill-starred second term sealed his party's fate in the next four national contests. Wilson's good fortune, of course, flowed not from an epic shift in voters' orientations (as in 1932) but from a temporary fracture among the long-dominant Republicans, which brought Democrats control of the White House and Congress. Four components of the new leadership's program had the potential to chill industrialists' hearts, though as might have been expected, their impact on manufacturing's many segments would prove widely differentiated. During the Wilson administration's first years, as promised and feared, Congress revised tariffs downward, installed a new national banking structure (the Federal Reserve System), updated and reframed antitrust law through the Clayton Act, and centralized regulation of commercial and industrial practices in the Federal Trade Commission. Little wonder that some manufacturers credited the 1913 recession to the anxieties and uncertainties that this political sea change heralded. As the federal Commission on Industrial Relations convened its Philadelphia hearings that year, reports of both widespread, deepening unemployment and overstressed public services and private charities haunted the testimonies of factory and social workers, business proprietors, and politicians.[1]

More troubles lay ahead, at least in the short term. The outbreak of European war deranged markets, prices, and prospects for manufacturers in fall 1914. Although exports represented a tiny proportion of America's industrial output, in sectors like machinery and machine tools they were significant and now significantly imperiled. On another count, war promised decisive interruptions in the flow of key imported materials for specialty production: exotic woods from European colonies, German aniline dyes for fashion textiles, Russian, Australian, and New Zealand wools for suitings and carpets, rare alloy elements essential to steelmaking, and platinum for fine jewelry.

Worse, not only did workers' immigration effectively cease, some employees departed to join military or revolutionary forces in their nations of origin. After the initial panic ebbed, orders from the warring powers swelled; but their volume and concentration in a few subsectors created fresh difficulties. Immense demand for weapons, ships, munitions, uniform cloths, and other military necessities in 1915–16 further shifted labor-force dynamics. Both domestic inflation and war industries' active worker recruiting pressed specialists to conserve their core forces by hiking wages, scrambling after war-related contracts, or both. America's entry into World War I augmented the strains. Government decrees directed flows of critical materials. The draft swept up men not engaged in essential, military-related production (and others who were). The excess-profits tax infuriated business by creating paperwork, demanding specific accounting practices, and shaving returns; and the rail system staggered into crisis when eastward flows overwhelmed carrying, unloading, and car-recycling capacities. DuPont and the steel corporations may have enjoyed this peculiar boom; but for thousands of other firms, it unsettled their webs of production relations.[2]

Russia's revolution soon effaced an array of czarist accounts-due, and domestic strikes for inflation-derived wage rises multiplied, while military intelligence and Bureau of Investigation agents trolled plants for subversives and slackers. The war's sudden end may have astonished manufacturers, but the cancellation of government contracts that followed surely disturbed them rather more. Suddenly, flexibility held a premium, for shifting back to civilian production was imperative. Then, for roughly eighteen months after the Armistice, inflation continued its surge, making every transaction a spot market deal that advantaged those with inventories, and pushing labor militancy to new extremes. When it came, the reversal was savage. The 1920–21 economic smash, now neutrally termed the "inventory depression," represented the most compact, radical price deflation in the republic's history. Retailers bled arterially. Manufacturers holding finished stock watched their net worths evaporate, and machinery orders lacked a pulse. Shortly, railroads' effective appeals to the Railway Labor Board for wage reductions, following the Interstate Commerce Commission's lowering of freight and passenger rates, triggered the bitter 1922 railway shopmen's strike. Ugly sequences.[3]

These lunges and lurches mark 1913–22 as American industry's convulsive decade, a period of monumental strains and uncertainties, of federal interventions and withdrawals, of ferocious labor and market conflicts, which in intensity and quickly shifting currents surpassed the erratic economic cycles of the preceding fifty years. Before long, the Great Depression and World War II would overshadow this stressful decade, being after all a longer era of crisis with extensive economic and institutional consequences. Yet, for our purposes, the convulsive decade—which I have here extended into the mid-1920s to capture some of its implications—locates historically a parting of the ways among the diverse segments of American specialty manufacturing. This final empirical section follows three cases that together document this divergence. In Cincinnati (and nationwide, through the National Machine Tool Builders' Association), toolmaking firms developed both novel technical designs and collective strategies to master wartime and postwar challenges, moving forward with confidence. At Providence, however, whereas the leading tool builders echoed Cincinnati's advances, the jewelry trades' nagging earlier troubles matured into a sectoral crisis from which there were few escapes. In Grand Rapids, by contrast, furniture firms allied in an influential trade association sustained their productive novelty and profits amid the war decade's constraints, then firmly resisted federal efforts to restrict their regional collaborations in the 1920s. The 1923 manufacturing census's snapshot of American industry late in this period will, in closing, complement earlier statistical assessments of specialty production.

## Cincinnati's Resilience

As the 1912 national election neared, *American Machinist* tried to calm those in the machinery community who "may be touched with suicidal mania" in contemplating a Democratic victory and another tariff revision. "The machine industry . . . is not going to be ruined, no matter who is elected or what policy is adopted," L. P. Alford editorialized. Still, the new year's economic slump did end the "prosperous times" for machine tool builders that Cincinnati Milling Machine head Frederick Geier had heralded in his 1911 NMTBA presidential address. Early in 1914, his successor, Brown and Sharpe's W. A. Viall, offered a grimmer picture: "[M]any, if not all of us, are working under subnormal conditions, and we are hoping from week to week that there may be a pronounced change for the better."[4]

Although demand had fallen off "rather drastically," tool makers put their idle facilities to use in designing, testing, and building new machine models. A prosperous 1912 had so crowded their shops that only three hundred new tools and accessories reached the market the next year; whereas a disappointing 1913 opened sufficient space for them to fashion over five hundred novel lathes, drills, chucks, and gearing systems for debuts in 1914, a record that stood for almost a decade. The best of these represented the "three essentials" in machine tool design (durability, productive capacity, adaptability); yet most already bore the influence of the industry's intimate relationship with automobile manufacturing. *Following* car makers, tool builders increased their use of alloy steels, redesigned gears for quiet running, and adapted low-friction bearings and force-pumped lubrication to reduce operating and maintenance problems. *For* car makers, they fashioned the first generation of "special," single-purpose machines to speed production of auto parts for mass assembly. The era when tool builders' most productive technical cross-fertilizations involved the railroads was closing. Yet, recognizing that auto firms' machine tool purchases had faded in the present slump, the NMTBA also promoted tool displays at New York's 1913 National Automobile Show.[5]

Tool builders weathered the tariff storm but regarded the progress of the antitrust Clayton bill uneasily, particularly those draft provisions outlawing both interfirm "agreements" and exclusive selling contracts that prevented dealers from carrying competing lines of lathes or millers. In machine tool practice, standardization of screws, bolts, and terminology had stalled, despite repeated association resolutions and scores of prescriptive articles in the trade press. Meanwhile, firms like Cincinnati Planer and Lodge and Shipley continued to refine systems for routing and cost finding in work processes—using Alexander Church's machine hour rate to manage the costing complexities that arose, as William Lodge put it, "because the work in every department is changing almost with every hour."[6] Bonus systems were less prevalent than their endless promotion might suggest. A 1914 survey showed them in place at only 83 of 285 tool-building shops (29 percent), affecting but one-fifth of all workers, the rest of whom earned their wages at piece or hourly rates.[7]

The recession struck Cincinnati tool makers differentially. G. A. Gray's 1914 planer sales were one-third of 1912's, and Cincinnati Shaper's output dropped 60 percent in the same period (both in constant dollars). Meanwhile, Geier's Cincinnati Milling Machine, a key auto industry supplier, set a regional benchmark in 1914 by topping $2 million in sales, a 34 percent real dollar advance in two years. A districtwide index of tool-building production slumped roughly a quarter, 1912–14, before nearly trebling the next year, once war-related orders flowed in. Tool prices remained fairly stable, 1911–15, their levels about a third higher than at the turn of the century. The war pressed Cincinnati machine builders' capacity to the limit but brought them stunning revenues as well. Adjusted for inflation, annual production at regional firms was nearly three times their 1914 output at unit prices that doubled by 1918; yet wages rose a modest 31 percent and materials costs 70 percent, 1914–18. Local employment trebled to over 9,000, outpacing the sector's national increase from 30,000 to 77,000 workers. At "The Mill" (CMM), sales sped past $6 million in the war's final year. Gray and Cincinnati Shaper hit the million mark for the first time in 1918, while Lodge and Shipley settled for $2 million in lathe sales. Thus were booked the profits and reserves that carried local builders through the ensuing deflation without a single failure.[8]

Gray's experience with planers offers a glance inside the machine tool boom. Essential for the "large work" that heavy munitions and naval construction entailed, planers were also crucial to metalworking processes in machinery manufacturing generally. Yet they were in terribly short supply during the war. As veteran trade commentator James See ("Chordal") explained during 1918: "[I]f you happen to want some big planing done in Chicago you might have to wait months for some busy big planer to find the time to do your job, and a much shorter process would be to buy the planer. But big planers, old or new, seem to be as scarce as hen's teeth, and appear to be made out of the precious metals . . . [W]ar calls for big guns, and big guns call for big lathes, and the making of big lathes calls for big planers, but neither big planers nor big planing is to be had."[9] G. A. Gray and Co. did its best to meet the emergency demand but, believing the boom to be short-term, declined to expand capacity. In a busy 1912, it had built 178 planers, only a shade above its 1897 total (167). In 1917, running overtime, Gray constructed just 197 of these "master machines." Little wonder that a wartime planer shortage existed, for Gray was one of fewer than a dozen American planer makers.[10]

G. A. Gray did not build the giant planers that Niles-Bement-Pond and William Sellers supplied to shipyards and armor plate fabricators, but it overlapped with them in offering middling and large tools with throats (through which the moving table passed the fixed workpiece

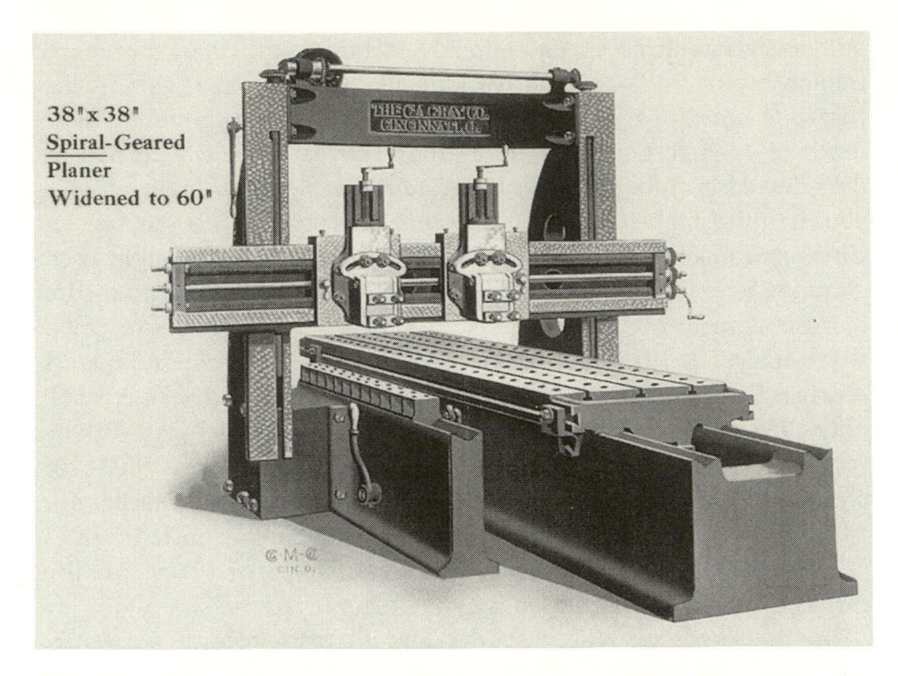


24. A large Gray planer, ca. 1915. This is a "special" with a throat widened to five feet, two cutting heads, and the optional spiral gearing Gray had adopted from Sellers in Philadelphia. The table, studded with holes for affixing the workpiece, moved back and forth underneath the cutting heads. Courtesy of Hagley Museum and Library.

across the metal cutting edges) from three to five feet square.[11] In this tool class, Gray's trade expanded in the war years. During both 1912 and 1917, Gray built planers in 44 combinations of throat and table sizes, no single model bringing in more than 13 percent of orders. Yet in 1912, over two-fifths of Gray's production was of small planers (throats under 30″), while but 16 percent represented large tools (over 40″). As "Chordal" noted, the war brought calls for "big planers"; hence in 1917, small machines accounted for under a fifth of Gray's output and big tools for well above a quarter. Midsize planers (throats 30″–38″) commanded 53 percent of orders that year, as contrasted with 40 percent in 1912. Gray's 1912 income amounted to $258,000 and five years later reached $892,000, yet that increase did not derive solely from the company's making more of its larger tools.[12]

Three factors figured in the firm's swelling revenues. First, whereas in prewar years about half the orders received called for standard tools, fewer than a fifth of 1917's sales involved base models. Each of the 161 special tools built that year possessed added features that escalated their prices. For example, a standard machine with a 30" throat and a 12′ table cost $3,250; but Gray billed $4,200 for one augmented with two "side heads"—29 percent more. A basic 36"/16′ model shipped to Wheeling Steel ran $2,972; whereas another, modified for three heads and an "extra heavy" frame cost a Cleveland firm $4,632—a 56 percent hike. Second, discounts to dealers disappeared entirely, replaced by add-ons to Gray's list prices that ranged from 5 to 35 percent, the highest charged to an order from the "Russian Commission" in New York amid the tumult of the revolution's early stages. Third, base prices moved steadily upward as the year progressed. Standard 24"/6′ planers Gray listed at $855 in 1912 cost clients $1,344 in January 1917, $1,605 in April, and $1,790 in November. These shifts, combined with the trend toward larger models, brought the firm unprecedented income in 1917 and 1918.[13]

As usual, Gray continued to rely on tool dealers for over four-fifths of its contracts (in dollar terms). Presumably wholesalers like Strong in Cleveland and Pittsburgh's Brown and Zortman, together accounting for 46 percent of all sales, replaced their vanished discounts by further marking up tool prices.[14] As "Chordal" implied, their frustrations and those of firms buying tools direct lay in another quarter—delivery delays. A standard 30"/10′ planer booked in early January 1917 hit the rails in just over 60 days, but Gray completed Ford's March order for an extra-heavy 36"/10′ model in mid-August, after 140 days. The lag soon lengthened. Westinghouse's request for four extra-heavy, three-head 36"/12′ planers was met seven months later with a late-October delivery; whereas a 36"/18′ extra heavy, booked in October, required ten months. Such delays led to cancellations of fifteen tool contracts in 1917, amounting to $110,000 in lost sales. Gray managers simply transferred several of these planers to waiting clients, while others could hardly have been begun when the revocations arrived, perhaps sent in response to the firm's fall 1917 terms: "delivery, nine months, not promised."[15]

The vast majority of Gray's clients, however, bided their time and welcomed their planers whenever shipped. Though most tools built in 1917 serviced war work in some fashion, the year's biggest orders directly reflected military needs—twelve midsize models for Dodge Brothers' Ordnance Department and thirteen more for the Army Engineers in France. On other fronts, General Electric purchased five planers for its Schenectady plant and twelve for its expanding Erie facility, including five of Gray's largest model. Westinghouse ordered seven machines sent to its Philadelphia works, while Carnegie Steel–Homestead, Bethlehem Ship, American Can, and three federal arsenals called for others. As planers were essential in making other machine tools ("big lathes call for big planers"), Gray also erected sixteen of them for midwestern tool builders, six for its own use, and six more for nearby LeBlond and Lodge and Shipley, who bought the year's largest products, five-footers with 40′ and 32′ tables, respectively.[16]

G. A. Gray booked its million in 1918, but the war's end brought the cancellations tool makers had long anticipated. As early as 1915, reflecting on the 1907 collapse, *American Machinist*'s editors worried that "the sudden close of the European war might easily throw another flood of cancellations upon the builders."[17] At Gray, that tide started running in September 1918, then swelled. By year's end, clients countermanded $307,000 in orders, three-quarters ($222,000) from contracts with the Allied Expeditionary Forces and Dodge Ordnance. Still, in the last third of 1918, Gray finished sixty planers, amounting to $299,000, a good year's income in prewar times. Although tool employment dropped 25 percent after war demand ended, the feared longer-term decline did not materialize. Gray's shipments in 1919 and 1920 totaled $1.6 million, Lodge and Shipley's, $2.6 million, and CMM's, a solid $10.3 million. The depression that gathered momentum in 1920 halved 1921's sales at both "The Mill" and Lodge and Shipley, while Gray surely had to draw on reserves to keep its workforce together. During 1921, its worst year between 1893 and the 1930s, G. A. Gray erected ten planers, netting less than $40,000 after deducting reinstalled dealer discounts (12–22 percent).[18]

Gray followed vice president August Marx's advice to fellow builders "in times of industrial storm." Responding to an *American Machinist* survey, Marx wrote, "Instead of disrupting a carefully built up engineering force, might it not pay to set your force to work on a constructive criticism of the present design of your product?" Such efforts could determine which features were commercially advantageous to users and which were "mere talking point[s]" for salesmen, or, through "more careful investigation of the design than was practical during the rush years," could help rework mechanical components, lower users' "upkeep," and cut a builder's production costs.[19] Three months after Marx's comments, Gray introduced its "Maximum Service Planer," with simplified, centralized controls grouped much like those on an auto dashboard, along with other features that enhanced safety, accuracy, and speed. By 1923, Gray brought two more new machine tools to market: the "Long Reach Planer" for heavy cutting (a thirty-seven-ton giant) and the specialty "Switch Planer" for heavy railway work. In 1925, this emphasis on novel designs yielded the innovative "Openside Planer" in four sizes and various table lengths, echoing the "massive knee and column" of CMM's heavy milling machines and discarding the square throat, "a marked departure . . . from previous designs of similar-purpose equipment." With first cost, machine hour calculations, and direct wages all included, the new planers, Gray argued, could reduce users' operating expenses by nearly 40 percent.[20]

Reinvesting war surpluses into new tool models and plant improvements was common at Cincinnati in the postwar period; but first, employers confronted a major organizing drive by the machinists' union in 1920. Earlier efforts to overcome area metalworking firms' opposition to union contracts had uniformly failed; now, expecting to build on wartime gains in other regions, the International Association of Machinists rallied for another try. During 1919 union members across the nation reportedly received appeals for dollar-a-month pledges to a fighting fund for an Ohio breakthrough. As tool builders sought fresh sources of scarce steel shafts and foundry castings that spring, organizers circulated through the district to recruit machinists smarting from the runaway cost of living, given their modest earnings gains since 1915 and, especially, plant owners' massive profits. The unionists' demands included a shortening of the workweek from forty-eight (or more) hours to forty-four, time and a half for overtime, a closed shop, and "a voice in the management" of firms. Encountering uniform rejections, the IAM called its strike in late April. One immediately ominous sign was that Geier's CMM workers voted against a walkout by nearly a nine-to-one margin. Likely more important to the conflict's outcome, however, were the ensuing trade collapse (which allowed firms with thin order books to hold fast) and the management of employers' collective policies during the strike by the National Metal Trades Association, metalworking labor's toughest antagonist.[21]

Geier's colleagues in the eastern Oakley complex largely avoided the turmoil that concentrated in the mill valley district west of downtown Cincinnati. Other local machine tool employers had pressed workers to sign the NMTA's "yellow dog" agreements, which affirmed that "loyal employees" would not join unions while in their employer's "service." By June, men who had taken this pledge, yet left workplaces at the strike's opening, started drifting back to positions in city plants. IAM advocates intensified picketing to balk this erosion, but soon manufacturers secured a court injunction that forbade "initiating . . . dealings, communications or interviews with any employee who has signed the non-union agreement," though others might be approached. The Ohio Superior Court also barred "threats, violence, abusive language, coercion and intimidation," undercutting all tactics but "peaceful persuasion" and dooming the strikers' cause. Soon after this finding, CMM cheerfully sponsored its annual August outing for workers and their families at the Cincinnati Zoo, where, after a "better babies contest," the "foundry huskies" defeated "Lil, the Zoo elephant," in a spirited tug-of-war. A core cadre of dedicated unionists held out through September, then folded their cards and returned to their jobs "under the identical open-shop and other conditions prevailing at the time the strike was called."[22]

One unanticipated strike outcome, however, may have been a general wage increase for Cincinnati toolmakers. Since the 1880s, their wages had been noticeably lower than what comparably skilled workers earned in eastern metal trades centers. In the wake of the strike that disparity ended, perhaps a manufacturers' gesture (despite slack markets) toward rewarding those who had proved "loyal" and encouraging similar attitudes among the rest.[23] Another factor, however, may be equally significant. For some years, higher-paying auto makers had been "robbing" toolmakers of "half-trained apprentices," skilled machinists, and even "superintendents of great value to the machine tool industry." From this angle, raising wage rates was arguably a strategy to retain present and potential talent. Either way, Cincinnati machine tool workers' earnings soared 71 percent, 1918–25, in constant 1914 dollars, thus powerfully augmenting their purchasing power and assuring labor peace through the next decade.[24]

Cincinnati machine tool firms chiefly used their wartime surpluses to renovate plants and equipment and develop new models. In 1921, the American Tool Works relocated into a five-story, custom-built, reinforced-concrete plant encompassing a quarter-million square feet of floor space, featuring a three-story, glass-roofed, open-court erecting shop and an employee cafeteria. Meanwhile Lodge and Shipley completed a one-floor, monitor-roofed addition, 50,000 square feet, with a 26-foot-high central bay for large lathe assembly. Like Gray, Cincinnati

Shaper and LeBlond focused on design innovations, Shaper bringing out the area's first metal-forming and shearing machines (1922–24) and LeBlond galvanizing the auto trade soon after with its automatic crankshaft lathe. As one analyst commented, "the firm that had produced perhaps twenty different sizes of engine lathes in 1910 now produced five times that amount of different type lathes, each designed to perform a single task on a specific part."[25] Thus did the advance of mass production in metalworking, particularly in the auto trades, stimulate further diversity in tool specialists' lines.

Cincinnati Milling Machine devoted its energies to novelty as well, but it also added a new class of tools by absorbing the Cincinnati Grinder Company in 1922. The next year, CMM introduced its "Centerless Grinder," which could finish "round metal shapes" to precision tolerances of two ten-thousandths of an inch, whereas "the previous limit had been about one thousandth." Earlier machines had held a workpiece between two centers, much as in a lathe, rotating it against the spinning abrasive wheel, but "the new process involved suspending the workpiece between two abrasive rolls, one of which ground the work while the other regulated the rotation." With sixteen-speed electric drive, CMM's grinder had "sufficient flexibility to be classed as a jobbing machine," *American Machinist* noted. In 1925, CMM announced two versions of a "Plain Cylindrical Grinder" that "differ[ed] materially from the usual design." Its increased rigidity eliminated most vibration while "simplified and centralized controls" facilitated precision work. The next autumn, the firm presented thirteen redesigned milling machines with dozens of improvements to "increase . . . delivered power," "maintain accuracy in service," assure lubrication, and ease "adjustment and maintenance." Their success was immediate and general. By the late 1920s, CMM again neared $10 million in annual sales, 30 percent of the entire district's tool output, and initiated its own "research laboratory." Although collective research lay beyond the differentiated Cincinnati trade's capabilities, this pattern of "successful new product development was the key factor enabling the local industry to maintain its position of machine tool leadership" nationally.[26]

New collective marketing tactics reinforced these technical advances. Though Cincinnati builders had commissioned shared industrial journal advertising and had jointly pursued plant-outfitting contracts in earlier decades, they faced marketing challenges more difficult than those which furniture makers had addressed through wholesale expositions. Their products weighed tons, demanded expert knowledge to demonstrate, and could hardly be hauled to a potential user's factory for a "look-see." Hence dealers who set up working examples of basic tool models in their urban agencies conducted the bulk of the trade's marketing. However, the price advantages of direct sales to users, which the war emphasized, combined with the demand crash during 1921, propelled tool builders into rethinking their sales approaches. Before the war, many companies had shipped sample tools to the "railway mechanical convention," usually held in Atlantic City. Some also took advantage of the Foundry and Machine Exhibition Company's services, arranging to display models at the American Foundrymen's meetings. Though few independent foundry owners would likely be purchasers, managers of integrated metalworking facilities that both cast and machined metal might be reached. The 1913 gesture toward the auto trades, showing tools at the New York car exposition, proved ineffectual. Manhattanites shopping for cars were simply the wrong target constituency.[27]

In the early 1920s, two new venues for marketing machine tools surfaced from quite opposite directions—Yale University and the steel heat-treating association. Yale's Sheffield Scientific School and the American Society of Mechanical Engineers' Connecticut branch sponsored their first exhibition in September 1922, noting the absence of "a concentrated effort to bring [machine tools] to the attention of the general public." By 1925, 30 tool builders and over 50 accessories makers offered their products for evaluation in New Haven. Midwestern firms accounted for three-fifths of the machine tool enterprises represented, 6 Cincinnati makers among them.[28] However, the inauguration of machinery exhibits at meetings of the American Society for Steel Treating proved more important. At the society's sixth convention (Cleveland, September 1924), nearly 200 firms "took space" for sales promotion. Thirty-five machine tool builders, including 8 Cincinnati firms, leased portions of a 25,000-square-foot tool section shared with 40 accessories makers. Why this attraction to the heat-treaters? A one-word answer may suffice: automobiles. Engineering and managerial personnel from the auto trades attended the steel treaters' meetings in droves, for that process was a key, rapidly developing element of automobile technologies. Some 40,000 individuals reportedly gathered at the 1924 meetings. Clearly there was no better point of access to the auto industry's yawning need for better tools. The next year, 53 cutting-tool builders were among the 180 exhibitors at the Cleveland festival. There 13 Cincinnati and 21 other midwestern tool companies presented their wares to another 40,000 potential buyers.[29]

Well, not exactly 40,000 *buyers*. The steel treaters opened their exhibitions to the curious public, which thronged the mechanical exhibits while ignoring the technical sessions. At the 1926 Chicago gathering, where total attendance neared 75,000, actual conference registrations numbered 8,500; and only among this latter group might machine tool purchasers be found. Reacting to the crush, to the difficulty of providing demonstrations amid seething crowds, and to the genuine sales potential that massing a host of tools in one spot could provide, the NMTBA announced prior to the Chicago exhibition that it would sponsor a national machine tool exposition in 1927. By September 1926, more than 50 tool firms had reserved advance spaces for the 1927 Cleveland show. The "exclusion of the general public" would be its core feature: "Admission will be by registration only. There will be no ticket distribution, no entrance fee, no crowds of mere sightseers." Cincinnati's R. K. LeBlond soon reflected: "The last two years have seen the greatest [technical] development in the history of machine tools. I think this was fully demonstrated at the Chicago show." Another builder added: "We doubt very much whether the users . . . appreciate the improvement in design and productive capacity made by the machine tool industry. An opportunity to visualize such improvements will be offered at the National Machine Tool Builders' Exhibition . . . and should greatly stimulate demand."[30]

In September 1927 over 100 machine tool enterprises, 14 from Cincinnati, plus 64 makers of accessories (motors, chucks, drills, cutters) crowded the stalls erected in Cleveland's Municipal Auditorium Annex, commanding 4,000 h.p. to drive their tools. Registrations topped 8,000; and "exhibitors agreed that at no previous exhibition had they been able to meet so many live prospects or to explain their machines so satisfactorily." Electric drive had become "so universal . . . that nowhere in the show was an overhead shaft visible." Builders unveiled scores of new models and, given the conjoined meeting of the Society of Automotive Engineers, caught the attention of their preferred audience, "the production men." Most important, "numerous sales were reported." The New Haven shows went on for a time but no longer held midwestern interest, whereas all the leading New England firms exhibited at the NMTBA's industry "congress." Forty-odd years after Grand Rapids furniture makers realized that drawing buyers to a central site could be a major marketing device for specialty goods, machine tool builders successfully duplicated that approach. Dependence on dealers might not be at an end, but the exhibitions assuredly reinforced "the present tendency . . . toward direct selling."[31]

Despite technical and marketing innovations, the mid-1920s were not solely times for rejoicing among machine tool producers. While the auto trades' "theft" of skilled machinists ebbed somewhat as toolmakers' wage rates ascended,[32] a perceived shortage of "all-round" workmen suitable for foremanship and managerial promotions remained. Thus, at Cincinnati, firms like LeBlond revised their apprenticeship courses to provide systematic training on different tools followed by a series of six-month stints in "the department[s] of the plant which use[d] the machines." Meanwhile, the University of Cincinnati initiated a new vocational program in which one or more foremen from the larger area plants learned enough about "the scope and meaning of foreman training work" to begin teaching the rudiments to promising candidates from their shops.

Less amenable to direct action, the upwardly revised tariff, which provoked retaliatory hikes abroad, effectively killed export markets for most machine tools, while German machine tool builders eagerly copied American design advances. Railroad and heavy machinery demand remained annoyingly flat. Cincinnati's expanded plants therefore routinely utilized only 60–70 percent of their capacity, but R. K. LeBlond brushed this aside: "The over-capacity of the machine tool industry is often referred to as a mistake on the part of builders. This is not altogether true. . . . The very nature of the industry requires that it should provide excess capacity to take care of peak demands." More positively, the 1921 smash had taught a salutary lesson to those makers inclined to build standard machines for stock. Three years later, the NMTBA's president explained that members' "present plan is to carry as small an inventory as possible . . . and not take the chance for loss attendant upon large stocks."[33]

Though discussion of standardization continued, so did widespread opposition. One builder explained: "The more difference we have in design, forgetting standardization, the greater effort is put forth by all thinkers along the lines of new things. Take this away and you haven't much left." Even the seemingly simple project of developing uniform screw specifications, advocated by Sellers in the 1860s, remained elusive. "For many years the thoughts of standardizing screws of all descriptions have been discussed back and forth, but nothing ever came of it." Another manufacturer supplied the logic behind tool makers' resistance: "[I]f it is carried to the extent that there is no difference between the articles or apparatus offered for sale by manufacturers, there is left only one basis of difference and that is selling price. This is . . . just as serious to the industry as stopping the progress of the art. . . . [Then] competition may become so keen as to be destructive. . . . It would seem that a logical solution of any such situation would then be the standardization of price as well as the qualities and characteristics of the product[, which] could be carried on only, if at all, under federal control. . . . a very radical departure from our present ideas of competition in industry."[34] Standardization of products *across* firms would generate no new demand for tools, builders believed, but would yield only price rivalry. Hence steps toward uniformity occurred *inside* firms, particularly through the building of "unit" components that could be installed in diverse models.

By 1925, tool production had rebounded to $176 million—in deflated dollars more than triple the depression trough—and Congress had revoked the despised excess profits tax. Two concerns, however, disturbed the industry's general optimism: (1) the Federal Trade Commission and Department of Justice's attack on sectoral efforts to regulate competition; and (2) automobile makers' presumptuous attitudes and, at times, predatory tactics in relations with tool builders. As Gray's records show, machine tool firms regularly exchanged price lists and technical details concerning their products as a means to avert price competition and focus sales on tool capacities and designs. Federal court judgments in the Maple Flooring, Cement, and Trenton Potteries cases appeared both to define this interfirm information sharing as unlawful under antitrust statutes and to bar exclusive sales agreements with "legitimate jobbers." The latter finding may have stimulated tool builders' interest in expositions and direct sales to users; but the former would undermine trade customs of long standing. A 1925 Supreme Court ruling doubtless relieved anxieties, for it held that the gathering and discussion of an industry's statistics on "production, stocks, unfilled orders, and average sales prices" were legal so long as this information was not used to fix common selling prices. As, given the industry's product diversity, there was no likelihood of price-fixing comparable to that in commodity hardwood flooring, the exchange of lists among competitors continued, as did regular reports to the NMTBA.[35]

The auto industry difficulty was more challenging, if less ambiguous. In a remarkable point-counterpoint exchange that *American Machinist* sponsored in October 1925, automobile makers articulated their dissatisfaction with tool builders' practices. Machine tool firms did not discuss new designs in advance with Detroit clients, failed to send their engineers for consultations, and did no research on metal cutting. Their basic tools had more speeds, feeds, and attachments than mass production required, while tool models and replacement parts were overpriced and often not instantly available when needed. Builders spent too much money on selling machines, incurring costs that increased prices, yet new machines often failed to perform up to specifications once installed in auto plants. Tool makers had not standardized features like work table heights and "tool holding and workholding devices," obstructing the integration of their machines into a "production line." Ultimately, "unless machine tool builders get together with the automotive industry more thoroughly," car makers threatened to build the tools themselves and to hire away skilled machine tool workers for the task.

Machine tool makers responded with acid courtesy. They declined to send engineers for collaborative designing because of "their experience in the few cases where this has been done. Designs for new machines and methods have been prepared and submitted, only to have them sent to other shops to be built, . . . leaving the designer nothing but the original tracing to show for his labor and expense." As for research, who had the deeper pockets to fund systematic inquiries into metal cutting—tool firms or auto corporations? If standard machines had more built-in flexibility options than car companies thought ideal, the complainers should realize that "though the automotive industry is the largest single customer, it does not use the bulk of machine tools made . . . and machines must be built to suit all customers." Certainly, auto firms desired special tools to match their production requirements, but this was a field with very limited demand. After all, "no automobile builder would design a car that he knew would be bought by only a hundred people at most." Regarding pricing, tool makers reminded critics that while "it is comparatively easy to spread develop[ment] and fixture charges of $100,000 over an output of fifty to one hundred thousand cars . . . , it is a very different matter to absorb a like amount in a few hundred machines without charging a price that may look like highway robbery." Equally frustrating, auto firms demanded tool prices that their accounting systems could amortize from a new machine's profits "in a single year," setting an "arbitrary and unfair" benchmark. The high cost of replacement components lay in the fact that "while automobile parts are frequently made in lots of ten thousand, the machine tool part is more apt to be made in lots of ten." Indeed, auto makers should examine their own practices in this regard. Common replacement parts for their vehicles, "if added together, average from 3½ to 5 times the cost of a completely assembled car."

Tool builders also flatly rejected the notion that their marketing costs were excessive, noting that selling expenses represented a higher proportion of car prices than of tool prices. Worse, auto firms "add much to [makers'] cost of selling" by demanding "tooling blueprints, layouts, time study and other engineering data." New machines often failed to perform as expected simply because they were regularly "forced" on foremen who had no role in their selection and because auto corporations skimped on training shop workers in their proper use. Every newly designed tool brought onto the plant floor was "as much an experiment as is a new motor or transmission in an automobile," a factor their critics seemed inclined to forget. On standardization, a tool builder responded bluntly. Though the auto trade had promulgated more standards than had any other industry, these were far from being "in universal use. The standards are used when it *pays* to do so and not otherwise. When the machine tool builder can be shown that it will pay him to adopt standards, he will do so." Last, if the car companies wished to build their own tools, let them attempt it. Given auto firms' persistent "antagonism" toward tool makers and the expensive "engineering and development work" required in creating special-purpose tools for them, builders expected that auto firms would soon learn that the "designing and development work must be paid for, no matter who builds the machines," and that their mass production efficiencies would prove useless in erecting "the comparatively limited number of these machines that can be used."[36]

In these exchanges, tool builders' recognition that they occupied a manufacturing world radically distinct from that of auto makers is paramount. They, not the car company spokesmen, again and again contrasted mass and batch production, standardizing and specialty strategies, and markets that did or did not grow through price-cutting competition. It is difficult to imagine that Detroit interests failed to understand the distinction, but the journal's editor thought so. "The automotive man can hardly appreciate the problems that confront a builder who knows that his market must, of necessity, be limited, [i]nstead of having a market that can be materially widened by a reduction in price." Auto makers appeared to regard machine tool builders as comparable to dependent tool-and-die shops or parts contractors over whom they could exercise authoritative influence. From the car companies' perspective, tool builders who rejected their demands showed themselves as not being "alive to [their] opportunities." If the builders would not "get together with the automotive industry," auto firms would shove them aside. Ultimately, in denying auto makers' claims to market power and technical authority, tool firms articulated their sovereignty in their field of expertise, offering cooperation with Detroit but declining to bow to its instructions.[37]

In 1924, a veteran German engineer visited thirty machine tool plants, fourteen tool-using factories, and the New Haven exhibition in order to draft a review of the industry's practice, thus providing an outside opinion that summarized a number of points discussed here. Dr. G. Schlesinger praised U.S. builders' design novelties and production systems, judging that "the standard of quality for machine tools set by leading American manufacturers is high. The best machinery and limit ga[u]ges are used in production." In the finishing of the "guiding surfaces" in machines of all sizes, hand-scraping remained a regular practice, "although with planing a high degree of accuracy could be obtained." Makers rejected a faster milling process for "finishing machine beds," here embracing planing, even at Cincinnati Milling Machine, which counted "the necessary loss of time . . . as more important than the possible cheapening of the product by milling." Small lot construction continued in place; "stock orders amounting to from 20–50 machines are seldom seen." Certainly, firms like CMM, LeBlond, Gray, Lodge and Shipley, and Brown and Sharpe had designed gauges and tools to assure parts interchangeability; and some had adopted "the progressive assembly method although they produce daily not more than 5–6 machines." Echoing John Richards' 1865 Brown and Sharpe visit, owners and managers had welcomed Schlesinger to their factories, furnished him with all information requested, and readily showed him "original drawings, results of experience of much value, special installations, and other valuable data."[38] While retaining such trade customs and continuing the batch production of specialty goods, machine tool builders by the mid-1920s had also embraced such elements of mass production techniques as fit their format, had designed a diverse new generation of machinery, and had begun to solve the puzzle of direct selling through exhibitions. Here was a specialty manufacturing sector in fine fettle, unawed by the automobile trades, facing the future with confidence. At Providence by 1925, Brown and Sharpe exemplified this stance, but Gorham Silver and the area jewelry trades assuredly did not.

## Prowess and Divergence in Providence

With 3,500 to 4,000 employees in prewar years, Brown and Sharpe was Providence's largest industrial establishment. President Henry Sharpe operated a resolutely open-shop enterprise regarded as the "backbone of the National Metal Trades Association." Vice president W. A. Viall, son of the works' manager in the 1890s, headed the NMTBA in 1914 and 1915. The 1913 recession had spurred hundreds of layoffs, but war orders began inundating Brown and Sharpe late the next year. Enlarging its staff to 5,500 by mid-1915, the company installed a night shift and paid time and a half for "overwork" beyond its regular 55-hour week. That summer, fresh from successful union drives in the munitions plants of Bridgeport, Connecticut, AFL Metal Trades organizers arrived in Providence to challenge the hard-line local NMTA branch, which effectively controlled the metalworking labor market through its central employment bureau. Beginning in July, Brown and Sharpe selectively raised the wages of key workers "with the expectation of allaying some of the uneasiness," then increased tensions by "discharg[ing] quite a number of men who had pronounced Union proclivities."[39]

On September 19, following a hasty membership campaign, AFL organizers and the Federation's Metal Trades Department leader presented demands for a 48-hour week, with no loss in pay, and union recognition. Brown and Sharpe rebuffed them and the strike commenced the next day after lunch. At least 2,800 shopmen deserted the plant, marched around it and then toward downtown before being "headed off" by police. The firm stonewalled all attempts to mediate, politely refusing the intercessions of two federal conciliators, ignoring Samuel Gompers when he arrived on the scene, and balking the state governor's effort to bring the parties together in his office. Early on, Brown and Sharpe pledged "a fight to the finish" and prevailed; for as

25. Brown and Sharp's plant in 1916. The original U-shaped building, ca. 1870, shown in plate 2, rests at right center behind the waterside five-story building, dwarfed by two generations of construction that expanded capacity. Courtesy of Hagley Museum and Library.

a disappointed conciliator wrote, "the outcome . . . is simply a question of endurance." By the close of the third week, when the company announced that it would no longer hold strikers' positions open, over half of the walkouts had returned, bringing the active force to 4,000. On October 30, strike chairman George Finnell unsuccessfully urged the remnant to give up the battle, for now 4,800 were at work. Brown and Sharpe would hire more each day; Finnell said, "and they will do with them the same as they have done with a whole lot of us, they will bring men in, and they will educate them. The Brown and Sharpe Manufacturing Company has been a school of education for a long time." The strike soon collapsed. Thereafter, "as quietly as possible," the firm rewarded the 2,500 workers who "stayed with us" by adding a week's earnings to their pay packets. Officials worried about the effect this loyalty bonus would have on the nearly equal number of returned strikers. Yet of the latter, Viall commented, "[W]hile they were disappointed, there was good sporting spirit shown among them, and they accepted their fate, many times making some very good jokes at their own expense." At least 500 others, permanently discharged and blacklisted, surely found nothing humorous in leaving Rhode Island to seek new jobs.[40]

Having crushed union organizing, Brown and Sharpe concentrated on war-related contracts and continued building its workforce—by late 1917 to 6,600 men and women (70 percent of whom subscribed over $260,000 to the Second Liberty Loan) and to 7,500 in April 1918, a level sustained through mid-1920. Though its postwar machine tool line included "milling, grinding, gear-cutting and screw machines" in "eighty-four sizes," the company hastened to bring out new models: an updated, column-and-knee milling machine in 1920, then an automatic control miller, a spiral-gear hobbing machine, and a "single-purpose" plain gear cutter, all in 1923. Its renowned "all-round" apprenticeship program expanded to 200 young men by 1921, offered a "dormitory" for 27 out-of-towners, and featured a co-op program for 25 more students at the city's technical high school. All earned wages that rose from $500 to $835 across four years, plus bonuses of 6 to 12 percent for shop work judged "good" or "excellent." Most remained with Brown and Sharpe for at least a year after completing their indentures. "[O]f those who leave to go with other firms, many return later. The management, including most of the heads of departments . . . are graduates from our training courses," Viall noted with satisfaction.[41]

Despite the collapse of orders and the resulting extensive layoffs in the 1921 contraction, Brown and Sharpe's corporate surplus for state tax purposes stood at $4.7 million two years later. Facing what it termed "a depression in business" in 1924, the company closed for a full month in August, rather than the usual two weeks, reopening with a full force September 1, but without the 10–25 percent wage reductions that other Rhode Island firms decreed. Employment gradually rebounded to above 6,000 by the late 1920s, while product diversity continued in full bloom. Although the firm's "Small Tools" catalog then offered 257 models of precision micrometers and "nearly 2,000" varieties of milling cutters, Brown and Sharpe invited clients needing "special tools and gauges in sizes not listed" to forward specifications for speedy estimates and timely delivery. One shop head noted earlier, in explaining the plant's detailed work-estimating and routing system, "Many of our gear orders are for special gears and are generally wanted by the customer about the time we get the order to make them. This means that the best possible date must be given." Last, in the 1920s, as sixty years earlier, Brown and Sharpe announced, "We are always ready and pleased to show our works to those who are interested in machine shop practice."[42]

At Brown and Sharpe, rigidity in defending proprietary prerogatives when challenged by labor complemented keeping the door ajar for visiting metalworking colleagues. A closely held corporation descended from the antebellum partnership of a craftsman and his first apprentice that led a fierce employers' association, Brown and Sharpe remained enmeshed in networks of skill, mutuality, and prowess, committed to product diversity and technical novelty even as it adopted or devised "modern" systems to manage operations on its thirty-two acres of shop floors. Surely a big business, it looked nothing like American Tobacco or Standard Oil, had much in common with great metalworking plants like Baldwin Locomotive and Cincinnati's midsize machine tool firms, yet contrasted starkly with Gorham Silver and its neighbors in the regional jewelry trades.[43]

By 1913, Gorham and its associated Silversmith's Company were in essence the personal fiefdom of Edward Holbrook, who had controlled the firm since 1894. Having acquired three midsize sterling silver firms before 1910, Holbrook moved in 1914 to purchase the Mount Vernon group, which operated three other enterprises in the trade. Through and after the war, Holbrook's "empire" included 40 percent of U.S. sterling production, with sales reaching above $10 million in 1917. By that time, however, its handcrafted flat- and hollowwares, ecclesiastical goods, and other specialties represented only a modest portion of total sales, for Gorham "invested heavily" in war production, building and equipping three new factories to mass produce cartridges, hand grenades (100,000/day capacity in 1918), and naval shells. Though it serviced the munitions needs of twelve nations, Gorham's experiment in military production "proved disastrous." As a company officer reflected later, "World War One dropped a bomb on us with unprofitable contracts, topped off by noncollectible debt[s] of the Russian government."

Holbrook died in 1919, the depression struck Gorham hard, and by 1922 "costs so far exceeded profits that the firm was effectively bankrupt." To meet the crisis, a new president tried to revolutionize marketing but succeeded only in wrecking it. Gorham had long sold its fine wares through upscale jewelers who held exclusive local retailing rights for part or all of the company's lines. In 1921, however, Franklin Taylor announced a new "open door" policy through which Gorham would welcome accounts with any interested retailer, including department stores, and invite the public to its regional wholesale showrooms for direct purchases. Within two years, a host of its elite retail clients dropped Gorham entirely (Towle Silver moved in to pick up the accounts); and Taylor's effort to "democratiz[e] the ownership of silverware" by selling to all outlets had failed emphatically. Management contracted with New York consultants Aldred and Co. for a full analysis of Gorham and its subsidiaries, aiming toward a "massive reorganization of production and marketing." Their 1924 report showed that the firm's accounting records had failed to deal with inflation (in real terms, predepression sales were 25 percent below those of 1910–13) and that its huge Elmwood plant was seriously underutilized, averaging a 1 percent return on investment. New president Edmund Mayo responded quickly; by 1927 he closed all but the largest of the scattered plants Gorham had acquired, consolidating their production at Elmwood while preserving their trade names and styles. He largely restored traditional marketing practices, sponsored advertising campaigns that again emphasized luxury, and created a personnel department to withdraw the authority for hire-and-fire decisions from foremen. Taking managerial guidance from consultants was Gorham's equivalent to machine tool builders' adoption of technical practices effective in auto production. Both represented the selective installation of approaches developed for and by routinizing corporations that seemed appropriate to specialty production contexts. Gorham continued to make thousands of distinct silver, brass, and bronze items, recovered its markets, "paid off its debts, and returned to profitability," taking full advantage of another boom in silverware during the late 1920s.[44]

In 1913, *Manufacturing Jeweler*, the voice of New England's producers, announced that "the decade from 1900 to 1910 was the most prosperous period in the history of the jewelry business." Fifteen years later, a correspondent fairly shouted, *"The jewelry industry is sick."* Jewelry's story into the mid-1920s maps familiar trade impasses and fluctuations across a landscape of stagnation and drift, at least in the Providence/Attleboro district.[45] Thus this discussion will proceed thematically through two periods, before and after 1918, in turn treating style and marketing, trade associations and the state, labor, production and technology, and training. In 1913–14, cheap jewelry fads continued to burst on the scene like fireworks, briefly brilliant, then rapidly fading to black. Beads of all varieties came into vogue just before the war, but the mesh bag rage ebbed before resuming in quite different postwar conditions. A style lull followed until 1917, when patriotic goods dominated the scene and Providence-area firms made flag lapel pins in stunning lots of ten thousand gross. That April one New York buyer, unable to secure deliveries, reportedly "visited a large ten-cent store, purchased a thousand gross or more at ten cents apiece and sold them for thirty-five cents" at his employers' department store. Style diversity in "military" lines and outwork in enameling ("manufacturers are turning to families for relief, as in years gone by") boomed for eighteen months, but all other novelties remained "dead" until the early 1920s. In 1918, a sanctimonious reporter commented on "baubles": "There is no profitable or patriotic place for them in a jeweler's stock during . . . war times." For several years thereafter, fashion dictated that jewelry "was comparatively little worn." In the novelty famine, producers relied on making "staples" in chains, religious goods, rings, and the like.[46]

The industry's marketing problems persisted. In "utter disregard of terms," jobbers and retailers routinely returned unsold seasonal goods for credit against invoices or ordered jewelry "on memorandum," a short-term consignment, then shipped it back many months later. Bills due in 60–120 days languished on makers' books for six months to a year, yet late remitters still took discounts offered for prompt payment. Horace Peck's 1913 study of trade terms and discounts in seventeen industries from hardware to varnish provided concrete evidence of jewelry manufacturers' indiscipline and lack of market leverage. Most sectors offered a 2 percent discount for payment in 10–30 days, uniform across all producing firms. Jewelry discounts started at 6 percent and ranged beyond 10, and this for remittance in four to eight months. "Credit men" from fifteen trades averred that they successfully compelled late payers to cover "net" billings after discount periods expired, and some collected interest on overdue accounts as well. Jewelry firms failed on both counts; indeed, bankruptcy records among jobbers and retailers showed that they had figured their books on the discounted value of orders. As usual, manufacturers applauded, then ignored, Peck's recommendations for uniform discounts and "a rigid rule" that net balances be collected from slow payers.

Once the European war opened, jobbers accentuated their predilection for forwarding "only sample orders" initially and purchasing "from hand to mouth" thereafter. As wartime materials costs escalated, Providence and Attleboro's low-end producers found fierce resistance to passing along rising expenses through rising prices, a consequence of retail price-laddering that fixed ten, twenty-five, and thirty-nine cents, for example, as retail jewelry ceilings. This squeeze informed proprietors' reluctance to deliver wage and piece-rate increases in these inflationary years, which stimulated both unionization threats and, more important, labor flight to other sectors. Local tradesmen later noted that between 1914 and 1920, as the cost of living more than doubled, they managed to advance their prices only 25 percent. The one positive note the war years brought to market relations derived from the diversion of regional die-cutters to munitions-related contracts. This shift reinforced the novelty slump, as tool and die firms were the crucial auxiliaries in a dis-integrated trade increasingly reliant on presswork. More significant, it encouraged a host of jewelry manufacturers to seek war production contracts for "small wares" in brass (e.g., buckles, buttons, insignia), shrinking the number of competitors in the low-end jewelry trade. As a result of this reorientation and of the liquidation of over a hundred area firms, 1915–18, makers briefly could secure prompt payments, even at times cash on delivery, for the staple and patriotic goods their clients ordered.[47]

Trade associations attempted to regulate competition and manage market relations in cheap jewelry before the war but soon found themselves cited by the FTC for antitrust violations. Groups of large national wholesalers and leading Providence-area manufacturers had agreed to blacklist those manufacturers who sold directly to retailers, department stores, and mail-order houses. No wholesale association member would purchase products from such firms. They also worked in tandem to exclude competing jobbers from the trade's "Rating Books" and "insisted that the [credit] classifications" of offending retailers and nonmember wholesalers "be changed and as a result the concerns were unable to buy goods from the manufacturers." This effort, if successful, would have reestablished the orderly flow of goods from makers through wholesalers to retailers that had existed thirty years earlier, thereby opening the way to uniform terms of trade and reduced price-cutting pressures. A federal court's 1914 decree barred all such collaborative attempts at trade governance. The mad marketplace scramble could not be controlled through such strategies.[48]

Soon thereafter, jewelry manufacturers entertained the idea of organizing the market from another direction, through style publicity. Perhaps by advertising in *Vogue, Harper's Bazaar,* and *Vanity Fair,* makers could both promote jewelry sales and, by closely monitoring women's fashions, set style lines that resonated with seasonal trends. Originating in Newark's high-end sector (hence the elite magazines chosen), this campaign found few adherents when presented to the New England Manufacturing Jewelers' and Silversmiths' Association (NEMJSA) in 1915. Doubting that their ultimate customers read *Vogue* regularly, Providence firms declined to forward their share of the $300,000 promoters sought. Newark manufacturers spent $10,000 on their own, but to little effect. Once the United States entered the war, NEMJSA and the other jewelry organizations could do little more than protest a series of federal actions that hamstrung the industry. First a treasury official announced that jewelry and other "luxury" trades were nonessential industries which could be suspended or deprived of materials, if necessary. Then in 1918, Congress passed a war revenue bill imposing a 10 percent jewelry tax, along with other luxury taxes; and the Council of National Defense forbade jewelers' use of platinum, gold, and finally brass. This last sequestering, just weeks before the war's end, was quickly revoked; but the jewelry tax remained in force through the mid-1920s, infuriating producers who watched impotently as Congress rescinded other luxury taxes.[49]

Nor were employers able to manage an increasingly discouraging labor situation. The Providence and Attleboro trade counted 21,400 workers in 1913, but only 15,400 remained two years later, "fewer jewelers at work than for any time since 1894." Two years' "dullness in the jewelry business had dispersed the skilled workers, who had left this vicinity, or [had] gone into other lines."[50] In this tightened market, labor turnover accelerated. "Two or three weeks work at a wage usually above the ability of the transient worker is the rule rather than the exception," one industry veteran griped in 1916. Convinced that they could afford only minimal rate increases, manufacturers virtually invited union organizers to the district. As before, recruitment efforts in Providence provided some excitement, ended by a brief, fruitless 1917 strike at Ostby and Barton, the city's largest jewelry firm. Labor's prospects initially seemed better in Attleboro, yet unionists again could make no dent in the employers, who were "contract bound amongst themselves to maintain . . . the open shop." There, after all but a few firms refused even to reply to the jewelry union's demands, a late-July strike call brought out roughly half the workforce. Within days the drift back to the shops commenced, crushing hopes for success. This upsurge led NEMJSA to collaborate with the MTA's employment bureau to prevent strikers from securing other metal trades positions, a rare moment of cross-sectoral alliance in the district. Predictably, however, jewelry manufacturers' solidarity against unionization did not imply any wider collaborations, particularly in the wake of the FTC case.[51]

Technological and labor process change was not quite absent from the industry in these years, but nearly so. Apart from recommendations to consider electric motors as an efficient substitute for shafts and belting, trade conventions offered no discussions of scientific management, time study, or advances in electroplating and die work. Still, the only significant technical innovation did effectively end outwork on mesh bags. Ernst Bek's patented automatic mesh machine, installed at his new factory near Newark in 1913, transformed the labor process by bringing all work stages inside the plant, much as power looms had done almost a century earlier. By 1917, Whiting and Davis erected a 50,000-square-foot, one-story Attleboro plant devoted to fashioning bags from mesh that streamed out of similar machines built in the company's metalworking shops. "The automatic mesh machines are wonders, and do everything but talk. . . . [S]o nicely are they adjusted that an imperfect or broken link or any defect in the weave automatically stops the machine, allowing the attendant to make repairs." Whiting and Davis employed 450 "hands," chiefly women, to assemble style lines. The Attleboro area's 9,000 outworkers, thousands more in Providence, and the women subcontractors who had dominated hand mesh making vanished from the industry and from reform agendas.[52]

The outmigration of labor to other districts and industries pressed home the unmet need to train a new generation of jewelry workers. Apprenticeship was moribund, but the Rhode Island School of Design and local high schools seemed promising venues. However, "[w]hen an attempt was made to establish a co-operative jewelry course in Providence, . . . it was found to be almost impossible to obtain students to take up the making of jewelry as a life-work." Instead, eight local shops sent nine employed "boys" to RISD for nine hours of daytime training per week; another firm commissioned an evening version of the course for fifteen of its young men, all the companies paying the students' eighteen-dollar-per-term tuition. By 1918, enrollment in the combined day and night programs had reached 120. Yet most manufacturers' remained reluctant to join the industrial education movement, "because of the possibility of the employee going to another shop . . . the owner of which benefits at the other's expense." The contrast with Brown and Sharpe's culture of apprenticeship could hardly be more vivid.[53]

If the region's jewelry sectors came out of the war era on shaky ground, nothing firmed their foundations by the mid-1920s. The labor force, which rested at 15,000 in 1918, drooped in the 1921 slump to just over 9,000 and rarely passed 11,000 across the next five years.[54] Attleboro's complement of firms slid from over 100 to 68 by 1925; and no new factory "apartment" buildings sprouted in Providence, though W. R. Cobb's heirs continued to supply findings to the trade, employing ten workers, much as forty years previously. Still, as predicted in 1918, Providence new starts arose in fair numbers, for men "from the bench" continued to enter the trade with small resources and large hopes, occupying spaces vacated by bankrupt tenants in the aging Manufacturers', Doran, and Bliss buildings. While appealing successfully to Congress for higher tariffs in 1921, NEMJSA's manager claimed that "eighty-five per cent of the jewelry manufacturers have risen from the bench by slow and hard work." He reserved his opinions of newcomers' penchant for pirating styles and cutting prices for other forums.[55]

Two years after the Armistice, a demand for novelties coursed through the New England districts. Interrupted by the depression, which in jewelry lasted through the close of 1922, it resumed with full force the following year. This "style craze" cost manufacturing jewelers dearly, for to catch the markets they had to shift away from war-era staples and commission new designs and dies to once again "snap out the style of the moment." Thereafter, each low-end maker fed, "primarily, a market for novelty or costume jewelry. His stock . . . must have a quick turnover, because novelties go in and out of style overnight." Yet producers' market positions provided little encouragement. With the depression came a raft of slow settlements and expired discounts taken (which nervous manufacturers tolerated) along with orders contingent on long credits (which they again granted). Hand-to-mouth buying resumed with assumptions of stock-production and immediate delivery on orders and the assurance of dead-loss inventories at seasons' ends. The renewed flow of cheap German and Austrian jewelry, despite an 85 percent ad valorem rate, triggered an import panic as well. Later Department of Commerce data showed that foreign goods accounted for well under 1 percent of retail jewelry sales in the United States, but New England producers spooked easily. Harvard researchers' discovery that the trade's retail branch had the lowest annual inventory turnover among ten categories of American stores only deepened the gloom. With mail-order firms' demand slipping, department stores and wholesalers pressing for price concessions, and retailers asleep, what were enterprising manufacturers to do?[56]

For some, again chiefly in the high-end trade, publicity campaigns to promote jewelry consumption, create style trends, and hike retail turnover rates seemed a sensible step. Yet now the cash gathering for magazine advertising carried million-dollar price tags. Retailers nationally and producers in the New York/Newark complex anted up, but most Providence-area companies backed away. An alternative tactic essayed in 1925 was a market-organizing style exposition—equivalent to those newly attempted in machine tools and long evident in furniture. This too was a disaster. Mounted in Boston as a regional display of fresh jewelry lines, the exhibition drew only seventeen Providence/Attleboro firms, plus forty-three other stall-holders, although over 90 percent of New England's output came from the Rhode Island/southern Massachusetts district. Worse, owing to the Providence cluster's indifference, exposition organizers canceled their 1926 show. Suspicion prevailed in the context of design theft and price-cutting, perhaps understandably.[57]

Jewelry trade associations floundered in the 1920s, as proposed mergers to build strength and save duplicate expenses failed to jell. NEMJSA had patriotically suspended its annual banquets during the war, reviving these festive occasions only in 1923, the first and only solid sales year through mid-decade. Equally telling, the Providence-based association opened a "used equipment bureau" in 1925, amid what president Ralph Stone described as "a very great depression" in the trade. Stone also offered members a "bureau of products," which would catalog their lines for buyers' reference at a single site—the organization's offices. Secrecy was assured: "no member and no officer of the association other than the manager . . . has access to the bureau's records." This too died in its infancy, as did Stone's proposed "styles committee," which he expected would organize "display[s] of merchandise" and, by subscribing "to the best fashion publications both foreign and domestic," would "notify [members] of new jewelry vogues."[58]

The defection of skilled workers to other occupations confirmed low-end jewelry firms' dependence on designing and die-cutting auxiliaries and their employment of low-skill pressworkers and assemblers. No technical advances beyond the mesh machine accompanied the continuing decay of skill in the labor process. The used machinery bureau alone suggests manufacturers' reluctance to invest in new equipment, as well as machine builders' disinterest in devising it. The multiple tensions between imagined huge markets for novelties, those markets' fragility, and the surety of design piracy entailed that current technologies be preserved, not that capital be risked on new machinery which would lie idle much of the time owing to the trade's volatile seasonality. Why make potentially dead stock more efficiently? Instead, travel light and pray for luck.

Appropriately, the enthusiasm for training demonstrated during the war decade ebbed steadily. When RISD proposed to replace its jewelry program's antiquated building with a new facility, costing $180,000, area manufacturers managed to gather only $15,000 for its equipment. Although Brown and Sharpe's supervisor of apprentices drew sixty jewelry managers and foremen to a 1924 lecture on his programs, no one in the troubled trade copied its model. Nevertheless, in response to complaints that its jewelry course drew "too few students," RISD instituted a "co-op" apprenticeship jewelry program in 1925, which involved sixteen full weeks of class instruction for each of three years, the balance of the time spent in shop work. Twenty students commenced that fall, sponsored by veteran jewelry firms, but the program dropped from sight as conditions worsened into and through the Great Depression. *Metal Industry*'s Providence correspondent summed up the trade's situation in 1925: "The jewelry branches have continued to drift aimlessly with comparatively little encouragement."[59]

Incapable of managing persistent problems of opportunism, design copying, interfirm suspicion, and price shaving under seasonal time pressures, defenseless against the flow of worker-entrepreneurs who fueled these "abuses," and legally barred from collective strategies to regulate competition and trade practices, the New England jewelry industry slid into a permanent malaise by the mid-1920s. Wholesale, department store, and chain store buyers ruled. When a national jewelry show was successfully mounted and repeated, it was Chicago wholesalers, not eastern producers, who sponsored and publicized it. Manufacturers could only rail against this age of "profitless prosperity," critique the "slow turnover" the trade's 15,000 retailers accepted, and beg for "genuine co-operation." Yet, as NEMJSA's manager moaned: "The pirating of ideas is not co-operation. . . . The regarding of merchandise for price, not for style, merit, or salability is not cooperation. . . . The inane cutting of prices is not co-operation." In the end, the association coded its surrender to buyers' market dominance in the form of its 1927 *Buyers' Guide*, "in which the manufacturers are listed according to the price range of the goods they manufacture, as five cents to twenty-five cents, twenty-five cents to one dollar," etc. In the years ahead, except for a handful of firms that specialized in niche or contract-bound goods, the region's jewelry makers would become auxiliaries to middlemen and their commissioned designers, bidding for orders to make styles "at a price."[60] Novelty would continue, to be sure, but the industry's "sickness" was chronic.

## Grand Rapids: Solidarity and Innovation

In April 1913, the Furniture Manufacturers' Association met to consider the shortening of Grand Rapids' summer factory hours. The previous year it had initiated fire and accident protection for members and now conducted both a financial and a factory review of new candidates. Its thirty-eight companies accounted for three-quarters of the city's ten thousand furniture laborers and regularly reported employment totals to the FMA, for they voted and paid dues and assessments on the basis of workforce size. At the April session, association members rejected the Board of Directors' recommendation, a Saturday half-holiday that would yield a fifty-hour week, and instead agreed on five days and forty-five hours—a step that reflected the current economic recession more than concern that workers should have summer Saturdays free for family fun. In August, the FMA announced a return to full time on September 1. As usual, local firms moved in unison on such minor, and most major, matters.

Three years later, facing a wartime drain of labor to shipbuilding and other trades, the FMA undertook a "canvass . . . of the wage situation," reopened its Labor Bureau, forswore "advertising locally for help," then funded recruitment in "the labor markets [of] other woodworking centers." Members also voted for a wage hike and set the procedures for its implementation ("no more than 25 per cent of the force should be raised in each week and raises should be made individually and not by groups or departments"). Meanwhile, the FMA Carloading Committee, displeased with the facilities the Michigan Central provided shippers, secured members' approval of a plan to construct their own warehouse and sidings, for whose costs they would be assessed pro rata. Equally important, in May 1916, the association issued a call "to furniture manufacturers of other cities to meet at Grand Rapids on June first, to consider and determine upon the uniform and unanimous raise in prices." Area firms thought that 20 percent on dining room pieces and 25 percent on bedroom lines was about right, but declined to fix the increases in advance of the gathering.

That fall, finding that some firms were not hewing to association directives, the board presented a sharp resolution to the membership: "Whereas there has been some misunderstanding as to the extent of the authority and discipline of the Association over its members, be it resolved THAT whenever this Association, as a whole, or its Board of Directors, shall establish a rule for the guidance of the members, any violation of the rule shall render it the duty of the Board of Directors to ask for the resignation of such violating members." Passed with but one dissent, the resolution led to the resignation two months later of the Raab Chair Company, which had deviated from the FMA's wage rule for carvers. Later, when a coal famine threatened in the terrible wartime winter of 1917–18, the FMA created a Fuel Service Department to "take charge of the fuel supply of its members," who then collectively purchased a Michigan coal mine to assure adequate power for uninterrupted production. Thus was trade governance practiced and manufacturers' solidarity fostered (and enforced) at Grand Rapids, where the trade association that crushed the 1911 strike functioned as the furniture industry's point of reference.[61]

The war years, with a few exceptions, proved relatively uneventful in Grand Rapids. Unlike Cincinnati's machine tool trade, which struggled to keep pace with demand, or Providence's jewelry sectors, which found patriotic opportunities mixed with disappointments, the city's furniture sector long stood aside from the war's swirl, if not from its effects, as the labor scramble suggests. Although the furniture business "went to smash" for a time after August 1914 and Grand Rapids' employment dropped 10 percent through 1915, a steady recovery to prewar levels followed. Combatants' orders for matériel included very little woodworking, but the indirect effects of war production in other sectors reached the furniture trades nonetheless. Munitions suppliers' prosperity fueled growing furniture sales from cheap to custom lines by 1917, as workers and managers used part of their rising incomes to buy new outfittings for their homes. That year F. Stuart Foote's Imperial Furniture paid its few shareholders a 40 percent dividend on $250,000 of common stock, followed by a 50 percent return in 1918.

Part of these profits derived from warplane construction, as Imperial joined thirteen other FMA firms to form a consortium that operated what was perhaps the nation's first "virtual" corporation. Finding little response to individual efforts at securing War Department orders for woodworking, the collective incorporated the Grand Rapids Airplane Company in November 1917 with Foote as vice president; and "the entire facilities of the 14 factories were put at [its] disposal." Snaring a contract to construct the British-designed Handley Page bomber, reputedly the war's largest aircraft, managers seconded to GRAC surveyed "all the factories, taking into consideration which . . . could manufacture the long parts most effectively; also which . . . could produce the most intricate parts as accurately as demanded by the government inspectors. This assignment was made upon a basis of factory output and was evenly divided among the 14 factories on the above basis." The associated firms committed a million square feet of workspace and two thousand employees, about a third of their capacity, to parts preparation and bomber assembly. "A large corps of experienced woodworkers" set about "designing and manufacturing special tools, jigs, and templates . . . made in this city" for repetition production. These more than halved the time required for fashioning components; hence the Signal Corps ordered duplicates "manufactured and shipped to England for assembly work there." Within ninety days of inking the contract, GRAC delivered "finished planes" and by fall 1918 had surpassed schedule requirements. The city's fine furniture makers thus put their industrial solidarity to new uses, showed their flexibility and capacity for precision manufacturing, and profited thereby.[62]

Other effects of the war surge were comparably cheering. Materials prices soared, of course, rising 15 percent in May 1916 alone; but FMA unity steeled producers to pass these advances along to dealers and consumers with hardly a pause. Indeed, despite price advances, Grand Rapids' 1916 market expositions drew 350 exhibitors and over 3,000 buyers, then a record. Wage hikes stanched the labor drain for a time, until America's war entry and Selective Service calls pressed area firms to hire a thousand women to fill places left vacant by volunteers and draftees. Mean earnings in 1918 reached $890, 43 percent above 1914's, and moved dramatically upward thereafter. Local manufacturers dismissed most women workers in 1919, rehired returning servicemen, and again raised wages to buy labor peace amid the strike surge that affected virtually all the other furniture centers.[63]

Strikes in Jamestown and Rockford and unanticipated postwar demand for styled lines turned 1919 into a "sold up" year for Grand Rapids furniture plants. Local firms joining the boom in styling phonograph cases had reduced space available for making case goods, but more than a dozen makers had used wartime revenues to erect "important and extensive additions," doubling the furniture center's 1914 capacity. As city firms sold direct to retail dealers eager for bedroom and dining sets, Grand Rapids' rounds of price advances percolated rapidly through the trade. Ample margins easily funded workers' increments (mean earnings reached $1,500/year in 1920) and provided surpluses for machinery purchases and dividends. At Imperial, Foote declared a 300 percent stock bonus, reflecting two expansions of his factories, raising its capitalization to $1 million, then paid hearty 15–20 percent annual dividends on that base throughout the 1920s. A later account argued that Imperial represented "a typical case."[64]

Meanwhile, the FMA completed its 125,000-square-foot warehouse (assessing members $20/worker to pay for it) and addressed the recognized shortage of skilled labor by opening both a carvers' and an upholsterers' school. It also sent small groups of foremen to Wisconsin for the Forest Product Laboratory's two-week courses on technical advances in plywood and kiln drying, then commissioned the University of Michigan to offer foremen training courses in the city. When the market slide materialized in 1920, the association installed an orderly series of stepwise wage and price reductions (e.g., for furniture, 15 percent on January 1, 1921, guaranteeing no further cuts for six months), which, if not uniformly copied elsewhere, reduced the depression's impact on Grand Rapids. In 1921, a reporter wrote that "very few of the Grand Rapids factories have been shut down, and on the average, they curtailed production less than factories in any other center." Unorganized workers took rate cuts without protest both because they evidently accepted manufacturers' pledge of restorations once markets recovered (a promise kept) and because, more than in any other furniture district, they owned their own homes, which made relocation unappealing. Buyers responded positively to the FMA's lowered price level, recognizing that it established a market "bottom" through the July openings. Retailers would have to look elsewhere for panic reductions of 40–50 percent. Thus in fall 1921, "when business began to pick up, the Grand Rapids factories were among the first to feel the impetus." By December, more of its plants were "working overtime than in any other center," despite the fact that prices had been advanced for the fall markets.[65]

The rebound, sustained by an extraordinary period in American home building, lasted through the late 1920s, when furniture retailers typically achieved five inventory turns yearly (jewelers achieved under two). Grand Rapids furniture employment reached 11,000 in 1923, passing Chicago, and neared 12,000 two years later, when city firms shipped goods worth a record $58.5 million. Styles had centered on English and French "period" designs since the fade-out of "mission" simplicity before the war, but a vogue for "colonial revivals" settled in after the depression. Period pieces continued to move steadily, whereas ventures into modernism drew few clients, except among office furnishers. Geographically, Grand Rapids' sales followed the nation's wealth—69 percent east of Chicago, 27 percent to the upper Midwest and Pacific coast, and only 4 percent to the South. Technical improvements continued to be incremental—larger twenty-four-spindle rough carving machines, faster wood planers, spray guns for applying varnishes, more general use of electric drive, and, based on Forest Product Laboratory research, more reliable kiln dryers for dimensional lumber and veneers. However, cost accounting remained elementary, factory cost plus add-ons for overhead and profit being most common. In 1921, the Grand Rapids trade journal hired an accountant to prepare a twelve-part series on proper costing, later published in book form; but the FMA never discussed the issue.[66]

Nationally, the Furniture Publicity Bureau secured acres of free promotion by supplying newspapers with copy concerning new styles and interior decoration, a far more effective strategy than anything jewelers had devised. The papers regularly printed its advice columns alongside retailers' display ads or compiled both into Sunday supplement sections. Renamed the American Homes Bureau in 1921, the agency argued the case for "better furniture" throughout the decade. At Grand Rapids, the FMA covertly funded (to the tune of $5,000) the retailers association's annual National Retail Furniture Institute, first held at the market expositions in 1925. The institutes updated visiting dealers' knowledge of accounting, inventory, display, and advertising practice and urged their support for the AHB's campaigns. Buyers' attendance at Grand Rapids market openings broke all records, rising above 5,000 in 1923 and 6,000 by 1928.[67]

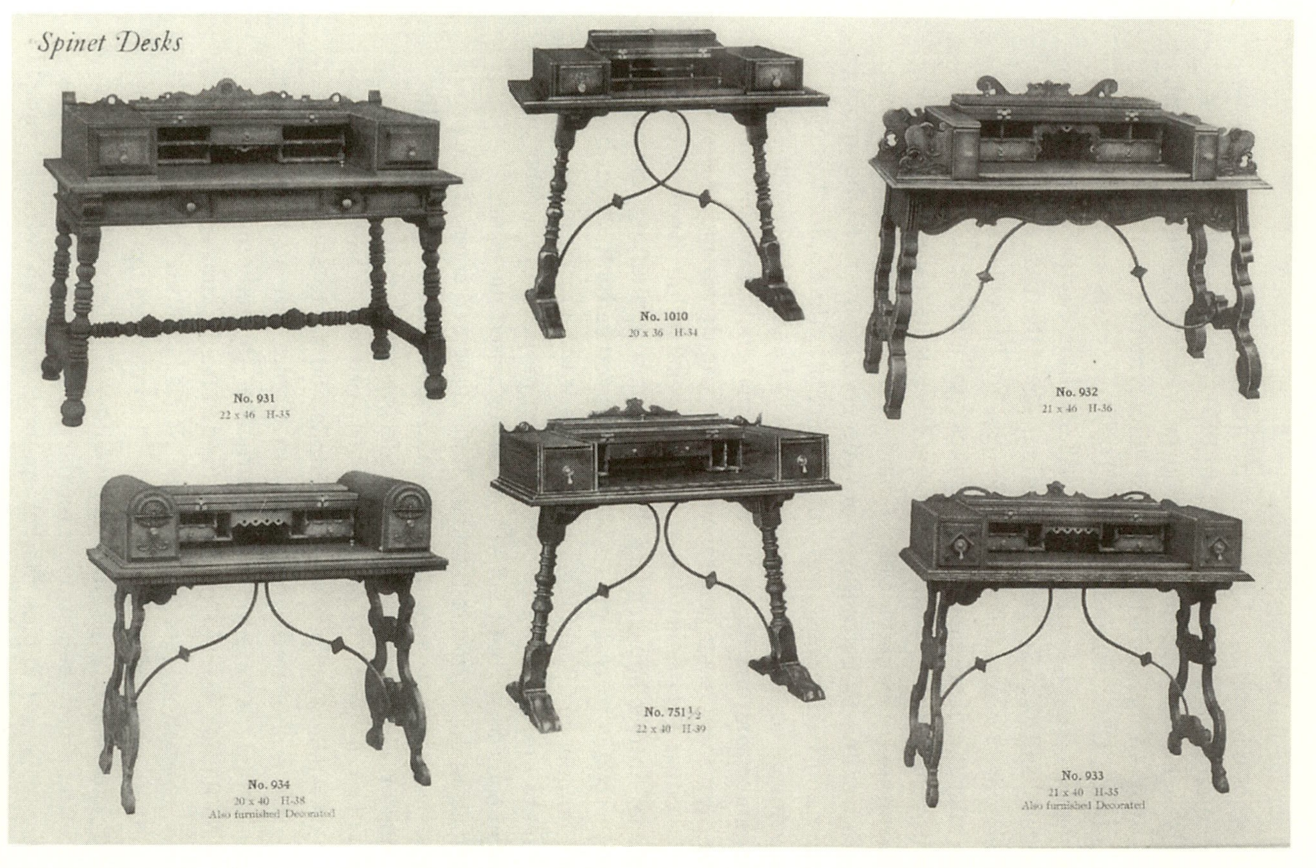


26. Six styles of small desks, aimed at women consumers, from the Grand Rapids Furniture Shops' 1925 catalog. Nos. 933 and 934, when "furnished decorated," had hand-painted trim in a variety of custom-ordered color combinations. Courtesy of Hagley Museum and Library.

The only blemish on this record of achievement and solidarity—increasingly acrimonious relations with the federal government—festered into a running sore by mid-decade. Furniture manufacturers had greeted the FTC with high hopes in 1913–14. Its initial publications on uniform cost-accounting systems drew praise as a potential means to eliminate innumerate proprietors' practice of selling below cost. Moreover, its early advocacy of retailers' education to honest advertising and sympathy for defending Grand Rapids' trade name against infringements reinforced the judgment that the FTC would combat unfair competition. However, the sharp inflation of furniture prices, ca. 1917–20, and a perception that manufacturers were victimizing both retailers and consumers changed the climate fundamentally. On two fronts, harsh exchanges between furniture makers and the FTC commenced in 1921. The "veneer" clash concerned what the FTC regarded as deceptive labeling. Its intervention challenged Grand Rapids' effort at business self-regulation. The "price-fixing" battle stemmed from perceptions of price gouging in the inflationary era and culminated in the indictment of nearly three hundred firms for antitrust violations in 1925. Good Republicans all, Grand Rapids producers alone fought back, infuriated at the Coolidge administration's assaults on trade governance practices that had so effectively and profitably stabilized a potentially chaotic marketplace. Though they resisted admitting it, FMA members had indeed colluded to regulate competition, driving it away from price and toward product qualities. That was not the point. At issue was their belief that they had the right to act collectively just as U.S. Steel acted at the Gary dinners—exercising leadership in setting trade "rules" to foster profitability without cutthroat competition.[68]

Stratospheric furniture prices in 1920 triggered private and government reactions that would entwine over the next seven years to distress the furniture trades. In 1921, a mysterious "Cincinnati lawyer," George Hawke, began circulating "carefully worded multigraph letters" to newspapers, manufacturers, and dealers accusing producers of defrauding the public by selling pieces made of veneered or stained cheap woods as real oak, mahogany, or walnut. Replying to a Grand Rapids inquiry, the FTC noted that Hawke "does not represent this commission," but the accusations hovered. Shortly, the Bureau of Labor Statistics published its price indexes for the 1913–20 period, which showed that no class of goods had so elevated in price as had "house furnishing goods," which rose 271 percent in seven years. This revelation brought forth U.S. Senate Resolution 127, calling for an FTC investigation of the trades. The association contested the government's findings, first by challenging the BLS data base and ultimately in a fiercely fought federal court trial.[69]

The "veneer controversy" had three sources. Since 1900, hardwoods had become more expensive and conserving materials represented sensible business practice. Fine furniture construction at Grand Rapids thus involved "built-up" designs that employed hardwood veneers as the outer layers of special "cabinet" plywoods for flat surfaces, though solid pieces were still used for legs, posts, and the like. Retailers understood that this technological fix kept prices within the middle class's reach and that it eliminated warping of table or cabinet tops, common when solid boards had been glued together. Next, during the inflationary years, some furniture makers commenced applying stains that gave a mahogany appearance to all-gum pieces. Retailers knew their woods and appreciated the price differences between such lines and "built-up" tables and dressers. Third, in this context, some retailers chose not to inform customers about the distinctions, advertising as "mahogany" pieces that were simply varnished in a mahogany shade.

This deceptive selling underlay Hawke's campaign. Manufacturers replied that dealers knew exactly what they were buying and that the trickery being practiced was in the stores, not at the factories. In 1922, the Associated Advertising Clubs opened an educational campaign about mislabeling of furniture, an effort producers supported. The next year, its representatives met with the National Council of Furniture Associations to point out the need for "changes in the nomenclature of furniture," to which the FMA responded in 1924 by framing a set of "Grand Rapids Rules" for description. The AAC and the three wood associations (mahogany, walnut, gum) approved them, but several other groups devised their own terms that spring. Then the trouble started.[70]

In an attempt to harmonize labeling, the FTC organized a September 1924 Trade Practice Submittal at Chicago. Commissioner Huston Thompson began by asserting that "the whole thing is absolutely voluntary . . . the industry is perfectly free to say that we do not care to go any further." After three hours, however, he sounded a warning: "if [the TPS] is not a success, then the manufacturers will have complaints issued against them."[71] The conference yielded pledges of cooperation but no immediate agreement on standard descriptions. Francis Campau, the FMA's counsel, believed that "until the FTC again took a hand, [producers and retailers] were making rapid progress. . . . in a short time an arrangement would have been reached to bring them all to a single frame of mind." FTC complaints did surface the following summer, but against New York retailers, not the manufacturers. Accused of defrauding the public through false labeling, Macy's and other stores attended a submittal on December 29, 1925 (to which producers were not invited), and adopted a set of trade descriptions that the FTC staff had prepared. The FMA objected promptly to the provision that "veneered surfaces were required to be so described." Consumers regarded "veneered" as indicating shoddy work, not realizing that fine furniture used cabinet laminates which were more expensive and durable than solid woods. Grand Rapids requested further hearings and discussions but was rebuffed. Instead, the FTC circulated its standards, demanding that manufacturers affirm them. As a body and alone in the industry, Grand Rapids firms refused. In July 1926, the commission filed complaints against twenty-five resisting Michigan furniture makers, charging them with "unfair methods of competition." When this failed to bring them to heel, the FTC issued "Cease and Desist" orders in 1928, which the FMA immediately appealed to the Sixth Circuit Court.[72]

The petitioners' brief challenged each of the commission's arguments and scored its rejecting further review of the 1925 submittal rules. Crucially, the Grand Rapids coalition attacked the FTC's shift from preventing monopoly to regulating obviously competitive industries, citing the *Sinclair Refining* decision that the FTC "has no general authority . . . to interfere with ordinary business methods or to *prescribe arbitrary standards* for those engaged in the conflict . . . called competition." Sixth Circuit judges wholly agreed. On June 28, 1930, they noted "that the finest of all modern furniture . . . [is] constructed of laminated wood," and that this practice "is substantially universal," hence in no way diverting trade to the accused. Also referencing *Sinclair*, the judges ruled that "the order of the commission [was] wholly unsupported by the evidence," and revoked it. Grand Rapids' furniture manufacturers stood "vindicated, but only at the end of a long, bitter, expensive litigation."[73]

The battle over pricing that the BLS's indexes set off in 1921 took almost as long to reach its finish line in 1928. Attacked for nearly quadrupling prices, 1913–20, Grand Rapids and trade association spokesmen devised an unusual countermove—critiquing the bureau's data sources and methods of analysis. They discovered that the BLS "house furnishings" index mixed price increases in "oak rocking chairs, hardwood kitchen chairs, [and] kitchen tables" with those for "white granite plates . . . glass tumblers, carving knives . . . and galvanized iron tubs." No one was assaulting the ceramics, glass, cutlery, or tub manufacturers, because to consumers "furnishings" meant furniture. W. H. Coye, a staff member at the National Alliance of Furniture Manufacturers (NAFM), a case goods group located in Grand Rapids, executed a statistical study that disaggregated furniture from other indexed goods. His research showed that average furniture prices increased 150 percent across the eight-year period (not 271 percent), twenty points less than the bureau's all-commodities index, and dropped in the depression to only 50 percent above their 1913 levels. Chastened, the bureau revised its classification in 1922, adding dining and living room furniture to its statistical group, a restructuring that instantly lowered the "housefurnishings index" by forty-two points. However, the FTC investigation of price gouging moved forward despite these revelations.[74]

Once again, Grand Rapids led the fight against federal charges, which accused associated case goods, chair, and refrigerator manufacturers of rigging prices and obstructing proper competition. The prehistory of these indictments is instructive. After FTC investigators had reviewed numerous firms' records, starting in 1922, the commission informed newspapers in April 1925 that indictments before a Chicago federal grand jury would be lodged "against all the furniture manufacturers of the country." The NAFM hired FMA counsel Francis Campau to gauge the situation. Traveling to Chicago, Campau met repeatedly with Roger Shale, the federal attorney responsible for the cases, who stated that the Department of Justice believed most firms had unthinkingly acted in restraint of trade. Shale offered a deal. If "a sufficient number of the manufacturers should agree to plead guilty" to indictments that "would touch the corporations only," no individuals would be charged. Arguing that the government's charges were baseless, Campau rejected this bargain, and on May 28 Shale's grand jury returned 269 company-centered indictments. Campau, in a meeting with nine lawyers for the defendants, asserted that forcing a trial was the only sensible response, but he was outvoted. Over the next two months, groups of refrigerator and furniture makers accepted fines to avoid battling Justice and the FTC, while a core of seventy-nine case goods companies, anchored by eleven Grand Rapids leaders, rejected any compromise.[75]

The Department of Justice turned up the heat by reindicting for criminal conspiracy over seventy firm owners in the resisting group, along with the NAFM and two of its officials. This generated a few defections, but sixty-six individuals held fast. In July 1925, alliance secretary Arthur Brown "refused to obey a court order to turn over [its] books and records," a maneuver that earned him a thirty-day sentence for contempt of court but temporarily obstructed the prosecution's access to damaging evidence. As a defense lawyer wrote, after reviewing the alliance's correspondence files: "Repeatedly, Mr. Brown in letters to manufacturers elsewhere informs them of action just taken by the Grand Rapids Association looking toward the raising of prices . . . and the like. Agreements as to hours of labor are also mentioned." Clearly, the NAFM was an extension of the FMA, a transmission belt for disseminating Grand Rapids policies. Though no member could be expelled for declining the NAFM's guidance, the government argued that its effective leadership from the war years forward yielded higher prices to consumers than pure price competition would have generated.[76]

The cost-finding consultations that Coye embodied represented a critical element in the NAFM's service. A former furniture manufacturer who prepared regular "cost and selling value bulletins," he visited members' plants to bring their cost systems into conformity with the alliance's model, derived from Grand Rapids practices. Working from a basis of regional (or "zone") labor, materials, power, and transport expenses, Coye showed owners how to develop "minimum selling prices" on their lines through reference to calculations on exemplary "articles of specified construction," to which firms added "a uniform 25 per cent for contingencies and profit." This mechanism established selling prices that guaranteed profitability if orders for styles could be booked. Coye's ingenious approach wonderfully fit specialty manufacturers' market and production environments. It was relatively simple, applicable to diverse designs, and fixed no uniform price for classes of furniture but rather uniform methods for arriving at wholesale prices. Companies calculated their own minima and could ask more or less than those figures, but in either case would have full knowledge of their potential gains and losses, as well as, through the NAFM's bulletins, information on typical costs drawn from Coye's research.

These efforts actualized commerce secretary Herbert Hoover's vision of associations as organizations for trade governance, management education, and efficiency promotion. The FTC and Justice regarded them as conspiracies to manipulate prices, evade competition, and defraud the public. The enduring irony, of course, is that such rationalizations of production, cost finding, and market strategy drew applause when conducted by large enterprises like General Electric and indictments when undertaken by alliances of midsize companies. As defense attorney Robert Golding observed, quoting Hoover in his opening statement, "Big business will take care of itself; the little manufacturers need associations."[77]

The disjuncture between two crucial concepts animated the nine-week Chicago trial (January–March 1927), the prosecution arguing that illegal "price manipulation" had occurred, the defense denying that "price fixing" took place. Both were correct. Associated producers had indeed moved price levels in tandem through percentage increases and decreases, although the best evidence came from the 1916–20 period, outside the three-year limit for culpable actions (July 1922–July 1925). However, Coye's work after 1922 *had* created accounting uniformities that tended to undermine price-based competition. The defense argued in reply that correspondence and bulletins introduced as evidence had been misinterpreted by the government and that no mechanism for enforcing guidelines or penalties for deviation existed. More usefully, the manufacturers' attorneys blurred the prosecution's rhetorical portrait of a "furniture trust" by asserting the impossibility of any central agency's fixing wholesale prices for the countless styles and variants brought to market. But their attempt to buttress this "economic defense" by having Harvard's Edward Gordon present "a mass of economic data" failed when the judge ruled that such testimony would be "immaterial, irrelevant, and incompetent." Baffled by the conceptual cross fire, the jurors could reach no unanimous verdict; after five days' deliberations, they reported themselves deadlocked and were discharged, ending the contest.[78]

A year later, when prosecutors informed the defendants that they planned to retry the case, the furniture operators' lawyers negotiated a settlement. The government would recommend that the court accept nolo contendere pleas from group members, who admitted no guilt. They paid fines totaling $100,400 rather than incur another round of "heavy attorney fees, court costs and other expenses." As six law firms spent twenty months in the initial defense, Campau earning $2,550 for one month's consulting, billings in the six-figure range could be anticipated for the second round. The settlement, attorney Golding averred, would save a number of firms from the edge of bankruptcy, but perhaps underlying the plea bargain was the hung jury's final vote in 1927: nine for conviction, three for acquittal.[79]

The extent to which furniture associations thereafter covertly continued practices the FTC abhorred cannot be documented from available records. FMA minutes became more terse, files more spare, in the wake of the prosecutions; but this points in no single direction. More significant was this local association's capacity (after decades spent building trust and discipline among members, plus valuable services for them) to generalize its approach on a national scale in the case goods furniture trade, at least for a time. In "better furniture," where style, finish, and construction made the sale, the FMA and NAFM worked carefully to reduce the role of price competition and to assure profits that would sustain both firms and communities. These specialty manufacturing collectives fought interventions by "ignorant" state regulators just as vigorously as they battled union "outsiders." However, the industrial politics of the 1920s, rooted in economic ideas that the courts would not permit even a Harvard professor to challenge, curtailed their capacity to manage competition.[80] Their approach to production and marketing was neither exhausted nor antique; repeatedly reconfigured, it sustained the district's furniture trades into the post–Cold War era.[81]

The tragedy here is that state interventions privileging imaginary consumers, informed by economic abstractions that enshrined price wars, suppressed specialty manufacturers' larger model for solidarity, self-regulation, and firm autonomy within communities of interest observing collectively legislated rules of business practice. Once installed as dogma, the key idea (that competition essentially and universally meant that pricing contests over standard goods rewarded "efficient" companies) devalued routines and rules that specialty producers had elaborated during several generations of "practical" manufacturing experience. As indicated above, some specialty sectors exhibited structural "flaws" that poised them for the economic drop, enhanced in the jewelry case when federal initiatives blocked private efforts to foster market ordering. Others, differently located in terms of production technologies, the dynamics of novelty, market and interfirm relations (including trust), and collective organization, cleared pathways toward a form of industrial capitalism that eliminated "ruinous competition" and rejected mergers, standard products, and corporate hierarchies. An inability to understand and tolerate economic diversity, a thin appreciation of sectoral differences in the framing of public policy, and the romance of mass production and cheapness as defining a ideal world of consumption eventually combined to undercut many specialists' capabilities, their strategies for collective agency, and, ultimately, their viability as creative components of America's manufacturing complex. Thus our national capacity for generating a profusion of goods to meet a diverse array of consumer tastes and industrial needs, having waxed between the 1860s and the 1920s, began to wane thereafter.

## A Statistical Supplement

In the mid-1920s, census officials constructed a long-term review of American manufacturing's course since 1899. In so doing, they outlined the nation's industrial structure, insofar as manufacturing censuses through 1923 documented it.[82] Their compilations make it possible to provide a closing complement to this study's introductory reprise of American industry in 1909, presented in tables 2–4, and the 1890 portrait tables 13–15 provided (tables 21–23). Nationally in 1923, specialty sectors contributed the same one-third of value added as had been recorded earlier ($8.55 billion of manufacturing's total $25.85 billion, both triple the 1909 figures). Printing, foundry and machine work, specialized railway shop construction, electrical production, along with women's clothing, fashion textiles, and furniture led the parade of specialty trades, which together employed 2.9 million workers, again a third of all manufacturing employment.[83] With a slightly expanded workforce, printing and publishing nearly tripled its 1909 totals to head the list (table 21). Just as automobile output exploded among bulk and mass production sectors, so too did electrical equipment soar in the specialty group, value added rising from $113 million to $744 million in fourteen years. Routinized manufacturing's substantial growth is also obvious and important; for table 23's sectors built their share of all value added from 32 to 40 percent by 1923, while doubling total employment.

Yet interestingly enough, mean value added per worker grew more rapidly in specialty sectors than in bulk and mass trades, increasing 143 percent versus 79 percent, respectively, to $2,970 and $3,348. One key to this differential may be that specialists' workforces expanded but 22 percent during the period, so that swelling value added spread

TABLE 21
Specialty Production Sectors, by Value Added and Employment, 1923

| | *Value Added (millions)* | *Employment (thousands)* |
|---|---|---|
| Printing and publishing | 1,537 | 281 |
| Foundries and machine shops | 1,400 | 449 |
| Railroad shops construction | 888 | 534 |
| Electrical machinery and supplies | 744 | 235 |
| Women's clothing | 597 | 133 |
| Woolen/worsted goods | 440 | 195 |
| Furniture | 434 | 168 |
| Hosiery and knit goods | 364 | 194 |
| Drugs, perfumes, etc. | 261 | 35 |
| Engines and machine tools | 239 | 82 |
| Dyeing and finishing textiles | 179 | 63 |
| Millinery and lace goods | 148 | 54 |
| Cutlery and tools | 135 | 45 |
| Shipbuilding | 128 | 62 |
| Marble work | 124 | 40 |
| Carpets and rugs | 102 | 35 |
| Locomotives | 91 | 31 |
| Textile machinery | 91 | 37 |
| Jewelry | 90 | 26 |
| Signs, advertising novelties, etc. | 87 | 26 |
| Scientific instruments, optical work | 80 | 25 |
| Fur goods | 72 | 14 |
| Men's furnishing goods | 66 | 28 |
| Pumps | 58 | 15 |
| Musical instruments | 57 | 22 |
| Paper goods, not elsewhere classified | 49 | 15 |
| Leather goods | 48 | 18 |
| Hats, fur/felt | 41 | 17 |
| Totals | 8,550 | 2,879 |

*Source*: Edmund Day and Woodlief Thomas, *The Growth of Manufactures, 1899 to 1923*, Census Monograph no. 8 (Washington, DC, 1928), tables 29 and 31.

*Note*: Average value added per worker—$2,970.

TABLE 22
Mixed Format Sectors, by Value Added and Employment, 1923

| | *Value Added (millions)* | *Employment (thousands)* |
|---|---|---|
| Cotton goods | 775 | 479 |
| Men's clothing | 594 | 195 |
| Boots and shoes | 474 | 225 |
| Auto parts | 449 | 164 |
| Pottery, brick, and clay products | 312 | 140 |
| Planing mill products and millwork | 291 | 103 |
| Silk goods | 282 | 125 |
| Glass | 196 | 73 |
| Brass products | 179 | 65 |
| Railway cars | 179 | 77 |
| Leather, including morocco | 167 | 60 |
| Stoves and furnaces | 165 | 50 |
| Steam fittings and accessories | 138 | 44 |
| Hardware | 130 | 50 |
| Shirts | 102 | 54 |
| Copper and sheet metal work | 95 | 29 |
| Stamped and enameled wares | 86 | 35 |
| Plumbing supplies | 81 | 27 |
| Tin wares | 80 | 31 |
| Phonographs | 62 | 20 |
| Corsets | 40 | 16 |
| Totals | 4,877 | 2,062 |

*Source*: See table 21.
*Note*: Average value added per worker—$2,367.

across a cohort of employees only moderately larger. Product novelty that underwrote profitable pricing, resistance to depression reductions, and some effects from technological change, improved work routing, and costing practices also likely contributed to these gains, though to what degree cannot be determined. On the other hand, price-competitive bulk commodity sectors surely served as a drag on the advances mass production trades achieved, for the former expanded output *extensively*, by adding workers and facilities in tandem, whereas the latter epitomized *intensive* development, the classic capital-labor substitution that, with standardization, built output much faster than employment. Thus both chemicals ($4,705) and autos

TABLE 23
Routinized Production Sectors, by Value Added and Employment, 1923

| | *Value Added (millions)* | *Employment (thousands)* |
|---|---|---|
| Food products and related | 2,282 | 626 |
| Iron and steel, crude and rolled | 1,845 | 608 |
| Chemicals and related | 1,741 | 370 |
| Automobiles | 1,015 | 241 |
| Lumber | 981 | 496 |
| Tobacco products | 528 | 146 |
| Tires and rubber goods | 370 | 108 |
| Paper and pulp products | 333 | 121 |
| Liquors and beverages | 247 | 31 |
| Cement | 163 | 35 |
| Nonferrous refining | 148 | 41 |
| Paper boxes | 120 | 57 |
| Manufactured ice | 118 | 27 |
| Photographic materials | 108 | 16 |
| Agricultural implements | 88 | 31 |
| Rubber boots and shoes | 86 | 29 |
| Cash registers | 82 | 15 |
| Wooden boxes | 69 | 39 |
| Mattresses and springs | 46 | 14 |
| Aluminum and products | 40 | 58 |
| Totals | 10,410 | 3,109 |

*Source*: See table 21.
*Note*: Average value added per worker—$3,348.

($4,212) stood well above table 23's mean value added per worker ($3,348), whereas lumber ($1,978) and boxes ($1,864) lagged badly. In short, mass production's cutting-edge productivity gains had not yet generalized throughout the national economy. Meanwhile, industrial specialists continued to ply their fields of expertise with their customary focus on diversity, novelty, and product competition, remaining as significant to creating value in American manufacturing as they had been before the war.

*Chapter 13*

# LOOKING AHEAD

ON MAY 31, 1926, the Philadelphia Sesqui-Centennial's opening ceremonies featured speeches by commerce secretary Herbert Hoover, city mayor Frederick Kendrick, and A. Philip Randolph, socialist head of the Pullman car porters union, among others. Promoters had spent millions on preparations and a massive Shriners' convention would follow the kickoff events. Yet unlike its illustrious predecessor a half-century earlier, the Sesqui-Centennial stumbled toward disaster. Its facilities lay incomplete well into the summer, and attendance totals ultimately rested a third below those of 1876. Even the weather was hostile, for rain fell on more than 100 of the fair's 184 days. The last international exposition in the Crystal Palace tradition, the Sesqui, perennially damp and ultimately bankrupt, accomplished little more than to clear the decks for the modernism, corporate pavilions, and futurist fantasies that animated Chicago's Century of Progress festival in 1933 and New York's epic World's Fair six years later.[1]

Though its best-remembered feature was the Dempsey-Tunney heavyweight championship fight, the Sesqui-Centennial also sponsored a "Congress of American Industry"—twenty-four sessions held over three September weeks at the Commercial Museum. Roughly twenty thousand individuals attended part or all of the congress and endured remarks by over fifty speakers, including General Electric's E. W. Rice, Westinghouse chairman Guy Tripp, John Edgerton of the National Association of Manufacturers, Baldwin Locomotive's Samuel Vauclain, and, as anomalous as Randolph at the opening, AFL president William Green. Among the speakers, only Dexter Kimball, a former journeyman machinist then heading Cornell's Sibley engineering school, carefully reflected on the 1876 fair:

> Before some of you were born, there was an exposition held here and the most remarkable machine was the great Corliss engine. . . . It could develop 1700 horsepower. Today we are building turbines with 55,000 horse-

power in a single unit and 75,000 in an assembled unit. . . . I do not know what the largest boring mill was at the time of the Centennial, but I do know there were some thirty feet in diameter. Today the largest machine can bore sixty feet. There is one in the General Electric Company at Schenectady.[2]

These immense power systems and machine tools embodied specialty manufacturers' capacity to transform technological ideas and scientific principles into precision artifacts for the infrastructures of American society. As Kimball spoke, GE and Westinghouse, classic "bridge" corporations, and other integrated anchors like Allis-Chalmers could look forward to profitably building hardware for the nation's continuing electrification. Such enterprises relied on product diversity. Despite the impetus for standardization and Hoover's promotion of simplification (the reduction of variety), a journal editor observed: "There are instances where the *multiplication of varieties*, as contrasted with simplification, has proven to be the key to successful business operation. Aside from the great electrical corporations, which are outstanding examples of this fact, there are many others." He explained, writing of GE, that "the production problems of the electrical industry, which arise in large measure from wide variety, complexity, and frequent change of design, are the problems that are most commonly encountered in the large majority of American shops and factories today." Although for a decade the two leaders had experimented with mass producing consumer electrical goods, none of GE's seventeen "principal works" was listed as making them by 1923. Indeed, its Schenectady complex represented "'the largest jobbing shop in the world' because of the bewildering variety of products and operations that are to be found in [its] 311 buildings." By contrast, Allis-Chalmers sought to speed the industrialization of agriculture by competing with Ford's tractors but found unwelcome challenges in creating assembly lines, for which its experience with customized power systems provided little guidance. Other anchors, however, struggled with secular tides that ran broadly against them.[3]

Baldwin's Samuel Vauclain, who served on Westinghouse's corporate board of directors, had 4,200 electric motors installed at his company's new Eddystone Works, built on a hundred acres twelve miles southwest of the firm's original city plants. Yet, however technologically current, these facilities, described as Vauclain's "personal monument," rarely operated at more than 30 percent of their 3,000 locomotives/year capacity. Beyond the millions Vauclain poured into this underutilized complex, at least four other factors precipitated Baldwin's gradual slide into bankruptcy (1935). The Hepburn Act and the Interstate Commerce Commission's resistance to rate increases narrowed railroads' profits and eventually thinned the revenues they committed to purchasing new motive power. Moreover, railroads' acquisition of bigger, more powerful locomotives, 1900–1907, had sharply reduced their engines' average ages while increasing their load capacities. That boom triggered Baldwin's Eddystone expansion, but the technical advances in locomotives implied that future orders would be fewer and more spread across time. Third, having become a public corporation before the war, Baldwin soothed its stockholders during the difficult 1920s by doubling dividends, paid not from profits but by depleting its reserves. Last, it diversified too little and too late, acquiring "other capital equipment firms making complementary products" only in 1929. Though prepared to erect steam locomotives in hundreds of designs, Baldwin found itself outpaced in the 1930s by General Motors' Electro-Motive division, which produced standard, more cost-efficient diesel engines that weakened railroads could either accept or do without. GM would "control innovation," ignore master mechanics' diverse specifications (to which Baldwin had long responded), and dominate this shrunken sector after 1940.[4]

Philadelphia's Cramp shipyards, another veteran anchor, closed in 1926, enfeebled by American shipbuilding's postwar collapse. Unfortunately, ship demand was even more volatile than that for machine tools or locomotives. The war's submarine menace had triggered an immense vessel-building program, but orders dwindled toward zero in the 1920s, particularly as federal officials sold to commercial users the surplus cargo vessels launched from Philadelphia's Hog Island and other yards. Naval downsizing, reflecting treaty agreements, entailed that warship contracts would also be scarce. Cramp, which from the 1890s had increasingly depended on U.S. Navy commissions for highly specialized ships, no longer could compete effectively for the few cargo and passenger vessels ordered through 1925. However, Camden's New York Ship stayed afloat, as did Sun Ship downriver at Chester. New York employed its up-to-date materials-handling and work-routing systems to build commercial ships economically, whereas Sun focused on constructing tankers for its parent oil company. Both firms endured through World War II, when Cramp briefly reopened, and for decades thereafter.[5]

Do these tales of stagnation and decline imply that production by integrated anchors had run its course by the mid-1920s, unless practiced by companies diversifying toward mass-market lines? Not at all. Pullman, the electrical giants, Gorham, and Steinway all devised paths to profitability while retaining their specialist capabilities. Moreover, as locomotive and ship sectors slumped, the American aircraft industry reproduced their reliance on skilled labor, precision, technical innovation, and links with networked subcontractors. Contemporary observers wrongly analogized planes to cars, expecting that mass production and standardization would soon result from the scramble among several hundred firms making largely customized aircraft in the 1920s. To be sure, the "Lindbergh boom" drew millions of investor dollars into the sector. However, the technical complexity of plane design and production combined with the Great Depression collapse of commercial demand to render moot any auto comparison, other than the parallel shrinkage of producers to a cluster of anchors (three in engines, thirteen in airframes, but no integrated leaders). During the thirties, governments here and abroad became the sector's only sizable customers, buying planes for military purposes, which made quality and performance crucial to competition. Until the U.S. mobilization in 1940, makers exported from a third to half of their aircraft, 1935–39, mirroring Baldwin's classic strategy in depressions.[6]

A Harvard Business School analysis of wartime plane "production techniques" showed that the industry's two divisions, airframes and engines, operated in 1940 with "batch" and "job-shop" approaches and used networks of associated specialists to build their extravagantly complex goods. The report judged that "simplicity, inexpensiveness, and flexibility" characterized the trade's tooling practices; hence "relatively skilled shop workers were required." Making parts in "small lots or batches" permitted "setup and capital costs to be absorbed in an efficient manner." In an aircraft machine shop, the "keynote was flexibility. It was usually made up of general purpose equipment [and] could handle a variety of production assignments, as well as produce experimental parts or assemblies." The war introduced mass demand, but planes remained diverse ("417 separate and distinct models") and stunningly elaborate. "Military airframes are made up of a vast number of parts; for instance the B-25 contains 165,000 separate parts, not counting 150,000 rivets or the engines, instruments, and other equipment. In contrast, there are roughly 5,000 parts in a medium-price automobile." Assembling the frame's sheet-metal components was technically intricate, and production planning necessitated "the laying out of templates on metal sheets by means of a process known as 'lofting' . . . adapted largely from the 'lofts' of the ship-building industry where somewhat similar problems are encountered." Midsize aircraft engines typically had 13,000 parts made from "about 1,400 individual designs . . . almost all calling for working to close tolerances and extremely high finish." These conditions underscored "the generally recognized fact that aircraft engines require very high standards of quality in manufacturing processes, compared with such mass production industries as the automobile industry."[7]

Contractors installed line assembly by 1944 despite numerous bottlenecks. Yet, given constant revisions to aircraft designs, they never fully achieved mass production. As the Harvard team explained: "The success of the mass production industries had been built in large measure on their ability to call a halt to changes prior to large scale output. . . . Obviously a rigid approach of this sort was impossible. . . . The fact that the aircraft industry was ultimately able to introduce a high degree of flexibility into . . . line production techniques in spite of change, constituted an outstanding contribution." Equally significant, analysts anticipated that the sector's older jobbing and batch approaches "probably will be well suited to their needs in the postwar period." The war experiment generated no mass production imperative for aircraft anchors, though they did fear that an overhang of product would threaten postwar markets, as in shipbuilding during the 1920s. New jet propulsion technologies solved the overhang problem through obsolescence, and though product diversity waned as cost and complexity escalated, precision, specialization, and technical innovation would define the sector long after the fighter and bomber factories closed.[8]

Networked specialist firms (and the industrial districts they generated) had a future as differentiated as that of their anchor colleagues, and in time they also witnessed the formation of successor flexible trades and districts. In the 1920s, specialist consumer goods sectors found themselves targeted by the "simplification" crusade's rhetoric, which tagged style diversity as evidence of inefficiency and "waste in industry." Companies might defend "the production of numerous varieties" as essential, because "[e]ach manufacturer is after business, and the more styles he shows the more business he gets"; but manufacturing's ideological climate now moved against them. Equally frustrating, in Philadelphia, styled-textile mills could not devise a strategy to reverse the market power shift that the postwar depression had cemented in place. After 1921, fabric buyers' hand-to-mouth purchasing became standard practice. Much as in jewelry, clients expected that manufacturers would make and hold seasonal inventories in advance of orders, then absorb losses on those designs that failed to hit market vogues. End of season "vultures" seeking cheap lots of dead stock plied a busy trade in the 1920s, while buyers' associations commenced systematically assembling rating guides that ranked mills according to their "cooperation" with the new terms of trade. As Philadelphia profits faded and area textile associations floundered, a cohort of southern staple-textile companies moved laterally into styled lines of household fabrics and hosiery to avoid a price squeeze that excess capacity and a higher, postwar wage floor had created in low-end cotton goods. Profitless prosperity, the advent of interregional competition, and the long depression drove scores of Philadelphia mills to the wall. Survivors executed the next war's fabric contracts, then closed within a few years, with only niche specialists continuing in any numbers by the 1950s.[9]

Together with the jewelry situation, this declension suggests that networked specialists in nondurable consumer goods trades could not surmount the changing dynamics of distribution which, by the 1920s, had constructed a permanent "buyers' market."[10] The state blocked jewelers' collaborative efforts to bring order out of chaos, whereas organized fabric buyers championed their punishing tactics as services to consumers. Those retailers that manufacturers could reach directly (department stores and chains) had established laddered price lines which offered little room for enhanced profits to makers. The consequence in both sectors was the collapse of technical and style novelty, a gradual shift of design capabilities outside manufacturing, a steady decay of district resources and capabilities, and an eager search for niches.

However, in durable specialty consumer goods (styled furniture), precision capital goods (machine tools), and the multivalent printing/publishing trades (as in New York), no such disaster ensued, though the Great Depression staggered them all. Both the Grand Rapids and New York districts regionalized production outside the district core, the latter especially into northern New Jersey; while firms in all three trades continued close, often face-to-face, contacts with clients and adopted, adapted, or initiated technological advances. In Michigan,

27. Knitting seamless hosiery at the Brown-Aberle mills during the 1920s. Seamless hose production, technologically a relatively simple process, rapidly fell to southern competition. Courtesy Pennsylvania State Archives (MG 219, Philadelphia Commercial Museum, Item #3168).

one key technical novelty was the 1920s integration of metalworking into furniture practice, particularly among office furnishings specialists who later became industry leaders (Stow and Davis, Herman Miller, Steelcase). Reportedly, this initiative capitalized on the depression's displacement of skilled Detroit metalworkers and their migrations to the Grand Rapids area, echoing Pullman's shedding of wood craftsmen twenty years earlier. In 1930, Cincinnati Milling Machine introduced the predecessor of current-day numerical control machine tools, the Hydrotel tracer, which "became popular for its reliability."

A quarter-century later, CMM worked with the air force and airframe manufacturers to develop and build a new generation of tape-controlled tools and "received the lion's share of orders." Still, Cincinnati's leading role in machine tools ebbed thereafter, even as CMM, renamed Milacron, grew to be the sector's largest enterprise. Meanwhile, printers rapidly assimilated novel photo-offset technologies, higher-speed presses, flexographics for packaging, and, eventually, computer-assisted design, echoing their first-mover alacrity in widely embracing electric drive during the 1890s. The co-location of leading publishers, advertising agencies, designers, and skilled technicians in New York's metropolitan area persisted, as did the sector's disaggregated networks and low levels of business concentration, at least into the 1980s. Though it is beyond the scope of this study to assess the long-run utility of the four industrial district categories outlined earlier, it does appear that post-1945 districts exhibited lower geographical compactness, plausibly a consequence of highway transportation and communications advances that reduced the salience of spatial propinquity to firm interactions and labor sheds. More important, specialists' vitality and attention to style, novelty, and technical opportunities continued to percolate through and mutate within key networked sectors.[11]

In addition, new specialty industrial districts rose to take the places of the fallen, some located in the United States despite an increasingly global economy. Though the German *Mittlestand* and northern Italian complexes for machinery, fashion fabrics, and other products have been much discussed, California's Silicon Valley, Massachusetts' high-tech Route 128 agglomeration, and Minnesota's Medical Alley have also begun to attract close attention, as might other localized American specialty trades. Custom work in close association with clients continues (even corporate giants have rediscovered the salience of client feedback), but demand/design trends toward servicing mass-market selling plans have been far more magnetic, reducing product diversity. The social relations of design and technical novelty, however, have remained chiefly a matter of personalized interactions, albeit in solving problems unimaginable to proprietors exchanging information at the Manufacturers' Club or furniture trade association.[12]

Jewelry making's descent into the miasma of hyperflexible and ruinously competitive relations resonated with the fiercely predatory practices of the fashion apparel trades. Then as now, the clothing industry had neither effective trade governance nor design cumulation, as did machine tools. Every new fabric or technical novelty fed either the bitterly contested and seemingly random production of styles or the rounds of price battling over cheap staple goods. To be sure, for a generation after the 1930s, clothing trade unions (International Ladies' Garment Workers, Amalgamated Clothing Workers) brought a measure of order to centers like New York, Philadelphia, and Rochester by negotiating and enforcing contract uniformities and taking wages out of competition. Yet, both before and after World War II, owners' relentless search for cost cutting pressed first staple and then fashion goods firms to locate new manufacturing facilities in nonunion rural and small-town areas, north and south.[13] Following the post-1945 reaffirmation of Parisian fashion leadership, "name" designers again ascended to transnational dominance of women's apparel trends. In later decades, designer fashions and their licensed reproductions for middling markets generalized to include U.S., British, German, and Italian "marques" and extended into menswear and children's lines, but this internationalization of styling little affected production pressures and competition for sales below the "couture" level. Indeed, the explosion of apparel production in developing nations (both in staples and fashion "knockoffs") created a dismal solution to cost and competition problems in mature industrial states, where apparel jobs vaporized by the hundreds of thousands.[14] Such specialists constructed terrains of endless novelty but stumbled into equally endless searches for short-term advantage and endless frustrations at their sectoral incapacity to manage markets, styles, and technical environments.

Auxiliaries marked out courses derived from the fates of their contractors' sectors.[15] The decline of Delaware Valley shipbuilding forced brass foundries and marine engine builders to reorient their capacities toward other clients, much as the erosion of Philadelphia's carpet, hosiery, upholstery, and wool/worsted sectors created dilemmas for spinners, dyeworks, and specialty machine makers. Elsewhere, auxiliaries moved with technological and product novelties to add new skills, such as electroplating at Grand Rapids or computer graphics in contemporary New York. The extent to which such capabilities went through cycles of corporate integration and dispersion deserves more detailed analysis. One scholarly team has underscored this point in researching the post-1945 Hollywood film industry. In the double-barreled crisis that television and antitrust judgments spawned, the integrated studio system broke down in the 1950s and 1960s. It was gradually replaced by "a highly disintegrated complex, involving a greater variety of film and video productions with most films made by independent production companies, more flexible contracts with writers, actors, producers, etc., and a flourishing of specialized shops for editing, special effects, recording, film processing, and so forth." Thirty years later a new phase of media mergers and concentrations has dawned, but industry leaders still treat each film as a custom product and work on the basis of project contracts involving scores of specialist auxiliaries, much as shipbuilders once did. To date they have unveiled no plans to re-create the huge back lots, long-term actor/writer/producer contracts, or other elements of the integrated studio system.[16]

On another front, early computers were custom and small batch products whose refinement involved extensive interactions with users, somewhat as locomotives did.[17] Yet the attainment of a version of mass production for chips and PCs in the 1970s and 1980s failed to generate market and technical stability, creating instead a "platform" for multiple, diverse software novelties, elaborate networks of core and auxiliary firms, and a shared agenda among engineers and designers for competitive technical advances, all of which have both energized and enervated the trade.[18] These brief comments, "updating" the course of specialty production patterns, are certainly preliminary and speculative. Indeed, another full-scale study would be necessary to tease out the peculiarities and permutations of industrial specialists through the Great Depression and World War II and into later decades, as well as the relevance of the frameworks sketched here to new industries like aircraft, high technology, and mass media. Assessing these later dynamics is a task for another day (decade?).

## Conclusion

Rather than viewing American industrialization after the Civil War as chiefly a story of increasingly rationalized and bureaucratic big businesses fashioning a "one best way" to mass produce and mass market standardized goods, this study has argued that a more complex and diverse process unfolded through the 1920s.[19] Custom and batch sectors drew upon standard materials turned out by giant corporations like U.S. Steel or Anaconda and in turn delivered power systems and machinery to them. Industrial specialists supplied growing consumer markets with a rich variety of stylish personal and household goods, supplementing the bulk-manufactured staples of everyday life. They also created supple networks of interfirm contracting and alliance, which in some measure duplicated spatially the quest for order that routinizing corporations' organizational innovations and mergers achieved inside firms. Making to order and selling from samples entailed their ready responsiveness to clients, market trends, and technological shifts. It also necessitated seeking efficiencies and economies based on skill, work routing, and variety, rather than simply on scale and speed. Just as oligopolizing sectors undertook to avert competition through price leadership or brand promotion of staples, so too did specialists struggle to manage competition by focusing selling on product quality, style, and differentiation. In both formats, some trades and firms managed these challenges better than others, and both routinized giants and specialty groups ran afoul of antitrust statutes in the effort. Overall, by spotlighting specialists' practices, accomplishments, and trials, this study has begun to frame a fuller account of American industrialization's dynamics and crosscurrents, thereby enlarging the meaning of the rubric "Second Industrial Revolution."

This book has also emphasized diversity among specialty manufacturing trades. A crucial cohort of very large enterprises delivered complex goods essential to the nation's transportation and power infrastructures. They were to Armour, Ford, and American Tobacco as chalk is to cheese, reminding us that big business was hardly homogeneous and that specialists cannot be simply categorized as small enterprises. Another sizable group of midsize firms designed and erected the machinery that made both mass and flexible production feasible. At Cincinnati, Worcester, and Providence, they did much more: organizing powerful trade associations, creating technical schools, revitalizing apprenticeships.[20] Without such industrial specialists, America's advance to global leadership in manufacturing would have been implausible; with them, it had coherence and cross-sectoral balance.

Equally, without specialty producers' mastery in furniture, printing/publishing, textiles, and the apparel and jewelry trades, this nation's status as a consumer society would have had a duller tone. Imagine a consumerism based on bulk-manufactured metal beds, standard chairs and sheets, work shirts, uniform blue serge suits, and plain bracelets. These specialists' vision of industrial capitalism centered on style diversity, ornament, and novelty. Though "efficiency experts" deplored their multiplication of varieties and modernist critics assailed their affection for elaborate designs, in responding to the vogues of their era they helped inaugurate an American society of vast consumer choice.

Although to most observers mass production defined American industrialization, those outside the conventional mainstream constructed the manufacturing capabilities pivotal to advancing technologies and the cultural artifacts that individualized consumption. In reminding us of historical complexity and contingency, this book has asked that an earlier, linear narrative be reframed to embrace their diversity of path and practice. Put simply, there were more critical players in industrialization than we may have realized. Many of those once thought peripheral constructed winning strategies in challenging markets and were far more central to building the nation's industrial base than has been recognized. If the old teleological and reductionist accounts are no longer tenable, building a new, inclusive narrative of American industrialization is now conceivable. American manufacturers did not define a uniform search for order, standardization, and corporate consolidation. Instead, order had different meanings from industry to industry, meanings that valorized sectoral technologies, social relations, and market tactics. Industrial specialists prized variety, prowess, and novelty as the keys to profitability and a blending of firm autonomy and judicious associationalism as strategies for growth and for managing competition. It is an open question whether their approaches remain instructive for firms facing the turn of another century, but their significance in America's industrial history is clear. At a minimum, specialists' distinctive alternative to the banality of "repetition work," price competition, and standard products is worth celebrating. In the best case, using their examples to think differently about our industrial past may provoke us to imagine diverse new venues that foster endless novelty.

# *NOTES*

## CHAPTER 1
## INTRODUCTION

1. For a fascinating example of the intersection of all three, see David Hackett Fischer, *Paul Revere's Ride* (New York, 1994).

2. Alfred Chandler, *The Visible Hand* (Cambridge, MA, 1977) and *Scale and Scope* (Cambridge, MA, 1990); David Hounshell, *From the American System to Mass Production* (Baltimore, 1984); Louis Galambos and Joseph Pratt, *The Rise of the Corporate Commonwealth* (New York, 1988); Susan Strasser, *Satisfaction Guaranteed* (New York, 1989).

3. Claude Fischer, *America Calling: A Social History of the Telephone to 1940* (Berkeley, CA, 1992); Susan Douglas, *Inventing American Broadcasting: 1899–1932* (Baltimore, 1987).

4. Sharon Strom, *Beyond the Typewriter* (Urbana, IL, 1992); William Leach, *Land of Desire* (New York, 1993).

5. Thomas Misa, *A Nation of Steel: The Making of Modern America, 1865–1925* (Baltimore, 1995).

6. Richard Tedlow, *New and Improved* (New York, 1990).

7. See Harry Braverman, *Labor and Monopoly Capital* (New York, 1974); David Montgomery, *The Fall of the House of Labor* (New York, 1987); and two collections edited by Stephen Wood, *The Degradation of Work?* (London, 1982) and *The Transformation of Work* (London, 1989).

8. Chandler, *Visible Hand;* Steven Sass, *The Pragmatic Imagination* (Philadelphia, 1982); Monte Calvert, *The Mechanical Engineer in America, 1830–1910* (Baltimore, 1967); David Noble, *America by Design* (New York, 1977), chap. 2.

9. Work in progress by Gerald Berk on Brandeis, managed competition, and the FTC should shed new light on this dynamic. See Berk, "Communities of Competitors: Open Price Associations and the American State, 1911–1929" (unpublished paper, 1995).

10. Michael Best, *The New Competition* (Cambridge, MA, 1990); Jonathan Zeitlin and Charles Sabel, eds., *A World of Possibilities* (Cambridge, UK, 1997), chap. 1.

11. Geoffrey Hodgson, *Economics and Institutions* (Philadelphia, 1988); Sharon Zukin and Paul DiMaggio, eds., *Structures of Capital* (New York, 1990); Michiel Schwarz and Michael Thompson, *Divided We Stand* (Philadelphia, 1990); Anthony Giddens, *Central Problems of Social Theory* (Berkeley, CA, 1979) and *The Constitution of Society* (Berkeley, CA, 1984). As regards industrialization, these works taken together indicate that we should expect to find multiple streams of intentionality and patterning, but neither a singular trajectory nor core logic. For a systematic attempt to model this contemporary diversity, see Michael Storper and Robert Salais, *Worlds of Production: The Action Frameworks of the Economy* (Cambridge, MA, 1997).

12. Maurice Levy-Leboyer, "The Quintessential Alfred Chandler," *Business History Review* 62 (1988): 519.

13. Daniel Nelson, *Managers and Workers* (Madison, WI, 1975), 7–8; Philip Burch, *The Managerial Revolution Reassessed* (Lexington, MA, 1972).

14. Philip Scranton, *Proprietary Capitalism* (New York, 1983) and *Figured Tapestry* (New York, 1989).

15. For a different four-part construction of the diversity of industrial practice, see Storper and Salais, *Worlds of Production*, chaps. 1, 2. Broadly, what I term specialty manufacturing, they regard as practice in the "interpersonal world"; bulk production to a degree corresponds with their "market world," and mass production to their "industrial world." In addition, they frame an "intellectual world" to take account of science-intensive products (e.g., bioengineering), which are not present in this study's time frame.

16. Gavin Wright, *Old South, New South* (New York, 1986), chaps. 5, 7.

17. This distinction suggests, in this context, the uselessness of the customary dichotomy between big and small business, which flattens key elements of differentiation among industrial enterprises.

18. Samuel Vauclain, *Steaming Up!* (New York, 1930), 146.

19. On the Ford transition, see Hounshell, *American System.*

20. Department of Commerce, Bureau of the Census, *Thirteenth Census of the United States,* vol. 8, *Manufactures, 1909* (Washington, DC, 1913). The 90 sectors displayed in the tables accounted for 92 percent of industrial value added, with 145 smaller sectors covering the residual. Of those, just under half (68) represented specialty work (artists' materials, billiard tables), whereas 57 were clearly bulk-oriented (bluing, wood screws) and 20 were mixed classes.

21. The absence of 1899 figures for foundry and machine shops is unfortunate here, but the sector's 1904–9 increase matched the national average at 35 percent. If over the entire ten-year span this sector grew at the national average rate (76 percent), this would reduce the overall specialists' gains only to 81 percent.

22. "Perhaps half" comes from adding to table 2's 33 percent one-half of the value added in the smallest sectors (4 percent) and a conservative one-third of that recorded for the mixed group (10 percent), for a sum of 47 percent. For the smallest sectors, see n. 20.

23. Frederick Scherer, *Industrial Pricing: Theory and Evidence* (Chicago, 1970), 131–33; Terry Burke, Angela Genn-Bush, and Brian Haines, *Competition in Theory and Practice* (London, 1988), 118–19.

24. Even Taylor and his boss at Philadelphia's Midvale Steel admitted this. See Daniel Nelson, *Frederick W. Taylor and the Rise of Scientific Management* (Madison, WI, 1980), 45.

25. For discussion, see Scranton, *Proprietary Capitalism* and *Figured Tapestry,* and Jonathan Zeitlin and Charles Sabel, "Historical Alternatives to Mass Production," *Past and Present,* no. 108 (1985): 133–76. The Zeitlin-Sabel position in this article has been much modified by subsequent research. For a restatement, see Zeitlin and Sabel, *World of Possibilities.* An extremely valuable reconsideration of the industrial district literature and concepts of trust is Udo Staber, "The Social Embeddedness of Industrial District Networks," in *Business Networks,* ed. Udo Staber, Norbert Schaefer, and Basu Sharma (Berlin, 1996), 148–74.

26. Philip Scranton, "Diversity in Diversity," *Business History Review* 57 (1991): 27–90.

27. Frank Comparato, *Books for the Millions* (Harrisburg, PA, 1971), 129–30.

28. This was a negotiated, contested relation, depending on buyers' and sellers' market power, and either's dependence on the other for sizable and continuing supplies and orders. In soft markets, leading wholesalers and direct-use clients pushed Brass Valley companies to make and store goods for their "call," with payment forthcoming only after requested lots were delivered. See Theodore Marbury, *Small Business in Brass Fabricating* (New York, 1956).

29. Mark Reutter, *Sparrows Point* (New York, 1988); Misa, *A Nation of Steel;* Harold Passer, *The Electrical Manufacturers, 1875–1900* (Cambridge, MA, 1953); Stephen Meyer, "Technology and the Workplace: Skilled and Production Workers at Allis-Chalmers," *Technology and Culture* 29 (1988): 839–64; Scranton, *Figured Tapestry,* 183–86; Roland Marchand, *Advertising the American Dream* (Berkeley, CA, 1986), 123–27.

## CHAPTER 2
## SPECIALTY MANUFACTURING TO 1876

1. "The Brown and Sharpe Manufacturing Co.," *Industry* 2 (1889–90): 38. Richards edited (and wrote most of) this San Francisco trade journal from 1888 through 1896.

2. Alfred Chandler, *The Visible Hand* (Cambridge, MA, 1977) and *Scale and Scope* (Cambridge, MA, 1990); David Hounshell, *From the American System to Mass Production, 1800–1932* (Baltimore,, 1984).

3. Brown and Sharpe Manufacturing Company, *Catalog and Price Lists* (Providence, 1880), quotation from unpaged frontispiece. The firm also supplied clients with micrometers, standard gauges, gear wheels, milling cutters, hair clippers (for people and horses), yarn reels, and calipers, and nickel-plated goods "to order."

4. Hounshell, *American System,* 75–82; W. A. Viall, "Report, January 25, 1917," Series C.2, Brown and Sharpe Papers, Rhode Island Historical Society, Provi-

dence. Clearly Brown and Sharpe's early contracts with Willcox and Gibbs gave the Providence firm a golden opportunity to develop its skills in precision gauge and tool work. What is important here is that the firm kept its main focus and built its growth trajectory on diverse tools, rather than gravitating toward mass production strategies. In 1913, for example, the machinery department processed 5,300 separate batch orders for components, most for 20–125 pieces. See L. D. Burlingame, "Investigation as to the Economy of Making Machine Parts . . . ," Series C.3, B&S Papers.

5. Unquestionably, American manufacturers were noted for novelty in product and practice before the war, as European visitors repeatedly remarked. However, those innovations were chiefly confined to refining systems for bulk production (as in New England textiles, shoes, or work clothing). The roots of specialty manufacturing are indeed evident by the 1850s in Atlantic coast urban centers, notably Philadelphia's fabric and machinery trades, but they sprouted and multiplied in the context of accelerating demand for infrastructure, producer, and consumer goods during the last third of the century. For styled textiles, see Philip Scranton, *Proprietary Capitalism* (New York, 1983), and for machinery, Ross Robertson, "Changing Production of Metalworking Machinery, 1860–1920," in National Bureau for Economic Research, *Output, Employment and Production in the United States after 1800* (New York, 1966), 479–95.

6. Lest this payment seem odd, it should be noted that when Lucian Sharpe commenced learning his craft with J. R. Brown in 1848, he entered into the "old type of indenture" in which his father provided Brown with fifty dollars a year plus a weekly board payment for a five-year apprenticeship, after which Lucian was taken in as a partner. See Henry Sharpe, *Joseph R. Brown, Mechanic, and the Beginning of Brown and Sharpe* (New York, 1949), 10–11.

7. "A Pioneer in the Woodworking Machinery Business," *Wood Craft* 15 (1911): 101–2.

8. For another example, see William J. Clark, "Reminiscences," in *History of the Bolt and Nut Industry in America*, ed. W. R. Wilbur (Cleveland, 1905), 360–61.

9. "Wm. B. Bement," *Industry* 8 (1895): 274; George Dickie, "John Richards," *Machinery* 12 (1900): 349–50.

10. On continuing expectations of plant visitations, see "The Egan Co's New Buildings," *The Wood-Worker* 9 (March 1890): 16. On screening visitors (including workers) for credentials, see *The Wood-Worker* 9 (April 1890): 22, (July 1890): 17; 10 (April 1891): 18, (December 1891): 15. J. T. Langdon ("Train Men for Your Business," ibid. 9 [February 1891]: 27) stressed that employers especially should visit other firms: "[The proprietor] is the one who wants to get out and hustle around among shops and find out what's going on in them, and if there is any one way which is an improvement on his style, he wants to adopt it and teach his men how to do it, and not wait for his men to show him how 'tis done." The prescriptive tone suggests that this was not done often enough. In a similar vein, proprietors and foremen who pretended knowledge where they had none were ready targets for "sharks" who peddled secret recipes for lubricants or alloy mixes or sold licenses to use worthless or imaginary patents. See Fred Colvin, *Sixty Years with Men and Machines* (New York, 1947), 154–55, and Charles Porter, *Engineering Reminiscences* (New York, 1908), 2–3.

11. Howell Harris has suggested in a personal communication that the "proper style of activist leadership" in late-nineteenth-century manufacturing may have been rooted in the Civil War military experiences common among proprietors and workers (Harris to author, September 14, 1992).

12. *The Wood-Worker* 10 (January 1891): 15.

13. Monte Calvert, "American Technology at World's Fairs" (M.A. thesis, University of Delaware, 1962); Thomas Pickering, "American Machinery at International Expositions," *Transactions of the American Society of Mechanical Engineers* 5 (1884): 113–23. In an appropriate reversal, European copyists routinely put Corliss or Brown and Sharpe labels on these goods, much as American styled-textile companies tried to pass off their products as English or French, each trading on the special cachet of either technical or fashion reputation. Such ploys were much less

likely to be found among purveyors of flour, nails, or kerosene.

14. Responding to this, some firms attached heavy weights to too-light tools in order to stabilize them. See Robert Woodbury, *History of the Grinding Machine* (Cambridge, MA, 1972), 62–63.

15. "Bement," 274–75.

16. Ibid. Consistent with his concern for quality and "good work," Bement regularly visited his firms' drafting rooms, passing "from one desk to another, perhaps without a word, even if he saw blunders and faults," but leaving a quiet "impression of approbation or dissent." His partner, James Dougherty, made his "course all over the foundry at certain hours each day, and whenever some difficulty arose." Bement detested "sham and bad fitting" and routinely "exchange[d] ideas" with other builders, including colleagues in Glasgow and Manchester (by means of "machine photographs"), for many years.

17. For vignettes on this process, see Joseph Roe, *English and American Tool Builders* (New Haven, 1916), and George Wing, "The History of the Cincinnati Machine Tool Industry" (D.B.A. diss., Indiana University, 1964).

18. See Dickie, "Richards," 350. Similarly, steam engine designer and manufacturer Charles Porter spent six years with British firms before returning to his native land in 1868 to expand his partnership works in New York City. See Porter, *Reminiscences*, chaps. 6–14. These and other life-course tales of postbellum engineers call our attention to the international character of the machine-building community. Richards' 1869 nuptials apparently represented a second marriage, as any son born to this couple could not have taken over his European operations in 1877.

19. "Pioneer," 101–2; *Industry*, passim; and Dickie, "Richards," 350.

20. Quoted from *Engineering*, Corliss obituary, *Scientific American* 57 (June 2, 1888): 34.

21. Ibid.; undated *Providence Journal* clipping, Box 10, Corliss Collection, MS 80:3, John Hay Library, Brown University; "Personal Characteristics of George Corliss," typescript, Box 10, Corliss Collection, 5; "Notes from the Builders' Iron Foundry," *Machinery* 3 (1897): 359–60.

22. Obituary, *Scientific American*; "Our Centennial," *Harper's Weekly*, May 27, 1876, in Corliss Centennial Scrapbook, Corliss Collection. This scrapbook contains clippings concerning Corliss's work as the Rhode Island delegate to the Centennial Commission, his installation of the engine for Machinery Hall, its workings, and reactions to it. The original is in the Hay Library at Brown; a microfilm copy is held at the Hagley Museum and Library, Wilmington, DE (cited hereafter as "Scrapbook"). For steam engine operations, see C. F. Swingle, *Twentieth Century Handbook for Steam Engineers* (Chicago, 1905), chap. 11, and for a full technical assessment of the Corliss innovations, see Louis Hunter, *Steam Power: A History of Industrial Power in the United States, 1780–1930* (Charlottesville, VA, 1985), chap. 5.

23. "Personal Characteristics," Box 10, Corliss Collection.

24. "George H. Corliss," Scrapbook, Corliss Collection. James Watt evidently used similar schemes for marketing his eighteenth-century improvements on the original Newcomen designs. See Embry Hitchcock, *My Fifty Years in Engineering* (Caldwell, ID, 1947), 63; Hunter, *Steam Power*, 265; and Robert Holding, *George Corliss of Rhode Island* (New York, 1945), 9–13.

25. John Pratt and Richard Zeckhauser, eds., *Principals and Agents: The Structure of Business* (Boston, 1985); Charles Sabel, "Studied Trust," in *Industrial Districts and Local Economic Regeneration*, ed. F. Pyke and W. Sengenberger (Geneva, 1992), 215–50.

26. Holding, *George Corliss*, 11.

27. "Personal Reminiscences," 4–5, and "Personal Characteristics," 5, Scrapbook, *Daily Graphic*, "Rhode Island at the Centennial," October 24, 1876, Corliss Collection.

28. Obituary, *Scientific American*; Hunter, *Steam Power*, 267 n. 27.

29. "Personal Reminiscences," 2–3, 6–12, Corliss Collection; Hunter, *Steam Power*, 283 n. 74, 298–99.

30. Pickering, "American Machinery," 114–15; Calvert, "American Technology," 133–34; William Corliss to Edward P. Allis & Co., January 7, 1868, Corliss Steam Engine Company Letterbook, 1867–68, 70, Box 2, Corliss Collection. Allis later re-

sponded to the rebuff by hiring away Corliss's manager, Edwin Reynolds, and his head draftsman, who took "the same positions in the Allis works at Milwaukee for the manufacture of the Corliss engine there" (Porter, *Reminiscences*, 247. Porter, a Corliss rival, was gleeful at this outcome). Hunter regarded this loss as balking a plausible engineering succession at Corliss and key to the decline of the firm after the founder's death in 1888 (*Steam Power*, 299–300).

31. "Personal Reminiscences," 1, 11; "Personal Characteristics," 1–4, Corliss Collection; Hitchcock, *My Fifty Years*, 60. Hitchcock, later dean of engineering at Ohio State University, was a Corliss draftsman for two years after graduating from Cornell's Sibley School.

32. "Personal Characteristics," 1–2; "Personal Recollections," 1; Broadbrim, "Centennial Letter No. 17," and "The Centennial Engine Builder," Providence *Sunday Telegram*, October 1, 1876, Scrapbook, Corliss Collection. "Carpet knight" refers to men who were replete with polished manners and parlor or boardroom savoir faire yet ignorant of production.

33. It would be valuable to know whether the Corliss Company made engines in standard sizes in small batches for inventory, so as to speed delivery when orders arrived and keep the workforce together in relatively slack periods. Hunter notes that Corliss made most engines to order (*Steam Power*, 265), so that auxiliary production may have served this bridging purpose. Brown and Sharpe did advertise in 1880 the on-hand availability of its basic screw-cutting machines and smaller machine tools (Brown and Sharpe, *Catalog*, at n. 3).

34. Ohio's Oberlin College was a focal point for antislavery sentiments after the 1834 secession of the entire senior class from Cincinnati's Lane Seminary (when its president forbade abolitionist activity) and their subsequent enrollment at Oberlin.

35. Arthur Cox and Thomas Malim, *Ferracute: The History of an American Enterprise*( Bridgeton, NJ, 1985), 7–10; *Ferracute: Eighty Years* (Bridgeton, NJ, 1943), chap. 1, unpaged.

36. The "Ferracute" name was a Smith fancy, imperfectly derived from Latin and Italian terms he thought represented "sharp iron," an apposite if etymologically dubious evocation of cutting-press work.

37. Cox and Malim, *Ferracute*, 11–21.

38. Roe, *English and American Tool Builders*, 268–69; Frederick Geier, *The Coming of the Machine Tool Age* (New York, 1949), 15; Wing, "Cincinnati Machine Tool Industry," 79–82. Sharpe apprenticed with J. R. Brown, who offered him a partnership "at a nominal figure" two years after Sharpe completed his five-year indenture. See also Donald Tulloch, *Worcester: City of Prosperity* (Worcester, MA, 1914), 247.

39. Thomas Savery Journals, 1864, 1881, Accession no. 291, Hagley Museum and Library, Wilmington, DE; Samuel Vauclain, *Steaming Up!* (New York, 1931), 48, 93–107.

40. Savery Journals, March to May, 1864, Hagley Library.

41. Savery Journals, April 1866 to December 1868, Hagley Library. In later years, Savery rarely mentions workers. Wages in 1875 ($173,000) accounted for 42 percent of the sales prices of orders shipped (December 31, 1876).

42. This format for pricing helps account for the practice of meeting competition by "taking the metal out" of machinery, as reducing weight immediately cut bid estimates. Of forty-eight machinery contracts for 1871 that Savery monitored, only five were for standard steam engines (Savery Journal, January 6–7, 1872). For by-the-pound pricing through World War II, see Henry Hess, "Foundry Costs," *Proceedings of the Engineers' Club of Philadelphia* 21 (January 1904): 29; and Lloyd Saville, "Price Determination in the Gray-Iron Foundry Industry" (Ph.D diss., Columbia University, 1950), 67–75.

43. Captain Henry Metcalfe, *The Cost of Manufactures and the Administration of Workshops* (New York, 1885). On pricing machinery by the pound, see p. 37. Metcalfe credits William Sellers' Philadelphia machine tool establishment with priority in refining workshop costing and labor process routing and oversight (24, 37, 74).

44. S. Paul Garner, *Evolution of Cost Accounting to 1925* (University, AL, 1954), 76–90; H. Thomas Johnson and Robert S. Kaplan, *Relevance Lost: The Rise and Fall of Managerial Accounting* (Boston, 1987), 20–

31, 34–38, 52–55. On Baldwin, see John K. Brown, *The Baldwin Locomotive Works* (Baltimore, 1995).

45. Savery Journals, January and September 1872, January and July 1873. One of the 1871 ship contracts had been secured only through a $6,500 kickback paid to the New York agent of the South American firm for which the *Rio Branco* was built. This sleazy deal left Savery seething.

46. Savery Journals, March 14, 1873, September 29, 1877.

47. Savery Journals, 1874 (esp. November entries), 1877–79.

48. Savery was periodically dispatched to deal with users' complaints. If Pusey & Jones' workmanship was at fault, the makers bore the costs of remedies, but if users' practices had caused failures, they covered all expenses including Savery's travel, meals, and lodging. Corliss too sent off skilled workers to deal with engine failures and needs for parts replacement (Hunter, *Steam Power*, 293).

49. For current discussion of interfirm networks, see N. Nohria and R. G. Eccles, eds., *Network and Organizations* (Boston, 1992), and Udo Staber, Norbert Schaefer, and Basu Sharma, eds., *Business Networks: Prospects for Regional Development* (Berlin, 1996).

50. Susan Strasser, *Satisfaction Guaranteed* (New York, 1989); Richard Tedlow, *New and Improved* (New York, 1990); Donald Hoke, *Ingenious Yankees* (New York, 1990).

51. Scranton, *Proprietary Capitalism*, chap. 4; Marta Sironen, *The History of American Furniture* (New York, 1936), chaps. 1–2.

52. For example, brass and bronze were smelted in lots of 150 to 300 pounds; scores of different formulas were used for the crucible mixtures, yielding metals whose composition "will vary and does vary more than most brass makers are willing to admit" (Bridgeport Brass Co., *Seven Centuries of Brass Making* [Bridgeport, CT, 1920], 26). Large batch production came only with the introduction of the electric furnace after 1910, yielding greater uniformity; yet dozens of compositions and grades continued to be made for diverse end uses. Similarly, yarns of different counts made from fibers of various qualities were spun on contract by firms whose frames had "change gears" to alter the size and tightness of the product; and builder's trim was run up from a broad range of woods (and grades) in an enormous number of patterns and dimensions (Scranton, *Proprietary Capitalism*, and personal communication from Herbert Gottlieb, Cornell University, November 1991).

53. That all were immigrants is not accidental. It is my impression that immigrant entrepreneurs and skilled workers played featured roles more commonly in specialty metals, styled textiles, and high-end furniture than in machinery building through the late nineteenth century.

54. "Fifty Years with the Disstons," *The Wood-Worker* 9 (December 1890): 25. In 1890, Bickley managed the firm's "long saw" shops.

55. Harry Silcox, *A Place to Live and Work: The Henry Disston Saw Works and the Tacony Community of Philadelphia* (University Park, PA, 1994), 1–4. See also W. D. Disston, H. W Disston, and William Smith, "The Disston History" (unpublished typescript, Hagley Library). For at least twenty-five years after starting to "melt" its own crucible steel, Disston continued to buy materials from Sheffield. For technical processes and an analysis of American crucible steel development, see Geoffrey Tweedale, *Sheffield Steel and America* (Cambridge, UK, 1987). Recruitment of Sheffield immigrants by U.S. firms was substantial in this era. Charlotte Erickson commented that "Sheffield cutlers were sought . . . because of their remarkable versatility, a product of long and careful training, which enabled them to shift from one job to another" (*American Industry and the European Immigrant, 1860–1885* [Cambridge, MA, 1957], 41).

56. Henry Disston married Mary Steelman (apt name!) in 1843, and the couple had nine children (ca. 1844–62). Three of the first five, born by 1851, died in infancy, but the four later children all survived, perhaps an indication of increased family wealth and improved housing that followed on the firm's prosperity. See Silcox, *A Place*, chap. 1, n. 17.

57. Ibid., 7–10.

58. On Cramp's wartime ship construction, see Thomas Heinrich, *Ships for the Seven Seas* (Baltimore, 1997), chaps. 1–2.

59. Tweedale points out that at this time Pittsburgh crucible output totaled a meager 20,000 tons annually, much of which was too poor in quality for tool uses (*Sheffield*, 23).

60. Silcox, *A Place*, 8. See also Silcox draft MS, chap. 2 (copy in author's possession).

61. Tweedale, *Sheffield*, 175. Disston kept only "certain hand saws . . . in stock and ready for sale" (Silcox, *A Place*, 10).

62. Tweedale, *Sheffield*, 148. Tweedale credits the Disston family's mechanical inventiveness in devising improvements in toothing, grinding, tempering, and hardening with making the firm "the world leader in sawmaking" by the end of the 1870s (148). Disston's principal rival, especially in large saws, was the New York family partnership R. Hoe and Co., started by another English immigrant twenty years before Henry Disston's 1840 beginnings. Hoe achieved greater prominence in the manufacture of printing presses but retained its sawmaking capacity into the twentieth century. See Stephen D. Tucker, "History of R. Hoe and Company, 1834–1885," *Proceedings of the American Antiquarian Society* 82 (1972): 351–453, and Frank Comparato, *Chronicles of Genius and Folly: R. Hoe and Co.* (Culver City, CA, 1979).

63. On this Philadelphia pattern, see Scranton, *Proprietary Capitalism*.

64. Daniel Robson, ed., *Manufacturers and Manufactories of Pennsylvania* (Philadelphia, 1875), 259. Rovings are ropelike tubes of straightened fibers ready for the twisting process that yields spun yarn for the loom.

65. Scranton, *Proprietary Capitalism;* Susan Levine, *Labor's True Woman: Carpet Weavers, Industrialization and Labor Reform in the Gilded Age* (Philadelphia, 1984), 26.

66. Along with at least one remarkable woman, Rebecca Pennock Lukens, a Quaker who operated a regional iron mill for nearly thirty years after her husband Charles's death. See Angel Kwollek-Folland, "Women's Business and the First Industrial Revolution, 1830–1880" (paper presented at the Delaware Seminar, Newark, DE, October 1995).

67. John C. Maule to Editor, *Philadelphia Record*, November 5, 1910; Robson, *Manufacturers*, 259.

68. Maule, *Philadelphia Record*.

69. Philip Scranton, "Build a Firm, Start Another: The Bromleys and Family Firm Entrepreneurship in the Philadelphia Region," *Business History* 35 (1993): 115–51.

70. Ibid.; Levine, *Labor's True Woman*, 30.

71. Philip Scranton, *Figured Tapestry* (New York, 1989) and "Build a Firm."

72. Sironen, *American Furniture*, chaps. 3–6; Page Talbott, "Philadelphia Furniture Makers and Manufacturers, 1850–1880," in *Victorian Furniture*, ed. Kenneth Ames (New York, 1982), 87–101 (published as vol. 8, nos. 3–4 of *Nineteenth Century*).

73. "Case goods" refers to the wood furniture classes listed above, distinguished from upholstered furniture (leather or fabric over a concealed frame). Chicago's packinghouses yielded vast quantities of hides that, locally tanned, underwrote its early supremacy in upholstered goods. Local firms easily made the shift to styled fabrics, and Chicago continued to lead that trade division into the 1920s. See Sharon Darling, *Chicago Furniture: Art, Craft, and Industry, 1833–1983* (New York, 1984).

74. Frank Ransom, *The City Built on Wood* (Grand Rapids, MI, 1955), 3–10. For midwestern industrialization, resources, and railways more generally, see William Cronon, *Nature's Metropolis: Chicago and the Great West* (New York, 1991).

75. Dwight Goss, *History of Grand Rapids and Its Industries*, vol. 2 (Chicago, 1906), 1041–42, 1048; Ransom, *City*, 10.

76. Goss, *History*, 1042–44; Ransom, *City*, 13–14. In 1870, mean employment at Grand Rapids furniture firms was thirty-five, and product value averaged $44,000. Widdicomb Brothers and Richards' capital in 1869 was only $12,000; hence it is doubtful that the 1873 figure represented "paid in" funds. See "Historical Notes on the John Widdicomb Co.," Widdicomb Papers, Collection no. 17, Grand Rapids Public Library, Box 1, File 2.

77. Clipping from Grand Rapids *Press*, December 27, 1927, Widdicomb Papers, Box 1, File 2.

78. Clipping from *Grand Rapids Daily Artisan and Record*, January 4, 1928, Widdicomb Papers, Box 1, File 2.

79. Ransom, *City*, 14; clipping from *Michigan Artisan*, May 10, 1909, Widdicomb Papers, Box 1, File 2.

80. Brian Page and Richard Walker, "From Settlement to Fordism: The Agro-Industrial Revolution in the American Midwest," *Economic Geography* 67 (1991): 281–315.

## CHAPTER 3
## INSTITUTIONS AND THE CONTEXT FOR SPECIALTY PRODUCTION

1. Louis Galambos, *Competition and Cooperation: The Emergence of a National Trade Association* (Baltimore, 1966), 20–21. Treasurers were effectively the CEOs of cotton corporations.

2. Bruce Sinclair, *Philadelphia's Philosopher Mechanics: A History of the Franklin Institute, 1824–1865* (Baltimore, 1974), 4, 31. The Rensselaer Institute, founded at Troy, New York, at roughly the same time, had quite different origins and goals. Later a premier engineering school, Rensselaer commenced as the project of a fifth-generation patroon and Harvard graduate, and initially focused on teaching science and its application "to agriculture, domestic economy, and the arts." Its first head was a botanist and geologist, an eminent student of Yale's Benjamin Silliman, drawn to Troy from Williams College. See Palmer Ricketts, *History of Rensselaer Polytechnic Institute, 1824–1934* (New York, 1934), 14–40.

3. For more on Baldwin's early work, see John Brown, *The Baldwin Locomotive Works* (Baltimore, 1995).

4. Sinclair, *Mechanics*, 33–116.

5. Ibid., 135–281, quotation from 281.

6. Ibid., 287–302. On the variety of technical education efforts undertaken in Philadelphia, see Nina Lerman, "From 'Useful Knowledge' to 'Habits of Industry': Gender, Race and Class in Nineteenth-Century Technical Education" (Ph.D. diss., University of Pennsylvania, 1993).

7. Sinclair, *Mechanics*, 303–11, quotation from 311.

8. Making a single, replacement, screw-threaded machine bolt and nut was not an especially demanding challenge for an experienced machinist, but it took far more time than one might at first imagine. Even after screw pitch standards had been established, a 1919 metalworking manual estimated an hour and three-quarters as a reasonable time for fashioning one hex-headed, 4½" threaded bolt and nut. See Robert Smith, *Advanced Machine Work*, 6th ed. (Boston, 1919), 3:42–43.

9. Sinclair, *Mechanics*, 313–25.

10. See Carlo Poni's contribution to *A World of Possibilities*, ed. Jonathan Zeitlin and Charles Sabel (Cambridge, UK, 1997), and Laurence Gross, *The Course of Industrial Decline: The Boott Cotton Mills of Lowell, Massachusetts, 1835–1955* (Baltimore, 1993).

11. Railway master mechanics' and master car builders' associations recommended adoption of the Sellers thread in the 1880s. Yet the process remained slow. In 1887, one observer noted that it "would now be hard to find a railroad anywhere on which it would not be asserted, at least, that the Sellers Standard was in use. Many more would assert that they use it than actually do use it, but that is natural" ("The Sellers . . . System of Screw Threads," *Journal of the Franklin Institute* 123 [1887]: 261–76, hereafter cited as *JFI*).

12. *JFI* 53 (1867): 3.

13. For the first section of Richards' work, see *JFI* 59 (1870): 305–13, and for his joining of the Arts and Manufactures Committee, 64 (1872): 144. See also Sinclair, *Mechanics*, 320, 322.

14. In the last years before its dismemberment, I had the opportunity to roam the stacks of the institute's library and discover priceless holdings of obscure trade journals. Many of these have been parceled out to historical institutions in the Philadelphia region, but most of the foreign-language journals, dating to the 1880s, became dumpster trash for want of new homes.

15. Louis McLane, *Documents Relative to the Manufactures of the United States*, 22d Cong., 1833, H Doc. 308, 1833, vol. 1, doc. 3, nos. 219–76.

16. Donald Tulloch, *Worcester: City of Prosperity* (Worcester, MA, 1914), 73; Charles Washburn, *Industrial Worcester* (Worcester, 1917), 293–97, 301. For a concise overview of the city's industrial development, see Roy Rosenzweig, *Eight Hours for What We Will* (New York, 1983), 11–16. For later, comparable patterns of space rental in Lancashire, see Andrew Marrison, "Indian Summer," in *The Lancashire Cotton Industry: A History since 1700*, ed. Mary Rose (Preston, UK, 1996), 245.

17. Joshua Chasan, "Civilizing Worcester" (Ph.D. diss., University of Pittsburgh, 1974), 164–80; Charles Buell, "The Workers of Worcester" (Ph.D. diss., New York University, 1974), 152–54; Washburn, *Industrial*, 302–20; Tulloch, *Worcester*, 213.

18. Washburn, *Industrial*, 142–52, 305–8; Chasan, "Civilizing," 216–26. Like Roebling and Disston, Washburn integrated backward as his trade developed and diversified, and he became distanced from machinery building. Ultimately, his firms (capitalized at $1.1 million in 1865 and $1.5 million in 1869) became bulk producers of standard wire as well as makers of "specialties": "galvanized steel wire cable for suspension bridges; phosphor-bronze and copper wire rope; transmission and standing wire rope; . . . hoisting . . . tiller . . . [and] switch rope; copper, iron and tinned sash cord wire; galvanized iron wire rope for ships' riggings, and galvanized crucible cast-steel wire rope for yacht's riggings" (Washburn, *Industrial*, 159). See also George Alden, "Technical Training at the Worcester Free Institute," *Transactions of the American Society of Mechanical Engineers* 6 (1885): 510–99, and more generally on early engineering and technical education, Monte Calvert, *The Mechanical Engineer in America, 1830–1910* (Baltimore, 1967). On Roebling practice, see Clifford Zink and Dorothy Hartman, *Spanning the Industrial Age: The John A. Roebling's Sons Company, Trenton, New Jersey, 1848–1974* (Trenton, NJ, 1992).

19. For a later review of curriculum and shop work, see *Machinery* 11 (1904–5): 517–20.

20. Chasan, "Civilizing," 226–29. Suggesting the substantive, rather than perfunctory, significance of production at the Washburn Shops is the saga of its hydraulic elevators. Devised and patented by Milton Higgins, once a draftsman for Washburn's firm and later the shops' superintendent, the elevators were built there "until 1896 when this department was bought out by a new company" and removed to another Worcester site. The shops' elevators were crafted for "buildings up to 27 stories" (Tulloch, *Worcester*, 89). See also Herbert Taylor, *Seventy Years of the Worcester Polytechnic Institute* (Worcester, MA, 1937), and *Machinery* 10 (1903–4): 578–81.

21. Alden, "Technical Training," quotation from Smith on 541; Calvert, *Mechanical Engineer*.

22. One of these, in the early 1930s, was my father, Clarence Scranton, who came to UC's co-op engineering program from Kansas, proceeded after graduation to work for the TVA and the Army Air Force engineers in World War II, and saw long service in specialty engineering and design at US Steel thereafter.

23. See "Techno in the Works," beginning at *Industry* 3 (1890–91): 74–75, and continuing in monthly installments for several years.

24. About them we know even less than about batch specialists, but as late as 1909 the census reported 40,000 enterprises in lumber, 24,000 bakeries, and roughly 15,000 cigar operations, groups whose employment had increased by 25 to 66 percent since 1899 (Bureau of the Census, *Thirteenth Census of the United States*, vol. 8, *Manufactures* [Washington, DC, 1913], 40). On cigars, see Patricia Cooper, *Once a Cigar Maker* (Urbana, IL, 1987), which addresses work relations, competition, and gender in midprice lines.

25. Tampa cigars were a notable exception, but the Tampa district represented the top-end specialty segment of the cigar trade and created skill-intensive, though obviously not styled, products. See Nancy Hewitt, "The Voice of Virile Labor," in *Work Engendered*, ed. Ava Baron (Ithaca, NY, 1991), 142–67.

26. Alexander Keyssar, *Out of Work* (New York, 1986).

27. Another example was Samuel Darling, a Brown and Sharpe partner after 1866, all of whose "activities were surrounded with considerable secrecy" (Henry Sharpe, *Joseph R. Brown, Mechanic, and the Beginnings of Brown and Sharpe* [New York, 1949], 22).

28. At Pusey and Jones, Savery's diaries show that the highest-paid machinists earned two-thirds more than the modal rate offered the largest number of journeymen.

29. David Leverenz, *Manhood and the American Renaissance* (Ithaca, NY, 1989); Peter Stearns, *Be a Man!* (New York, 1979).

30. Philip Scranton, "Where's Poppa: The Manhood Question in American Manufacturing, 1870–1930," *Business History* (UK), forthcoming. For documenta-

tion of celebrations of competence, see David Shayt, "Machine Climbers," *American Heritage of Industry and Technology* 4 (Fall 1988): 56–61, and Philip Scranton and Walter Licht, *Work Sights* (Philadelphia, 1986), 32, 223.

31. These relations also generated elaborate shop jokes and the customary victimization of new "hands" and apprentices. For these, see John Richards, *Mechanical Humor* and *Industrial Anecdotes* (San Francisco, n.d.). For a British view, see David Collinson, "'Engineering Humor': Masculinity, Joking, and Conflict in Shop-Floor Relations," *Organizational Studies* 9 (1988): 181–99. For a fascinating study of masculinity as a racially charged and imperializing concept in this era, see Gail Bederman, *Manliness and Civilization* (Chicago, 1995). Attention to constructions of manhood among manufacturers and industrial workers has thus far been rare in the literature on the social dynamics of defining and enacting maleness.

32. See Scott Sandage, "Deadbeats, Drunkards, and Dreamers: A Cultural History of Failure" (Ph.D. diss., Rutgers University, 1995).

33. This was the 1880s pattern for worker-entrepreneurs launched from William Lodge's machine tool shops in Cincinnati. See George Wing, "The History of the Cincinnati Machine Tool Industry" (D.B.A. diss., Indiana University, 1964), and Joseph Roe, *English and American Tool Builders* (New Haven, 1916), 270–73.

34. Philip Scranton, *Proprietary Capitalism* (New York, 1983). This pattern repeated in jewelry, silk weaving, and full-fashioned hosiery across the next sixty years. All were relatively low entry-cost trades. Such displacements may have been one source of the multiplication of jobbing machine shops as well, but there are only fragmentary indications on this count. Eager distributors in the 1870s and later were frequently themselves manned by wholesalers' salesmen dismissed in economic slumps. No similar phenomenon has come to my attention in specialty producers' goods. The vogue among service professionals shed by public and nonprofit companies for starting their own firms in the late 1980s and early 1990s echoes this desperate entrepreneurship, though perhaps in an economic climate that offers better prospects.

35. Frank Comparato, *Books for the Millions* (Harrisburg, PA, 1971), 128–29; Scranton, *Proprietary Capitalism;* Louis Hunter, *Steam Power* (Charlottesville, VA, 1985), 270–76.

36. It appears that export-oriented bulk producers were willing to sacrifice to tariff reciprocity any sector in which they were not involved, most of these being specialty production realms. In any event, it was not until the 1930s that detailed reciprocity deals were executed effectively; and with World War II intervening, their consequences became entangled with the postwar GATT embroilments. For an insightful assessment of this fracturing's political consequences in the 1930s, see Thomas Ferguson, "Industrial Conflict and the Coming of the New Deal," in *The Rise and Fall of the New Deal Order*, ed. Steve Fraser and Gary Gerstle (Princeton, 1989), 3–31.

37. In textiles at least, the mess at Lowell neatly contrasted with the busy-ness of fabric firms in flexible Philadelphia. See Scranton, *Proprietary Capitalism*, chap. 8.

38. David Tyler, *The American Clyde* (Newark, DE, 1958); Scranton, *Proprietary Capitalism*, table 8.1.

## CHAPTER 4
## THE 1876 EXPOSITION AND PHILADELPHIA MANUFACTURING

1. For reviews of the literature on industrial districts, see Udo Staber, Norbert Schaefer, and Basu Scharma, eds., *Business Networks: Prospects for Regional Development* (Berlin, 1996), chaps. 1, 8, 9.

2. See Philip Scranton, "Conceptualizing Pennsylvania's Industrializations," *Pennsylvania History* 61 (1994): 6–17.

3. At some sites, auxiliaries also worked extensively with expanding mass production corporations, notably Detroit's tool-and-die shops associated with automobile manufacturers.

4. They appear to have been more frequent in the East than elsewhere and include the collection of glove-making specialists at Gloversville, New York, the three isolated machine tool companies in Springfield, Vermont, and a handful of textile firms that dominated small towns, as Cheyney Silk did in Connecticut. Similarly, a number of cities reported one or

two billiard table makers and other niche enterprises. These are not considered in this study.

5. Walter Peterson, *An Industrial Heritage: Allis-Chalmers Corporation* (Milwaukee, WI, 1976), 25.

6. There had been an "unsuccessful" New York effort in 1853–54 to duplicate the 1851 British Crystal Palace exposition. Financial difficulties brought about the seizure of exhibits by creditors, wrecking American prospects for world's fairs for a generation (Monte Calvert, "American Technology at World's Fairs" [M.A. thesis, University of Delaware, 1962]).

7. John Kouwenhoven, *Made in America: The Arts in Modern Civilization* (Garden City, NY, 1949), 113.

8. Translated for and quoted in the *New York Evening Post*, November 16, 1876, Clippings Scrapbook, Corliss Collection, John Hay Library, Brown University.

9. See, for example, *Reports on the Philadelphia International Exhibition of 1876*, 3 vols. (London, 1877), in which American machinery and systems for repetition production receive more fulsome praise than do specialists' consumer goods. For an American view of the exhibits, see New York Tribune, *Extra No. 25: Letters about the Exhibition* (New York, 1876), and Phillip Sandhurst, *The Great Centennial Exhibition Critically Described* (Philadelphia, 1876).

10. At the 1878 Paris Exhibition, American pottery and styled fabrics were scorned, though displays of machinery, printing equipment, saddlery and harness, and one piece of wood carving in the French style earned good marks. However, one reviewer thought it "most probable that [the carving] work was produced in Paris, or at least by French workmen, and that it ought therefore really to be in the French section." See The Society of Arts, *Artisan Report on the Paris Universal Exhibition of 1878* (London, 1879), quotation from 213, and National Association of Wool Manufacturers, *Bulletin* 11 (1881): 269.

11. Frank Ransom, *The City Built on Wood* (Grand Rapids, MI, 1955), 17.

12. *Furniture Manufacturer and Artisan* 62 (1911): 94.

13. *Report of the Industrial Commission on the Relations of Capital and Labor . . .*, vol. 7 (Washington, DC, 1901), 458.

14. Ibid., 14:276, 278.

15. Calvert, "American Technology," 149, 157–62. The *Journal* noted, as did newspapers, that "thousands of tons" of material for exhibits remained on railway sidings at the formal opening in May.

16. Thomas Savery Journals, 1875–76, Accession no. 291, Archives, Hagley Museum and Library. According to Savery's November 23, 1875, entry, his firm realized a $9,200 gross profit on the Machinery Hall contract despite being bargained down from its initial bid.

17. Savery Journals, 1876–77. Yearly sales account summaries can be found in entries for December 31, 1876, and December 26, 1877.

18. Arthur Cox and Thomas Malim, *Ferracute: The History of an American Enterprise* (Bridgeton, NJ, 1985), 23–25. For the text of Smith's medal citation, see U.S. Centennial Commission, *Reports and Awards, Group XXI* (Philadelphia, 1877), 11.

19. *Providence Journal*, November 15, 1875; *Philadelphia Daily Evening Telegraph*, April 6, 1876; *Philadelphia Times*, April 11, 1876; *Scientific American Supplement*, May 6, 1876; *Philadelphia Evening Bulletin*, May 11, 1876; in Clippings Scrapbook, Corliss Collection.

20. *New York Examiner and Chronicle*, June 8, 1876, and *New York Times*, June 5, 1876, Clippings Scrapbook, Corliss Collection.

21. Charles Porter, *Engineering Reminiscences* (New York, 1908), 248–50. For technical details on Corliss engines, see Thomas Hawley, *American Steam Engines*, 2d ed. (Boston, 1914), chap. 9, and Louis Hunter, *Steam Power* (Charlottesville, VA, 1985), chap. 5.

22. Corliss's publicity coup was expensive. The firm secured a $79,000 contract for shafting, steam pipes, and the like from the commission, but its total costs for the above, plus transportation, erection, and disassembly, ran $41,000 above this. The cost of fashioning the engine itself is not evident, but it was not sold until April 1880, when George Pullman purchased it for his Chicago "Palace Car" works. "Expenditures incident to the . . . Centennial Exhibition," Box 4, Folder 14, and *Boston Daily Advertiser*, April 27, 1880, Clippings Scrapbook, Corliss Collection.

23. Philip Scranton, *Proprietary Capitalism* (New York, 1983), 304. Adult women represented another 40 percent, with youths of both sexes being the residual.

24. Promissory notes were a routine component of business in this era. They represented collateralized pledges to repay short-term commercial loans within a network of trade familiars, for whom the "first name," the borrower, was regarded as a good risk given his reputation and set of accounts collectible. Such notes bore interest and could be circulated to the lender's own creditors, if deemed sound, ultimately being presented to the issuer at the due date for redemption with interest accrued. In a personalized trade collective, these bits of paper were central to the provision of working capital.

25. Derived from sectoral data on Philadelphia in Department of the Interior, Census Office, *Report on the Manufactures of the United States at the Tenth Census* (Washington, DC, 1883), 421–24. Printing, like textiles, had a network structure that included, in addition to printers and publishers, separate bookbinders, engravers, lithographers, and stereo- and electrotypers.

26. Scranton, *Proprietary Capitalism*, 396–404.

27. Ibid., chap. 10; Philip Scranton, *Figured Tapestry* (New York, 1989), chap. 2.

28. Leslie Miller, the art school principal, testified to the Industrial Commission that it "was established to perpetuate the lessons of the Centennial Exhibition" (*Report of the Industrial Commission*, vol. 14 [Washington, DC, 1901], 218).

29. Scranton, *Proprietary Capitalism*, 405–12.

30. To clarify, fabric makers might sell on a basis of 2 percent discount with payment received within thirty days, the net charge being effective thereafter. Late payments that deducted such discounts were a constant annoyance, for though they might each be small, their accumulated totals from fifty or a hundred clients could undermine seasonal profits.

31. Inspection of issues of *American Fashion Review* and *Clothing Gazette* in the late 1880s reveals extensive advertising by Philadelphia specialists. These serials are held at the Library of Congress.

32. Unfortunately, William Leach's *Land of Desire* (New York, 1993), an acclaimed, recent study of American department stores, scarcely mentions the purchasing dynamics that underlay the provision of fashionable goods.

33. Scranton, *Figured Tapestry*, chap. 3.

34. For more on manufacturers' institution building, see Philip Scranton, "Webs of Productive Association in American Industrialization" (paper presented at the European Social Science History Conference, Leeuwenhoek, the Netherlands, May 1996).

35. Scranton, *Figured Tapestry*, tables 3.3 and 3.4; *Report . . . Tenth Census*, 421–24; and Department of the Interior, Census Office, *Report on Manufacturing Industries in the United States at the Eleventh Census*, pt. 2, *Statistics of Cities* (Washington, DC, 1895), 434–53.

36. Michael Storper and Richard Walker, *The Capitalist Imperative: Territory, Technology and Industrial Growth* (Oxford, 1989), esp. chaps. 1, 2.

37. *Report . . . Eleventh Census*, pt. 2, 434–53. By contrast, Cleveland and Pittsburgh had fewer than half as many sectors, and at the latter site basic iron and steel contributed 35 percent of all value added.

38. It might be worth here noting that Philadelphia's largest steel plant, the Midvale Company, bought and remelted pig iron and Bessemer steel to make alloys for cannon, crankshafts, and the like, rather than producing tonnage steel and iron.

39. Centennial Commission, *Reports and Awards: Group XXI*, 14–18. Pratt and Whitney of Hartford commanded two pages of comparable praise. For illustrations and further discussion of Sellers' entries, see Sandhurst, *Great Centennial*, 338–59. The recently discovered collection of nearly a thousand Sellers machinery drawings (ca. 1840–1895) at the Franklin Institute shows the persistence of general machine building at the firm into the 1890s.

40. See "Discussion" following H. L. Gantt, "Recent Progress in the Manufacture of Steel Castings," *Transactions of the American Society of Mechanical Engineers* 15 (1894): 268–69 (hereafter, *Transactions ASME*), for Taylor's account of his early days at Midvale. Taylor dated his arrival to 1877, but Daniel Nelson gives 1878 as his starting date (*Managers and Workers* [Madison, WI, 1975], 56, and *Frederick W.*

*Taylor and the Rise of Scientific Management* [Madison, WI, 1980], 29).

41. This sketch intentionally elides several complexities that beset textile specialists: variations in character of wool, cotton, or silk, inconstancies in dyeing, effects of heat and humidity on product qualities, yarns that were "ratty" or "knobby" in the loom, and so forth. Nevertheless, these were child's play compared to the complexities of specialty metalwork.

42. For a comprehensive discussion of metal molding, see Howell Harris, "Little Drops of Water, Little Grains of Sand" (unpublished paper, 1992, in author's possession), and "The Rocky Road to Mass Production: Explaining Technological Conservatism in the U.S. Foundry Industry," Research Seminar Paper no. 36, Hagley Museum and Library, 1996.

43. On the challenges of making and working iron and steel, see Thomas Misa, *A Nation of Steel* (Baltimore, 1995).

44. The Towne engineering school at the University of Pennsylvania derived from a single bequest. The donor's son had by 1870 relocated to Connecticut where he prospered at Stamford's Yale and Towne lock company. The city's other engineering school, Drexel Institute, owed its origins to financial capital.

45. John K. Brown, *Baldwin Locomotive Works* (Baltimore, 1995); "The Manufacture of Locomotives and Railroad Machinery," in Charles H. Fitch, *Report on the Manufactures of Interchangeable Mechanism*, 44–59, bound with *Report . . . Tenth Census*. See also *Industry* 1 (1888–89): 12, for more on Baldwin.

46. For an example of such refusals, see *Industry* 2 (1889–90): 99. For the significance of special jobs in "educat[ing]" skilled workers, see scientific instrument manufacturer John Gray's testimony to the Industrial Commission (*Report*, 14:212).

47. James See, "Standards," *Transactions ASME* 10 (1889): 542–75, quotations from 544, 549, and Smith's discussion, 570. James See was better known as "Chordal," under which pseudonym he contributed trenchant critiques and commentaries on machinery practice to trade journals.

48. Oberlin Smith, "Experimental Mechanics," *Transactions ASME* 2 (1881): 55–69; "Nomenclature of Machine Details," ibid., 358–69; "Inventory Valuation of Machinery Plant," ibid., 7 (1886): 433–39.

49. Charles Harrah, manager of Midvale by 1886 and later its president, averred that the state, railroads, and shipbuilders were the firm's "only clients" (*Report of the Industrial Commission*, 14:355).

50. Nelson, *Frederick W. Taylor*, 31–46. The differential piece rate created a two-tier system of payment by results. Workers achieving the daily output goal earned a much higher rate than those who failed to reach the quota. This provoked considerable resistance among machinists, but Taylor evidently fired the objectors and used the shortfall penalty to force greater production from the rest. See also Charles Wrege and Ronald Greenwood, *Frederick W. Taylor, the Father of Scientific Management: Myth and Reality* (Homewood, IL, 1991), 31–62.

51. Henry Metcalfe, "The Shop-Order System of Accounts," *Transactions ASME* 7 (1886): 440–88, quotations from 442, 481, 482. Metcalfe's comprehensive *The Cost of Manufactures* (New York, 1885) had been published the preceding year. For discussion of Metcalfe's significance in cost accounting, see S. Paul Garner, *Evolution of Cost Accounting to 1925* (University, AL, 1954), chaps. 2, 11.

52. Metcalfe dealt with indirect costs, often called "burden," by distributing general expenses to each department "according to the most probable hypothesis," then dividing the sum by the number of man-days of work executed there yearly. That figure was added to each product's departmental day-work labor charges, and the total was summed with materials to yield the item's "gross cost." When expected selling expenses and profits were tacked on, a market price that fully covered exact expenses could be derived for any reorders.

53. "Discussion" following Metcalfe, "Shop-Order System," Taylor comments at 475 and 485.

54. Ibid., 486. Emphasis in original.

55. Thurston, a Providence native, soon directed Cornell's Sibley College of Engineering and had authored the era's definitive steam engine history, first published in 1878. See Monte Calvert, *The Mechanical Engineer in America* (Baltimore, 1967), 45–57, 96–97, and Robert Thurston, *A History*

*of the Growth of the Steam Engine*, 4th ed. (New York, 1897).

56. Frederick Hutton, *A History of the American Society of Mechanical Engineers from 1880 to 1915* (New York, 1915), 4–20, 78, 92; Bruce Sinclair, *A Centennial History of ASME* (Toronto, 1980), chap. 2. For broader context, see Edwin Layton, *The Revolt of the Engineers* (Cleveland, 1971), and Calvert, *Mechanical Engineer*.

57. Sinclair, *Centennial History*, chap. 2. Philadelphia's Engineer's Club predated the ASME formation, being another product of the Centennial Exhibition. Its proceedings commenced in 1879. The numbers of trade journals exploded in the 1880s in considerable measure because of the 1879 postal act, which offered cheap, second-class rates to magazines, specifically mentioning those "devoted to literature, the sciences, arts or some special industry." See James Wood, *Magazines in the United States* (New York, 1949), 100.

58. Charles Porter, *Engineering Reminiscences* (New York, 1908), 173–298.

59. Ibid., 310.

60. Ibid., 313–24.

61. Ibid., 333–34.

62. Samuel Vauclain, *Steaming Up!* (New York, 1930); Harry Silcox, "Disston and Tacony" (unpublished paper, 1990); Philip Scranton, "Build a Firm, Start Another," *Business History* 35 (1993): 115–51; Philip Scranton and Walter Licht, *Work Sights: Industrial Philadelphia, 1890–1950* (Philadelphia, 1986).

## CHAPTER 5
## PROVIDENCE AND NEW YORK: JEWELRY, SILVERWARE, AND PRINTING

1. David Landes, *Revolution in Time* (Cambridge, MA, 1985); Michael Harrold, *American Watchmaking: A Technical History of the American Watch Industry, 1850–1930* (Columbia, PA, 1981).

2. Charles Venable, "Gorham and Tiffany" (Ph.D. diss., Boston College, 1993)—revised as *Silver in America, 1840–1940: A Century of Splendor* (New York, 1994); Oliver Carsten, "Work and the Lodge: Working Class Sociability in Meriden and New Britain, Connecticut, 1850–1940" (Ph.D. diss., University of Michigan, 1981), 78–81, 87–92; U.S. Tariff Commission, *Silverware* (Washington, DC, 1927); George Gibb, *The Whitesmiths of Taunton* (Cambridge, MA, 1946), chaps. 9–13.

3. Similar auxiliaries appeared in Newark and New York. See *New Jersey Commerce and Finance* 2 (1906): 145; *Manufacturing Jeweler* 1 (1884–85): 335, 397; 2(1885–86): 268; 14(1894): 12–16, 324 (hereafter cited as *MJ*).

4. Department of the Interior, Census Office, *Report on Manufactures of the United States at the Tenth Census* (Washington, DC, 1883), 428.

5. *Rhode Island State Census, 1885* (Providence, 1887), 599; Arthur Cox and Thomas Malim, *Ferracute: The History of an American Enterprise* (Bridgeton, NJ, 1985), 21. Chain and clasp work and attaching goods to cards for retail display and packing were tasked to women workers at this time, but polishing remained a job for skilled men into the early 1900s.

6. *MJ* 4 (1887–88): 492.

7. *MJ* 2 (1885–6): 161, 170, quoted matter from 228; 3 (1887–88): 221; 5 (1889–90): 294.

8. See Maxine Berg, *The Age of Manufactures, 1700–1820* (London, 1985); and for a useful U.K.-U.S. comparison, Kenneth Sokoloff and David Dollar, "Agricultural Seasonality and the Organization of Manufacturing during Early Industrialization: The Contrast between Britain and the United States" (unpublished paper, 1991).

9. *The Garment Manufacturer* 8 (February 1902): 19–20, (June 1902): 35–36; B. M. Selekman, Henriette Walter, and W. J. Couper, *The Clothing and Textile Industries in New York and Its Environs*, Regional Plan of New York, Monographs 7–9 (New York, 1925).

10. One could, of course, order an item for duplication through a proxy buyer, but delivery would take six weeks to two months, and only then could the copying process commence. Hasty sketches were the method of choice, but their accuracy was dubious and they could reveal nothing of the internal construction details. Copyists pretending to be retail buyers could closely examine a sample's structure but would be escorted out if they commenced measuring dimensions.

11. On textiles, see Philip Scranton, *Figured Tapestry* (New York, 1989), and for

furniture, Philip Scranton, "Manufacturing Diversity," *Technology and Culture* 35 (1994): 476–505.

12. Jewelers' low entry costs might again be noted here. By 1885, rental space in purpose-built jewelry factories was easily leased by the room with or without power. A foot press and a Brown and Sharpe jeweler's lathe would run about $200, less if purchased used, and most skilled workers would have a tool kit already. The turnover of new starts stimulated used machinery dealer start-ups. Both rental factory lofts and used textile machinery were available in Philadelphia, but textile operations took far more room and starting with fewer than ten looms or five hundred spindles had become a dubious proposition. At Grand Rapids and elsewhere, furniture production was space-demanding and shared quarters unusual. New firms thus had to find an empty mill or have a rudimentary factory built, install power, and secure machinery from makers, for active secondary markets were then rare.

13. Small firms might stretch to afford the $100 estimated cost of two two-week selling trips to New York, but keeping a full-time salesman on the road demanded about $1,500 annually for travel expenses and a like sum for commissions (*MJ* 3 [1886–87]: 174; 9 [1891]: 1363).

14. *MJ* 5 (1888–89): 193, 581, 587; 6 (1889–90): 115, 264; 7 (1890): 70.

15. *MJ* 9 (1891): 1364, 1407; 10 (1892): 283.

16. *MJ* 1 (1884–85): 173, 216; 2 (1885–86): 215; 5 (1888–89): 307, 540, 581, 587: 6 (1889–90): 115, 222; 11 (1892): 253; 25 (1899): 92; 29 (1909): 510.

17. *Report on Manufactures ... Tenth Census*, 428; Department of the Interior, Census Office, *Report on Manufacturing Industries in the United States at the Eleventh Census*, pt. 2, *Statistics of Cities* (Washington, DC, 1895), 470–77; William Lathrop, *The Brass Industry* (Mount Carmel, CT, 1926), 154–57.

18. *Report ... Eleventh Census*, pt. 2, 470–77.

19. Ibid., 474–77. Worsteds' surge had begun as a postbellum consequence of the Morrill Tariff when Massachusetts corporations sent abroad for the required wool combing and spinning machinery and a number of British proprietors emigrated to Philadelphia to slip under the duty barrier. Its continued expansion paralleled the rise in white-collar employment and the replacement of woolen staples in men's suits and overcoats and women's jackets and skirts by conservatively styled worsteds. See *The Arlington Mills* (Boston, 1891); Arthur Cole, *The American Wool Manufacture* (Cambridge, MA, 1926), vol. 2; and Paul Cherington, *The Wool Industry* (Chicago, 1916).

20. Calculated from sectoral figures in *Report ... Eleventh Census*, pt. 2, 470–77. That worsteds' growth was financed by substantial incursion of debt is indicated by the sector's interest expense in 1890 of $450,000, versus $56,000 in cottons, $60,000 in jewelry, and $68,000 in metalworking.

21. *Report ... Tenth Census*, 13; *Report ... Eleventh Census*, pt. 1, *Totals for States and Industries*, 42–43. The larger firms included Gorham and Tiffany, Meriden Britannia, Towle, Whiting, Schieble, Wallace, Knowles, and Reed and Barton. For the last, see Gibb, *Whitesmiths of Taunton*.

22. *Report ... Eleventh Census*, pt. 2, 474–77; Edward Holbrook Papers re Silversmith's Company, File 1.9, Gorham Collection, John Hay Library, Brown University.

23. Quoted from the French Ministry of Commerce's report on the silver exhibits at the Columbian Exposition, in Charles Carpenter, *Gorham Silver, 1831–1981* (New York, 1982), 206.

24. Carpenter, *Gorham*, 22–31; *MJ* 5 (1888–89): 140; 8 (1890–91): 264; 9 (1891): 306–22.

25. Hollowware includes all varieties of bowls, creamers, tea- and coffeepots, compotes, urns, pitchers, goblets, cups, and tankards, while flatware includes knives, forks, and spoons.

26. Carpenter, *Gorham*, 40–47.

27. Metal spinning is a machine technique for shaping a flat metal blank into a bowl on a specially designed lathe. For further details, see *Machinery* 16 (1909–10): 519–23, 606–9.

28. The technical advances, however, did not reduce the centrality of handwork to every later production step.

29. Carpenter, *Gorham*, 42–62.

30. "Employees of the Gorham Manufacturing Co.," ca. 1891, File I.9, Gorham Collection.

31. The first of these was for the *Maine* in 1891, the last for the cruiser *Long Beach* seventy years later (Carpenter, *Gorham*, 177–78).

32. Ibid., 66–146, esp. 90, 116; *MJ* 7 (1890): 334. A large proportion of the Gorham photographic prints and plates has been preserved in the collections at the Hay Library, Brown University.

33. The 1890 catalog offered fifty pages of tea sets and forty pages of spoon variants, for example. *MJ* 8 (1890–91): 826–28.

34. *MJ* 3 (1886–87): 336, 378, 390, 411; Carpenter, *Gorham*, 136–40, 201–3; Gorham Manufacturing Company, *Description of the Works* (Providence, 1892).

35. Gorham, *Description*, 44–45.

36. An in-house memorandum from the early 1890s notes that of the 1,377 workers employed at the huge Elmwood plant, about half were "American and educated in our own shops," roughly 40 percent were English immigrants, "many of them long residents in this country," and the rest more recent arrivals from Germany, France, and Sweden. Swedish immigrants gradually became an important skilled group also at Brown and Sharpe, numbering 642 in a workforce of 5,000 by 1915. See "Employees of the Gorham Manufacturing Co.," File I.9, Gorham Collection, and "Labor Relations, 1915–18," Series C.1, Brown and Sharpe Collection, Rhode Island Historical Society, Providence.

37. See letters from Towle and Wallace to Holbrook in File I.9, Gorham Collection, which suggest that the grouping was concerned only with sterling flatware, not holloware or plated silver.

38. Even if producers moved their per-ounce price points together as silver fluctuated, the long deflationary trend meant that old stock made from higher-cost materials might have to be sold years later at figures that left little or no profit. An 1891 balance sheet gives Gorham's materials and merchandise on hand as $1.8 million but does not break this down into raw stocks and finished goods inventories. See *The Statistical History of the United States* (Stamford, CT, 1965), 371, and File I.9, Gorham Collection.

39. There was no special return to large-scale operations in silverware, for firms like Towle and Wallace did quite well with workforces of a few hundred. Gorham's size stemmed from its multiplication of activities, its extension beyond tableware into many fields of ornamental goods, etc., and beyond silver to bronze, copper, plate, and decorated glass (Carpenter, *Gorham*, passim).

40. See Carsten, "Work and the Lodge," chap. 3.

41. Carpenter, *Gorham*; Carsten, "Work and the Lodge." Advertised brand names in silverware and Britannia were in place in the 1850s. Indeed, that competition on reputation terms was established early is indicated by the many clones of Hartford's Rogers Brothers silverplate company (C. Rogers and Bros., Rogers and Bros.), one new start going "so far as to invite a cigar dealer from Hartford, named William Rogers, to become a stockholder—so that it could use his name" (Carsten, "Work and the Lodge," 77). It should also be noted that silverware was a far smaller market than jewelry. In 1880, output was $2.2 million, plus $8.6 million in plated and Britannia ware (the latter made from alloys of nickel, zinc, and copper), versus $22 million for jewelry. By 1890, silverware values reached $5.8 million, plate and Britannia, $11.5 million; but jewelry neared $35 million (*Report . . . Tenth Census*, 11, 13; *Report . . . Eleventh Census*, pt. 1, 40–43).

42. Carpenter, *Gorham*; Carsten, "Work and the Lodge"; Tariff Commission, *Silverware*.

43. For example, up-market New York–area firms wanted all goods using gold to be marked with a karat rating to ensure quality and focus competition on style. Providence and some Newark firms resisted, for they were masters of low gold work that echoed the style of luxury items at far lower price ranges, and argued that retailers knew what they were getting for their money. The New Yorkers' complaint was, however, that consumers could not know what they were buying, that retailers sold low gold, plated, or filled goods at elevated prices, and that the discovery of this knavery damaged the reputation of all gold jewelry. It took legislation to resolve the matter, but it was erratically enforced.

44. Sean Wilentz, *Chants Democratic* (New York, 1984).

45. Patricia Malon, "The Growth of Manufacturing in Manhattan, 1860–1990"

(Ph.D. diss., Columbia University, 1981). Malon's work focuses on clothing, metal-working, and printing sectors. See also Iver Bernstein, *The New York City Draft Riots* (New York, 1990). Bernstein observes that Roach's iron plant, with 1,200 workers by the late 1840s "was at the time the largest metalworking and machine shop in the country" (168).

46. John Tebbel, *A History of Book Publishing in the United States* 4 vols. (New York, 1972–81), condensed as *Between Covers* (New York, 1987). For company histories, see Tebbel's bibliography in the latter, pp. 469–72. For a mid-twentieth-century assessment of New York's regional printing trades, see W. Eric Gustavson, "Printing and Publishing," in *Made in New York*, ed. Max Hall (Cambridge, MA, 1959), 135–239.

47. For a sparkling manifesto that identifies issues for scholarship and an excellent bibliography, see William Pretzer, "The Quest for Autonomy and Discipline: Labor and Technology in the Book Trades," in *Needs and Opportunities in the History of the Book*, ed. David Hall and John Hench (Worcester, MA, 1987), 13–59.

48. Ava Baron, "An 'Other' Side of Gender Antagonism at Work: Men, Boys, and the Demasculinization of Printers' Work, 1830–1920," in *Work Engendered*, ed. idem (Ithaca, NY, 1991), 47–69; and "Questions of Gender: Demasculinization and Deskilling in the U.S. Printing Trade, 1830–1915," *Gender and History* 1 (1989): 178–99. The classic labor study is Seymour Lipset, Martin Trow, and James Coleman, *Union Democracy* (New York, 1956), which treats the typographers union. More recently, Robert Max Jackson's *The Formation of Craft Labor Markets* (New York, 1984), gives considerable attention to printing workers in chaps. 11 and 12.

49. See Victor Strauss, *The Printing Industry* (New York, 1967), 781–82, for a representative bibliography, and the biannual journal *Printing History* for occasional articles that move beyond antiquarianism. One interesting focus of such work, however, has been the design of books and typefaces. See Stanley Morrison, *Four Centuries of Fine Printing* (New York, 1949), and Daniel Updike, et al., *Updike: American Printer* (New York, 1947).

50. The sizable section of the trade that unionized (120,000 strong by 1910) affiliated with the AFL and remained relatively uncombative (despite bursts of strike activity in the 1880s, 1906, and 1919) and focused on efforts at shared authority. See Emily C. Brown, *Joint Industrial Control in the Book and Job Printing Industry*, Bureau of Labor Statistics Bulletin no. 481 (Washington, DC, 1928).

51. Though in rural areas they did persist at least through the end of the century. See, for example, Irvin Cobb, *Stickfuls: Compositions of a Newspaper Minion* (New York, 1923), chaps. 1–3. A recent study of the Philadelphia trades suggests that the division between printing and publishing may have commenced by the 1830s. See Rosalind Remer, *Printers and Men of Capital: Philadelphia Book Publishers in the New Republic* (Philadelphia, 1996).

52. Tebbel, *Book Publishing*, 4:454.

53. James Wood, *Magazines in the United States* (New York, 1949), 75–79, 90–98; Tebbel, *Between Covers*, 161–64.

54. Frank Comparato, *Books for the Millions* (Harrisburg, PA, 1971), 130. Among the exceptions were Boston's Little, Brown and Houghton Mifflin, along with a number of religious and foreign-language firms.

55. "The Last Yiddish Linotype in America," *Wall Street Journal*, March 5, 1992, A-12. Urban dailies have received the most attention, but in 1890 they represented a tenth of New York City's periodical press (50 of 495). Rough estimates based on census figures suggest that their revenues in 1890 were about $15 million, or 30 percent of the city's newspaper and periodicals output, a sizable but not overwhelming proportion.

56. It should be noted that outside the major cities, newspapers continued to mix job work with publishing the paper. As late as 1931, the United States supported 2,400 daily newspapers and nearly 12,000 weeklies, only 15 of which had circulations above 300,000 (David Gustavson, *The Importance of the Printing Industry* [Pittsburgh, 1931], 6, 17). It also appears that in the first wave of New York's newspaper expansion, ca. 1820–50, editors and publishers contracted with independent print shop owners before securing their own presses (Charlotte Morgan, *The Origin and History of the New York Employing Printers' Association* [New York, 1930], chap. 3). Newspaper mergers developed

only after 1950 (Elizabeth Neiva, "Chain Building: The Consolidation of the American Newspaper Industry, 1953–1980," *Business History Review* 70 [1996]: 1–42).

57. Irene Tichenour, "Theodore Low deVinne (1828–1914): Dean of American Printers" (Ph.D. diss., Columbia University, 1982), 9–10. Bernstein suggests that these characteristics were common in other sectors of New York City manufacturing (*Draft Riots*, 164–65).

58. Quoted in Tichenour, "DeVinne," 81–82. For a summary of Tichenour's dissertation, see her "Theodore deVinne: Unlikely Leader," *Printing History* 11, no. 1 (1989): 17–26.

59. Malon, "Growth," 153; Tichenour, "DeVinne," 13; and A. F. Hinrichs, *The Printing Industry in New York and Its Environs*, Regional Plan of New York, Monograph 6 (New York, 1924), 18–19.

60. Tichenour, "DeVinne," 13.

61. Tebbel, *Between Covers*, 107, 146–49; Raymond Shove, *Cheap Book Production in the United States, 1870 to 1891* (Urbana, IL, 1937).

62. By 1890, advertising revenue matched income from sales of papers. See *Report . . . Eleventh Census*, pt. 3, 650, 656.

63. Alfred M. Lee, *The Daily Newspaper in America* (New York, 1937), 263–65. Into the 1890s New York papers sold roughly three-quarters of their product through middlemen and struggled unsuccessfully to distribute directly. Weeklies and journals relied far more on subscriptions than on impulse sales. Still, the importance of advertising to their revenues was for the most part equally immense, for otherwise their one- or two-dollar annual subscription charges would have been ruinous.

64. See Baron works cited in n. 48.

65. *Report . . . Eleventh Census*, pt. 3, 658.

66. For details on the technics of these processes, see Michael Winship, "Printing from Plates in the Nineteenth Century United States," *Printing History* 5, no. 2(1983): 15–26, and Victor Strauss, ed., *The Lithographers Manual* (New York, 1958), chap. 1.

67. *Report . . . Eleventh Census*, pt. 3, 654. The reports detailed not only materials and labor costs, but also interest payments, rent, power expense, and, as "sundries," charges for all "necessary and important expenditures," including "commissions on advertising, telegrams, associated press privileges, postage, and amounts paid for composition and presswork when done by contract."

68. National comparisons derived from subsectoral data presented at *Report . . . Eleventh Census*, pt. 1, 36–45.

69. These issues led to the formation in the 1890s among leading firms of a New York collusive group (the Inner Circle) (Morgan, *New York Employing Printers*, 89–93). For the complexities of competition in quality and service, see *Inland Printer* 50 (1912–13): 692–93; 51 (1913): 279; and 52 (1913–14): 58–60, 215–18, 377–81, 537–39, and U.S. Senate, *Book Paper Industry*, 65th Cong., 1st sess., S. Doc. 79 (Washington, DC, 1917), 18–20.

70. Lee, *Daily Newspaper*, 101–3; *Report . . . Eleventh Census*, pt. 3, 649; Judith McGaw, *Most Wonderful Machine* (Princeton, NJ, 1987).

71. Lee, *Daily Newspaper*, 118–22; Comparato, *Books for the Millions*; Frank Comparato, *Chronicles of Genius and Folly: R. Hoe and Company and the Printing Press* (Culver City, CA, 1979); Strauss, *Lithographers Manual*, chap. 1; Winship, "Printing from Plates."

72. Warren Devine, Jr., "The Printing Industry as a Leader in Electrification, 1883–1930," *Printing History* 7, no. 2 (1985): 27–36.

73. The Pearl Street Station, which supplied power for lighting only, commenced operating in 1882. See Francis Jehl, *Menlo Park Reminiscences*, vol. 3 (Dearborn, MI, 1941), for an extensive account of its planning and activation. Bright illumination obviously had real value for printers.

74. George Stevens, *New York Typographical Union No. 6* (Albany, NY, 1913), 41–202. The "em" was a flexible measure of work accomplished by typesetters. Much like weavers' scales calibrated to varied yarn sizes, it attempted to register and equilibrate payment for the efforts needed to set tiny type and larger faces that filled pages faster. As Walter Conkey explained, "The smaller the size of type the larger the number of ems there are in the page" (W. B. Conkey, *What a Businessman Should Know about Printing and Book Making* [Hammond, IN, 1928], 18). Setting type would go swiftly or slower depending on the point size of the letters used,

but payment by the em equalized the compensation of those condemned to doing minuscule work and others favored with larger, page-filling faces. According to Lee (*Daily Newspaper*, 122), an "average" compositor would set a thousand ems of type an hour, making typical weekly earnings at $0.32/thousand ems of about $19. The em linked earnings to output, replacing hourly or day's pay in shops with diversified printing tasks. Once Linotype and Monotype machines replaced hand setting for "straight" work, all in the same face and point, ems were relevant more to job work than to newspaper or book text setting.

75. Stevens, *New York Typographical*, 234.

76. Ran Mendel, "Cooperative Unionism and the Development of Job Control in New York Printing Trades 1886–1898," *Labor History* 32 (1991): 354–75, and Jackson, *Formation*.

77. Leona Powell, *The History of the United Typothetae of America* (Chicago, 1926), 9–19. Mendel indicates that a compromise settlement was achieved, which established the "first comprehensive schedule of wages and work rules" but not shorter hours or the closed shop ("Cooperative Unionism," 373). These would come unevenly in the 1898–1914 era.

78. Mendel, "Cooperative Unionism," 371. That shop foremen were regularly union members, and often officers, reinforced the trade's relatively placid labor relations, Mendel suggests.

79. These figures exclude principals and office staff, and are derived from *Report . . . Eleventh Census*, pt. 1, 36–43, and pt. 3, 130–45.

80. It may well be that employers' repeated complaints about feeble profits stemmed in part from their factoring in anticipated equipment and plant expenditures and in part from their general ignorance of cost and capital accounting, an issue that became central to trade debates once economic depression arrived. Or they could have been shamming.

81. Prominent among these activists were Sidney Hillman of the Amalgamated Clothing Workers and Emil Rieve of the Hosiery Workers. See Steven Fraser, *Labor Will Rule* (New York, 1991), and Scranton, *Figured Tapestry*, chaps. 6, 7.

## CHAPTER 6
## MIDWESTERN SPECIALISTS: CINCINNATI TOOLS AND GRAND RAPIDS FURNITURE

1. Brian Page and Richard Walker, "From Settlement to Fordism: The Agro-Industrial Revolution in the American Midwest," *Economic Geography* 67 (1991): 281–315.

2. See Philip Scranton, "Multiple Industrializations: Patterns of Manufacturing Deployment in the American Midwest," (paper presented at the Business History Conference, Glasgow, 1997), tables 1–12.

3. Leading sectors contributed 1 percent or more of urban production in the relevant census year. Cincinnati's fourteen leaders cumulatively represented over 70 percent of all manufacturing by value, and 68 percent of value added.

4. Derived from sectoral rosters given in Department of the Interior, Census Bureau, *Report on the Manufactures of the United States at the Tenth Census* (Washington, DC, 1883), 421–24, using the same 1 percent criterion for Philadelphia.

5. George Wing, "The History of the Cincinnati Machine Tool Industry" (D.B.A. diss., Indiana University, 1964), chaps. 2, 3; Steven Ross, *Workers on the Edge* (New York, 1985); Writers Program, WPA-Ohio, *Cincinnati: A Guide to the Queen City and Its Neighbors* (Cincinnati, 1943).

6. Wing, "Cincinnati Machine Tool Industry," 33–52.

7. Over $9 million of liquor and beer production values represented the addition of federal alcohol taxes. The wages differential reflects in large part the substantial involvement of female workers in apparel making and the almost exclusively male employment in the metal trades.

8. John Steptoe, George Gray, Frederick Geier, R. K. LeBlond, Fred Holtz, and William Lodge (in Wing, "Cincinnati Machine Tool Industry," 62–85), as well as Henry Bickford, William Gang, William Osterlein, William Barker, T. P. Egan, and Fred Dietz (in Joseph Roe, *English and American Tool Builders*, [New Haven, CT, 1916], 268–73).

9. Wing, "Cincinnati Machine Tool Industry," 62–85, quotation from 72; Roe, *Tool Builders*, 268–69.

10. "The Machine Tool Progress of Cincinnati during the Past Twenty-Five Years," *American Machinist* 26 (1902): 1622.

11. Roe, *Tool Builders*, 269–70. Clearly, the management of competition was one of the most critical and farseeing elements of the Lodge-Davis sales center. The company, according to a 1907 article cited by Wing, handled the entire output of five tool makers and secured "large orders" for a more sizable group, clearly accounting for the majority of the twenty or so tool-building companies active in the early 1890s (Wing, "Cincinnati Machine Tool Industry," 82–83, 108).

12. William Lodge, "A Brief History of the Cincinnati Machine Tool Industry," *Cincinnati Magazine* 1 (November 1909): 8–9. On the significance of trust in economic relations, see Mark Granovetter, "Problems of Explanation in Economic Sociology," in *Networks and Organizations*, ed. N. Norhria and R. Eccles (Boston, 1985), 25–56; Charles Sabel, "Studied Trust," in *Industrial Districts and Local Economic Regeneration*, ed. F. Pyke and W. Sengenberger (Geneva, 1992), 215–50; and L. Zucker, "Production of Trust," *Research in Organizational Behavior* 8 (1986): 53–111.

13. Roe, *Tool Builders*, 271; Wing, "Cincinnati Machine Tool Industry," 83, 86–87. These charts, documenting the multiplication of firms over time, are the visual inverses of postmerger genealogies showing the amalgamation of once-independent companies into corporate giants. For a companion New England machine tool "family" genealogy, see *American Machinist* 59 (1923): 1–4.

14. Wing, "Cincinnati Machine Tool Industry," 83–84; Brown and Sharpe Papers, Parts Production Ledger, Rhode Island Historical Society, Providence.

15. An early metalworking proprietor, Miles Greenwood, was a major sponsor of the institute and its president, 1847–54. Steptoe and others emerged from his shops, but it was not until after 1900 that Worcester-like curricular development was attempted. The expositions were the brainchild of a local paint manufacturer. Metal tradesmen did exhibit, but by the later 1880s they valued local industrial boosterism less than Lodge's maneuvers toward penetrating national markets, and they made no determined effort to rescue or revive the expositions after their 1888 termination. See Geoffrey Giglierano and Deborah Overman, *The Bicentennial Guide to Greater Cincinnati* (Cincinnati, 1988), 66, 221–22, and Philip Speiss, "The Cincinnati Industrial Expositions, 1870–1888" (M.A. thesis, University of Delaware, 1970).

16. Wing's estimates of output growth from 1880 to the early 1890s show an increase in sales from $250,000 to ca. $1.8 million ("Cincinnati Machine Tool Industry," 108). In real 1890 dollars, this would represent about an eightfold increase in a decade, given the deflation of wholesale prices during the 1880s. See Department of Commerce, *Historical Statistics of the United States* (Washington, DC, 1976), Series E-1 and E-13.

17. Frank Ransom, *The City Built on Wood* (Grand Rapids, MI, 1955). Though the significance of smaller cities in industrialization has been often slighted, geographers have repeatedly urged closer attention to their roles within larger city-systems as a conceptual step beyond the individual urban case studies favored by historians. See Page and Walker, "From Settlement to Fordism," 302–4, and works cited therein by Allan Pred, David Meyer, and Edward Muller.

18. The tables include only those sectors contributing 1 percent or more to total product value. Woodworking's proportion of employment and value added would rise above 50 percent were minor trades like coffin making, wood carving, barrels, pianos, et al., included. Most other leading sectors fed local consumption (brewing, printing, flour and baking, confectionery, clothing, and tobacco). Transforming wood was the center of interregional "exports."

19. See J. E. Land, *Historical and Descriptive Review of the Industries of Grand Rapids, 1882* (Grand Rapids, MI, 1882), for firm profiles.

20. In value-added terms, leading bulk versus batch production sectors' shares of city industry totals were: at Cleveland, 33 versus 27 percent in 1880; at Cincinnati, 31 versus 27 percent; and in Chicago, 35 versus 21 percent; but in Grand Rapids 17 versus 43 percent in 1880 and 14 versus 51 percent in 1890. See Scranton, "Multiple Industrializations," table 13.

21. Ibid., tables 5 and 8; Sharon Darling, *Chicago Furniture: Art, Craft and Industry, 1833–1983* (New York, 1984), 45.

22. Woodworking technology advanced dramatically in the second half of the nineteenth century, a process that John Richards' *Treatise on the Construction and Operation of Wood-Working Machines* (London, 1872) summarized in midstream. Yet as Michael Ettema has explained, "the greatest changes came in those operations which were already inexpensive," i.e., planing, shaping, rough carving, sanding. Further, "the degree to which machinery was capable of reducing labor costs in furniture manufacturing was inversely proportional to the total cost of the product" ("Technological Innovation and Design Economics in American Furniture Manufacture of the Nineteenth Century" [M.A. thesis, University of Delaware, 1981], 12, 27).

23. It must be noted that the number of firms doubled, 1889–90, but this in no way alters the huge increase in sunk capital. At Chicago, capital trebled, 1880–90, while workforces in furniture rose only 60 percent, arguably a better ratio than at Grand Rapids. See Scranton, "Multiple Industrializations," tables 7 and 8.

24. Ransom, *City*, 31. For example, a five-piece bedroom suite might be offered in walnut, oak, or mahogany, flat or mirror varnished, a total of thirty piece-options, and in three grades of ornamentation each at differing prices, yielding ninety options. "Case goods" refers to wooden furniture for bedrooms, dining rooms, parlors, halls, and libraries (i.e., dressers, sideboards, china cabinets, dining tables), as distinct from chairs, kitchen furniture, or upholstered products.

25. Traveling salesmen were not displaced by the expositions, however. After courting visiting buyers, they ventured out after each show, armed with information about the best-selling styles that helped them chase supplemental orders from stay-at-homes. This process makes a recent observation by Charles Sabel apposite: "Economic historians usually know, even if economists do not, that markets do not arise spontaneously" ("Comment," on J. B. de Long, "Did Morgan's Men Add Value?" in *Inside the Business Enterprise: Historical Perspectives on the Use of Information*, ed. Peter Temin [Chicago, 1991], 234).

26. Ransom, *City*, 22–23, 47.

27. They may well have agreed with workers' claim that as much production could be achieved in the shorter day as in the longer, but could not carry the point with their colleagues.

28. Ransom, *City*, 41–43, 47–48. The NFMA soon collapsed, for it was "composed of manufacturers in all branches of the industry," makers of parlor, kitchen, rattan, office, and school furniture, as well as Grand Rapids' specialty, case goods for households' hallways, libraries, dining rooms, and bedrooms. These divisions found they "had practically no interests in common," and formed national subsectoral associations that Arthur White later judged as, "without an exception, successful" (*Furniture Manufacturer and Artisan* 63 [1911]: 636).

29. Dwight Goss, *History of Grand Rapids and Its Industries* (Chicago, 1906), 1054–55. The patented caster had an inserted metal sleeve into which the caster's vertical post fitted.

30. Jeffrey Kleiman, "The Great Strike: Religion, Labor, and Reform in Grand Rapids, Michigan, 1890–1916" (Ph.D. diss., Michigan State University, 1985), chap. 1.

31. Ransom, *City*, 38; Goss, *History*, 1043. Goss gave the Widdicomb capital figure as of its 1873 incorporation, whereas Ransom, who had access to now-unavailable firm records, provided the percentage return information but did not note whether stated capital had been increased during the 1880s.

32. Essential delays for glue and varnish drying made rushing the work a disastrous tactic, and fitting, finishing, and polishing demanded careful attention if the solidity and quality of the goods was to be assured.

33. On the basis of average gain in output, sectors producing $10 million in 1890 products would be yielding $22–23 million in 1909, which the $25 million cutoff figure closely approximates.

34. All industrial sectors enjoyed expanding demand in this period, and a number of batch and bulk trades had serious tariff protection. To assert that larger markets or tariff walls insulated batch firms' inefficient practices from the discipline of a price-governed market is to assume just that abstract homogeneity of economic relations that this study ques-

tions. Such firms prospered by realizing that differentiated demand entailed differing constructions of efficiency, by recognizing that there were behaviorally relevant boundaries to any successful technique in production or marketing.

35. Stephen Meyer, "Technology and the Workplace: Skilled and Production Workers at Allis-Chalmers, 1900–1941," *Technology and Culture* 29 (1988): 839–64.

36. Gorham grew by differentiating its capacity *extensively* into multiple markets—plated ware along one vector and statuary, ecclesiastical goods, and architectural ornament on another—as did piano builders who created multiple lines of uprights and parlor and concert grands, variously finished and ornamented, echoing Ettema's furniture makers. See Charles Carpenter, *Gorham Silver, 1831–1981* (New York, 1982); Craig Roell, *The Piano in America* (Chapel Hill, NC, 1989), and Ettema, "Technological Innovation."

37. Some anchors, like Baldwin Locomotive and Gorham, strove to be utterly self-reliant, making virtually all components in-house. Others, like shipyards and heavy machine works, relied on outside companies to execute subcontracts for ship plate, brass work, castings, and other components.

38. Jay Galbraith, *Designing Complex Organizations* (Reading, MA, 1973); Paul Milgrom and John Roberts, "The Economics of Modern Manufacturing: Technology, Strategy, and Organization," *American Economic Review* 80 (1990): 511–28.

39. Naomi Lamoreaux, *The Great Merger Movement in American Business, 1895–1904* (New York, 1985).

40. The Gorham papers show active collaboration among silverware firms in the 1880s, and Savery's journals document bid jockeying among shipbuilders and papermaking machinery builders.

41. In this period such activities were most prominent in metalworking, the ASME being the key institution, and less prevalent stepwise in printing, furniture, and styled textiles. On cost-and-profit in shipbuilding, see testimony of Charles Cramp, *Report of the Industrial Commission* (Washington, DC, 1901), 14:406.

42. Mushroom starts in job printing had a somewhat different character, for they stole clients from established firms by quoting very low estimates, then often failed to meet quality expectations and/or make money and went broke, making room for replacement upstarts. The Typothetae quite early on undertook to disseminate information on sensible estimating, as did trade journals like *Inland Printer*. These beginners usually handled only low-end work (tags, labels, simple broadsides, and calendars), for they had not the machinery for large-run or complex printing.

43. This was the original pattern at Baltimore and midwestern cities, though all had a custom sector that decayed. In Philadelphia and Rochester, New York, conservatively styled "semistaples" appeared as a compromise for midmarket men's and women's wear. Both formats could make use of specialized sewing machinery, large lot sizes, and factory supervision to achieve efficiencies. See Philip Scranton, "Apparel Arts and the U.S. Men's Clothing Industry," in *Apparel Arts: Fashion Is the News*, ed. Gianinni Malossi (Milan, 1989), and idem, "Market Structure and Firm Size in the Apparel Trades, Philadelphia, 1890–1930," *Textile History* 25 (1994): 243–73.

44. This problem also faced New York job printers doing contract work for large corporations in the new century.

45. One of the best examples of this comes from the several silk dyeworks of Paterson, New Jersey, which rapidly grew far larger in the late nineteenth century than many of the local throwsters and weaving mills, thereby establishing local market power that discouraged new entrants. See Philip Scranton, ed., *Silk City* (Newark, 1985).

46. These were local schedules and customs, however, and trade journals published occasional articles documenting widely varying rates in various cities.

47. Batch production's expansion was directly derived from market development, technical change, and related forces, but auxiliaries were second-order beneficiaries of this dynamic.

48. Jane Jacobs, *The Economy of Cities* (New York, 1969), 85. For greater detail, see Barbara Tilson, ed., *Made in Birmingham: Design and Industry, 1889–1989* (Birmingham, UK, 1989).

49. Jacobs, *Economy*, 87, 89.

50. Jane Jacobs, *Cities and the Wealth of Nations* (New York, 1984), 40–41.

51. Eventually, leading Grand Rapids firms shifted from wood to metal furniture, from household to office provision, and thereby devised a new means to profitability in the twentieth century.

## CHAPTER 7
## CHICAGO AND GRAND RAPIDS: PALACE CARS AND FURNITURE

1. For fair visitors as chiefly middle-class and for Pullman, see James Gilbert, *Perfect Cities: Chicago's Utopias of 1893* (Chicago, 1991), 21–22, 60–61, 131–68. On the stockyards, see Siegfried Gideon, *Mechanization Takes Command* (New York, 1948), 213–46, and Louise Wade, *Chicago's Pride* (Urbana, IL, 1987). For a regionalized view of the city's course, see William Cronon, *Nature's Metropolis: Chicago and the Great West* (New York, 1991).

2. Gilbert, *Perfect Cities*, 16, 57; Cronon, *Nature's Metropolis*, 137–47, 234–35, 335–36. Cronon shows how the refrigerated car not only transformed the dressed beef trade but also created a huge ice-gathering and shipping industry in Wisconsin into the 1890s.

3. Sharon Darling, *Chicago Furniture: Art, Craft and Industry, 1833–1983* (New York, 1984), 26–30. On the lumber trade, see Cronon, *Nature's Metropolis*, chap. 4.

4. Darling, *Furniture*, 34–40, 46–48; Jane Jacobs, *The Economy of Cities* (New York, 1969), 153–54.

5. Gilbert, *Perfect Cities*, 16; Department of the Interior, Census Office, *Report on the Manufactures of the United States at the Tenth Census* (Washington, DC, 1883), 391–93; Department of the Interior, Census Office, *Report on the Manufacturing Industries of the United States at the Eleventh Census*, pt. 2, *Statistics of Cities* (Washington, DC, 1895), 130–45; U.S. Department of Commerce, Bureau of the Census, *Historical Statistics of the United States* (Washington, DC, 1957), ser. E-1, "All Commodities." Farm products and food prices fell less rapidly than the overall average, whereas textiles, fuels, home furnishings, and metal goods prices dropped as much as 30 percent in the 1880s (ser. E-2, 3, 5, 6, 7, 10). In nominal dollars average industrial worker income moved from $434 to $539 annually. In real terms, with added purchasing power taken into account, the latter figure was $657, a 51 percent increase.

6. If, to take account of these shifts among the top four bulk and batch sectors contributing 1 percent or more to citywide industrial output, we review their reports for 1880 and 1890, the bulk group's share moves from 51 to 50 percent, and the specialists from 10 to 18 percent (calculated from census tables cited in n. 5).

7. To clarify, value added in manufacture represents the difference between product values and the costs of input materials for production. It can be viewed as a pool of funds from which labor and other expenses (fuel, insurance, et al.) must be paid, leaving a residual gross profit that may be applied to loans, banked as company reserves, sent off as dividends to shareholders, distributed as bonuses or profit sharing, or pocketed by proprietors. Bulk sectors, despite their high product value totals, added less value in production than did specialists. In 1890 Chicago, 21 percent of product value represented value added for the four leading bulk sectors, versus 58 percent for the top specialty sectors. As 1890 census tables provide data on most expense categories, rough estimates of profitability can be made—for the bulk group about 10 percent versus 24 percent in batch sectors. There is, of course, no guarantee that owners actually reaped net profits at these levels, but the difference remains suggestive. The raw Chicago figures for value added and workforce in 1880 are: bulk, $14.3 million and 12,530; specialty, $11.8 and 14,969. For 1890: bulk, $56.7 million and 38,077; specialty, $49.3 and 35,907.

8. At the 1 percent output share level, six specialist sectors accounted for 51,000 jobs, 27 percent of value added, and 18 percent of product value in 1890, versus 43,000 jobs, 31 percent of value added, and 50 percent of output for nine bulk sectors in that year.

9. Gilbert, *Perfect Cities*, 136; Darling, *Furniture*, 135.

10. Stanley Buder, *Pullman* (New York, 1967), 11–12, 147–201; Jack Santino, *Miles*

*of Smiles, Years of Struggle* (Urbana, IL, 1991), 7–8.

11. Almont Lindsey, *The Pullman Strike* (Chicago, 1942), 24; Buder, *Pullman*, 12–17. Lindsey indicates that these early cars cost $24,000 each to construct (22).

12. Buder, *Pullman*, 17–24, 27, 59, 134–35, 223; Lindsey, *Pullman Strike*, 24; Gilbert, *Perfect Cities*, 150. Buder notes that three-quarters of the workforce were skilled mechanics (78). If so, they averaged $735 in 1893 and laborers $245. Chicago annexed the town and Hyde Park Township, of which it was a part, in 1889.

13. Lindsey, *Pullman Strike*, 26–27, 93; Buder, *Pullman*, 134, 138–39, 148–51. Pullman workers had at least two other fundamental grievances that fed into the strike dynamic: rents in the company town averaged 20 percent or more above those in adjacent neighborhoods for comparable homes, and Pullman permitted no resident to buy a house in the town, a contrast with Disston's approach. By 1893, Lindsey indicates, 17 percent of Pullman workers did own homes, but none in the planned town (91).

14. There were no evictions during the strike, but probably a thousand workers left Pullman's shops permanently, some blacklisted. The press widely attacked the founder for his "pigheadedness," and Chicago society reformers who had once admired his planned industrial town vilified him. See Gilbert, *Perfect Cities*, 163; Buder, *Pullman*, 196–98; and *Engineering Magazine* 11 (1896): 63.

15. School trustees delayed more than a decade in acquiring a site, "allow[ing] the bequest to grow." The institution opened to receive its first 106 students only in 1915. It operated until 1950 (Buder, *Pullman*, 211, 223).

16. The move to steel displaced a host of woodworkers. Pullman undertook retraining efforts, but they were unsuccessful (ibid., 219).

17. Ibid., 210–23. Homes sold at from $1,400 to $6,000, one-sixth down, the rest in monthly payments, prices being set at a hundred times then-current monthly rents. Robert Lincoln, the murdered president's son, a Pullman board member since 1880, was a prominent Chicago lawyer and George Pullman's closest adviser in 1893. Company profits reached $8 million in 1901 and $10 million in 1904 (T. J. Zimmerman, "One Man Power vs. System," *System* 7 [1905]: 242–52).

18. Buder, *Pullman*, 217. Interstate Commerce Commission authority to regulate sleeping car rates was confirmed in federal courts in 1910, after being written into the 1906 Hepburn Act.

19. Crane was impressively diversified in plumbing-related goods. Its product lines in 1905 included "stationary, marine, and locomotive . . . safety valves, drainage fittings, extra heavy brass and iron valves and fittings, hydraulic [and] Ferrosteel flanged fittings and valves, ammonia fittings, steam traps, steam and oil separators, malleable and Ferrosteel companion flanges, electrically and hydraulically operated and steam actuated valves and . . . flat band fittings" ("Fiftieth Anniversary of the Crane Co.," *Machinery* 11 [1904–5]: 620). For Richard Crane's personalized management style, see *Factory* 2 (1908–9): 93, and for his critique of contemporary technical schools, *Wood Craft* 6 (1907): 168. The shipbuilders focused on lake vessels up to 270′, many double-bottomed; the Chicago yard was one firm in a Great Lakes merger of 1899 (*Engineering Magazine* 19 [1900]: 674–81). See also W. J. Chalmers' testimony in *Report of the Industrial Commission*, vol. 8, *Chicago Labor Disputes of 1900* (Washington, DC, 1901), 7–19.

20. For lumber, see Cronon, *Nature's Metropolis*, chap. 4, and for packers' by-products, Darling, *Furniture*, 46. Extremely odd novelty furniture for the western trade also incorporated steer horns in this era.

21. *The Wood-Worker* 10 (August 1891): 26 (hereafter cited as *TWW*); Darling, *Furniture*, 45. One woodworking employment estimate accounted for about 15 percent of industrial jobs in the city at the time, but its size (28,000) suggests that it included all woodworkers, those in millwork, cooperage, and box making, as well as furniture operatives. The 1890 census gave furniture employment as 7,500 or 4 percent of Chicago's industrial workers, and the 1900 tally counted 8,600. See *Eleventh Census* citation from n. 5 and U.S. Census Office, *Census Reports: Twelfth Census of the United States—Manufactures*, pt. 2, *States and Territories* (Washington, DC, 1902), 182–83.

22. Darling, *Furniture*, 53, 55–57; *TWW* 15 (June 1896): 28. Skilled woodworkers seem generally to have earned less than their brethren in the metal trades, but Chicago rates were the highest in the Midwest, 10–15 percent above those in Grand Rapids or Milwaukee.

23. Darling, *Furniture*, 51, 54, 58, 63, 73. Because of the diversity of products, the minima approach was appealing. The Chamber Suite Association, for example, "made a minimum price" for varieties of case work. "A dresser 20 inches deep and 40 inches wide, with a 20 × 30 [back] plate, would be so much, in elm, ash, oak, bird's-eye [maple], birch and mahogany. A firm could make the price as much more as it pleased by adding work, carving, etc." The complexity of such schedules hampered their effectiveness, and the CSA failed by 1902. See *TWW* 21 (August 1902): 31.

24. *TWW* 14 (August 1895): 28–29; Darling, *Furniture*, 59–62. In 1900, four monthly and three weekly furniture journals were printed in Chicago, as were hundreds of catalogs filled with engraved or halftone plates. Thus was the furniture trade a stimulus to the city's printing and publishing sector.

25. Darling, *Furniture*, 48, 66–68, 113, 115.

26. In addition, a substantial woodworking machinery sales, service, and repair sector intersected with the furniture trade's expansion, including the Machinery Exchange on North Canal Street, which rebuilt used machinery and offered several lines of makers' new machine samples. Based on trade journal ads, woodworking machinery centers were at Gardner, Massachusetts, Cincinnati, Grand Rapids, and Indianapolis, though a handful of builders did locate in Chicago (*TWW* 21 (July 1902): 40).

27. Darling, *Furniture*, 124–29.

28. *Furniture Manufacturer and Artisan* 34 (July 1926): 61–62; 38 (October 1929): 41–42 (hereafter cited as *FMA*).

29. In the 1870s and 1880s, progressive shops installed powered planers, band saws, molders, drills (increasingly with multiple spindles), dowel makers, and sanders, but gluing, assembly, varnishing, ornamentation, and polishing remained largely handwork. The necessary times for parts assembly, for the drying of glues and varnish, and for repairs, touch-ups, and packing constituted structural obstacles to any dramatic improvement of throughput well into the new century. For fuller details, see Michael Ettema, "Technological Innovation and Design Economics in American Furniture Manufacture of the Nineteenth Century" (M.A. thesis, University of Delaware, 1981).

30. *TWW* 21 (June 1902): 29; Darling, *Furniture*, 49–50, 72. Karpen bought the patents to A. Freschl's tufter and sold "shop rights" to other firms adopting it. It authorized the Union Embossing Machine Co. of Indianapolis to build its multiple-spindle carvers. For rotary veneer work, see *TWW* 20 (July 1901): 34–35, and for veneer practice, the monthly series commencing in *Wood Craft* 6 (1906): 17 (hereafter cited as *WC*). Veneer output in the United States increased 50 percent 1907–10, using 478 million board feet of lumber in 1910 (*WC* 16 [1911–12]: 138).

31. Darling, *Furniture*, 50. Unlike printers, furniture plants seem to have been quite slow in adopting electric drive. The first extended descriptive series of articles I have encountered starts at *WC* 9 (1908): 1–3, continuing into 1909. As late as 1925, only three woodworking firms in Chicago were counted among the 272 companies using more than a half-million kWh annually (Harold Platt, *The Electric City* [Chicago, 1991], 220).

32. Darling, *Furniture*, 51.

33. For early pieces, see *TWW* 11 (August 1892): 28; 12 (January 1894): 26; and 16 (October 1897): 31. An extended debate opened in 1904 with the series "Figuring Cost in a Furniture Factory," *TWW* 23 (March 1904): 32–33—other sections in subsequent months. See also *Factory* 1 (1907–8): 25–27, 48; 4 (1910–11): 22–23.

34. *FMA* 1 (1911): 45.

35. *TWW* 20 (January 1902): 31.

36. *FMA* 1 (1911): 47.

37. *Factory* 4 (1910–11): 261. Emphasis in original.

38. The first published accounting guide for the trade is Frank Timken, *Accounting in the Furniture and Woodworking Industries* (Chicago, 1915). Timken assumes that "the product of manufacturers in [these] industries . . . is, as a rule, either entirely composed of staples, . . . or the proportion of specialties is so negligible a quantity that, in the formulation of a sys-

tem of factory accounting, we cannot afford to consider it a really important factor" (38). His text is thus more ideological than practical or helpful, a harbinger of things to come.

39. One other difficulty challenged local firms. About 1910, the major mail-order houses in Chicago began acquiring or erecting furniture factories to serve their clientele, integrating backward into production in the cheaper lines that had provided fill-in orders between styles seasons for area firms (Darling, *Furniture,* 63). No serious damage seems to have eventuated from this shift by Sears, Ward, and Hartman.

40. For 1899 data, see *Twelfth Census* figures cited in n. 21. For 1909 totals, see Department of Commerce and Labor, Bureau of the Census, *Thirteenth Census of the United States,* vol. 9, *Manufactures, 1909, Reports by States, with Statistics for Principal Cities* (Washington, DC, 1912), 296–97. Nominal dollar figures for 1909 are deflated using ser. E-23, House Furnishings, *Historical Statistics of the United States,* as cited in n. 5.

41. Darling, *Furniture,* 63–64.

42. Frank Ransom, *A City Built on Wood* (Grand Rapids, MI, 1955), 57; Jeffrey Kleiman, "The Great Strike: Religion, Labor, and Reform in Grand Rapids, Michigan, 1890–1916" (Ph.D. diss., Michigan State University, 1985), 4, 9–11; *WC* 4 (1905–6): 101; "Notes on the Grand Rapids Furniture Market," Henry Masten Papers, Collection 10, Box 1, Folder 4, Grand Rapids Public Library. Klingman showed his business acumen by setting up a separate company to purchase out-of-town samples at the market's close and operating a year-round retail outlet for their sale on the first floor of one of the halls. One reason for the multiplication of banks appears to be a Michigan statute forbidding any bank to provide loans above 10 percent of its capital to any single client. Hence subsets of the furniture group created "their own" banks as trade and working capital needs expanded. See John Widdicomb to J. P. Uptegrove, March 9, 1901, Widdicomb Papers, Collection 17, Box 5, Folder 28, GRPL (hereafter, WP-B [number], F [number]).

43. Ransom, *City,* 27. Hompe later became vice president of Royal Furniture (Kleiman, "Great Strike," 20).

44. *TWW* 14 (December 1895): 28. Beside shows in Chicago and New York, an exposition craze in the mid-1890s produced copies in Minneapolis, Cincinnati, Rockford (Illinois), and Rochester and Jamestown (New York). All but Jamestown soon faded away. See *TWW*'s monthly furniture columns for 1895–96.

45. Kleiman, "Great Strike," 4–5; *WC* 4 (1905–6): 99–105; 7 (1907): 91–92; "Employees of Furniture Industry—1910, 1915, 1916," Furniture Manufacturers' Association Papers, Collection 9, Box 9, Folder 3, GRPL. Office furniture firms remained batch specialists, however, as Macey's sectionals were fashioned in a dozen sizes, four woods, with doors having writing desks, plain or leaded glass, "and in sundry other styles or finishes" (*WC* 4:104). Gunn's desks in oak were finished in twenty-two different shades, and each took four weeks in the finishing stages that followed assembly (*WC* 7:92). Gunn produced roughly a hundred desk designs, along with bookcases and cabinets.

46. F. L. Furbish Papers, Sales Ledger, 1889–90, Collection 19, Grand Rapids Public Museum. See also Furbish's Letter Copy Book, 1881–85, in the same collection.

47. Kleiman, "Great Strike," chaps. 1–2; John Widdicomb Letter Copybooks, 1899–1908, WP-B 5–7 (1898–1908); Michigan Bureau of Labor and Industrial Statistics, *Seventeenth Annual Report* (Lansing, 1900), 73–78; author's interview with Richard Harms, GRPL archivist, August 27, 1990. The furniture freight cars, by agreement with the roads, were labeled for return *empty* to Grand Rapids, no small indicator of the trade's collective influence. The trademark group, Grand Rapids Furniture Association, was separate from the renamed Furniture Employers Association. The former's founding statements can be found in B 27, F 93 of the Widdicomb papers, as can a complaint to the association regarding poaching of a skilled worker (JW to Walter Drew, August 19, 1905, B 5, F 35).

48. There were many variations of these historical cost and rules-of-thumb procedures. At Grand Rapids in the late 1890s, some firms figured materials and labor in making samples, then doubled

this total to "make" a wholesale price (Ransom, *City*, 31).

49. On Pittsburgh-basing in steel and analogous schemes in linseed oil, hardwood lumber, and cement, see Frank Fetter, *The Masquerade of Monopoly* (New York, 1931).

50. H. E. Scholle to JW, March 29, 1899, WP-B 5, F 27. It should be noted that price-fixing of goods was undertaken by makers of relatively staple furniture, notably among school goods producers whose "trust" was broken up by federal intervention in 1907 (Darling, *Furniture*, 63–64).

51. Transcript of a *Grand Rapids Herald* article, January 28, 1908, WP-B 1, F 1. Like many family splits, the Widdicomb feud was kept alive through many tiny unpleasantries. Mail for JWCo. was often mistakenly delivered to the original WFCo., where William Widdicomb supposedly held it for a few days before handing it back to the mailman for proper delivery. JW, for his part, wrote snide letters to colleagues, critiquing WFCo.'s annual statements, which he received as a minority shareholder.

52. Ibid.; also, JW to R. G. Dun & Co., March 29, 1908, B 7, F 37; and *Grand Rapids Press* clipping, August 31, 1928, B 1, F 1.

53. JW to Dun, March 28, 1908, WP-B 7, F 37; JW to W. E. Corey, February 13, 1905, B 6, F 35.

54. *Herald* article, and "Fine Furniture for Over a Century," undated clipping, WP-B 1, F 1. Ralph stayed with JWCo. as head designer for the next fifty-three years.

55. JW to H. E. Scholle, May 1, 1900; JW to R. J. Horner & Co., July 5, 1900, WP-B 5, F 28; JW to F. I. Billings, December 4, 1901, B 6, F 31; JW to Hamilton Salmon & Co., April 28, 1904, B 6, F 34; JW to A. W. Litschgi, July 24, 1909, B 7, F 38. In Philadelphia JWCo. sold through the Strawbridge and Clothier department store, but it ignored Marshall Field and Macy to focus on the top-end trade in Chicago and New York.

56. Michigan Bureau of Labor, *Seventeenth Report*, 87; JW to Kinsella Co., May 31, 1900; JW to Jacques Kahn, June 5, 1900, WP-B 5, F 28.

57. JW to Jacques Kahn, May 6, 1901, WP-B 5, F 28.

58. Mahogany, oak, birch, maple, bird's-eye maple, and "toona," an exotic wood JW had renamed and repeatedly chided others for spelling "tuna." Toona was a market failure, accounting for less than 2 percent of JWCo. production in late 1899.

59. The cut order book does not include sample making, but exposition display sets were started in September after design approvals, if Widdicomb followed general trade practice.

60. Fashionable case goods makers designed furniture sets as room-filling groups, which in chamber styles included five to eleven pieces, any of which could be ordered singly or as a bundle, depending on retailers' estimation of the size of clients' bedrooms. In February 1911, the firm set in motion 84 cuts on 80 style numbers to produce 1,009 pieces, the mean cut size being 12 (Cutting Order Records, 1911, WP-B 16, F 3).

61. Calculated from Cutting Order Records, 1899–1902, WP-B 16, F 3. No style was cut more than four times in the half year, and the total number of wood + pattern variants was 86 for bureaus and 81 for chiffoniers. The median cut for both products was 10 units. These were typical arrays. At the smaller Century Furniture Co., a maker of tables, wooden and upholstered chairs, and sofas, production of 9,430 units in 18 months during 1906–7 involved 290 styles, 653 cuts, and a mean cut of 14. See Century Furniture Shop Cost Book, 1906–7, Archives, Grand Rapids Public Museum.

62. JW to Kahn, December 7, 1900, and January 19, 1901; JW to Jardine, April 2, 1901, WP-B 5, F 28; JW to J. H. Badger, WP-B 6, F 31.

63. JW attempted to secure bank support for a preferred stock issue in order to end a constant cycling of commercial paper for working capital, without giving up ownership. See JW to Uptegrove, May 19, 1900, and JW to Kahn, July 3, 1900, WP-B 5, F 28. On the merger boomlet, see JW to Robert Jardine, April 2, 1901; JW to H. A. Marston, April 4, 1901; JW to Uptegrove, April 22, 1901; WP-B 5, F 28. Marston, the merger promoter, tried again in 1902, with no success (see Furniture Manufacturers' Association Papers, Collection #17, Box 1, Folders 3 and 4, GRPL).

64. JW to Uptegrove, August 15, 1901, WP-B 6, F 31; JW to B. Cannon and Co., December 22, 1902, B 6, F 32; JW to

Leighton Pine, May 25, 1904, B 6, F 34; JW to O. B. Starkwather, May 6, 1908; B 7, F 37. No records survive regarding the origins of the Singer contract.

65. JW to Singer Sewing Machine Co., February 10, 1902; JW to Uptegrove, May 1 and 8, 1902; JW to Kahn, June 7, 1902; WP-B 6, F 31; W. P. Williams to W. Stridiron, April 17, 1902 (copy to JW), B 27, F 95; JW to Singer, January 6, 1903; JW to M. B. Pine, February 11, 1903; JW to W. G. Smith, March 6, 1903; JW to Uptegrove, May 30 and August 22, 1903; JW to National Sewing Machine Co., June 19, 1903; JW to Domestic Sewing Machine Co., December 24, 1903, B 6, F 32; JW to Uptegrove, March 29, 1904, B 6, F 34. Trying to sweeten South Bend managers, Widdicomb shipped two of them a bureau with mirror and a chiffonier late in 1903. By February 1904, shipments to Elizabethport were once again authorized (JW to W. M. Weld, November 5, 1903, B 6, F 32; JW to Pine, February 4, 1904; JW to Uptegrove, February 10, 1904, B 6, F 34).

66. JW to Domestic, May 14, July 9, December 8 and 20, 1904; JW to National, October 7, 1904; JW to Leighton Pine, May 25, 1904, WP-B 6, F 34; Pine to JW, June 15, 1904, B 6, F 33; JW to Pine, February 7, 1905; JW to Domestic, March 13 and 25, April 1, June 22, 1905; B 6, F 35; Charlotte Property Certificate, June 19, 1906; National to JW, August 17, 1906, B 6, F 33. On profits, see JW to Kahn, August 9, 1902, B 6, F 31; JW to Uptegrove, September 19, 1903 (B 6, F 32) and February 10, 1904 (B 6, F 34); JW to T. J. O'Brien, February 2, 1907, B 7, F 36. In 1901–2, the firm kept accounts on a July-to-June year but shifted by 1906 to calendar years.

67. Grand Rapids Furniture Manufacturers Employment Association to JW, September 22, 1903, and June 13, 1906, WP-B 6, F 33. Once the Charlotte plant was closed, total employment moved downward to about 400.

68. *Herald* article, WP-B 1, F 1; JW to R. G. Dun and Co., March 28, 1908, B 7, F 37. Long-term debt in 1897 was $40,500; in 1907, $38,500.

69. JW to Horner and Co., June 29, 1907, WP-B 7, F 36; JW to Dun, March 28, 1908; *The Furniture Journal* 27 (July 10, 1907): 106.

70. JW to Max Adler, February 4, 1907, WP-B 7, F 36; JW to O. B. Starkwather, April 8, May 6, 8, 11, and 23, 1908; JW to Sears, Department 1, June 6, 1908, B 7, F 37; JW to Starkwather, April 23, 1909; JW to Sears, Treasurer's Office, August 25, 1909, B 7, F 38; JW to Adler, September 23, December 15 and 29, 1909, January 8, 1910, JW to James Arthur, October 2, 1909; JW to Julius Rosenwald, October 7, 1909, B 7, F 39. Premium houses offered inexpensive furniture and housewares as prizes when quantities of their other goods were purchased. The 1909 Sears suites were style number 1796; in 1900, Widdicomb's style register had just reached 400. Kitchen cabinet lot sizes ranged from 200 to 500 units, much larger batches than styled line cuts, yet a far cry from Singer-scale production.

71. *Herald* and *Press* articles, WP-B 1, F 1; Wage Increase Notice, January 5, 1903, B 27, F 96; Frank Bockus to JW, August 31, 1901, B 5, F 30; Dr. C. E. Patterson to JW, November 14, 1901, B 27, F 94; JW to H. J. Klomprens, April 28, 1904; JW to Dickema and Kollen, August 27, 1904, B 6, F 34; JW to H. S. Jordan, November 27, 1907, B 7, F 36; JW to B. W. Brinthall, May 9, 1908; JW for M. B. Pine, June 30, 1908, B 7, F 37.

72. Ransom, *City*, 41–43, 64; *Furniture Trade Review* 16 (December 1895): 43; Thomas Kidd, Amalgamated Woodworkers, to JW, March 8 and 9, 1900, WP-B 5, F 27, JW to Drew, August 19, 1905, B 5, F 35; On employers' hostility, see Sanford Jacoby, "American Exceptionalism Revisited," in *From Masters to Managers*, ed. idem (New York, 1991), 173–200.

73. Kleiman, "Great Strike," 97–103.

74. Ibid., 103–4; *FMA* 62 (1911): 211; Ransom, *City*, 64; R. W. Irwin, "Story of the Grand Rapids Strike: Address at the National Association of Furniture Manufacturers' Semi-Annual Meeting, December 1911" (Indianapolis, 1912), 12 (excerpted at *FMA* 63 [1911]: 622–25). The investigation provision was added at the city furniture association's suggestion.

75. *FMA* 62 (1911): 211; Irwin, "Story," 16–17.

76. Kleiman, "Great. Strike," 107–25; Ransom, *City*, 65–66; Irwin, "Story," 18–21. Kleiman highlights ethnic/religious divisions among workers as contributing to the failure. Dutch Reformed employees responded to their clergy's denunciations of unions as ungodly and stayed at work or returned, whereas Polish Catholic

workers accepted their bishop's invocation of Pope Leo XIII's *Rerum Novarum* as supporting unionization. As one-third of Grand Rapids' 112,000 residents (as of the 1910 census) were Dutch and but 9 percent Polish, this split mattered (W. Jett Lauck, "Furniture Manufacturing Centers [and the] United States Census," *FMA* 64 [1912]: 94–95).

77. Campau to William H. Gay, May 23, 1912; Furniture Manufacturers' Association Papers, B 9, F 18; D. H. Brown to Fred Tobey, November 11, 1911; Tobey to Brown, November 21, 1911; and Guild statements, B 1, F 5, *FMA* 63 (1911): 352–53. On later, limited gains, see Ransom, *City*, 47–60.

78. Board Minutes, July 1912–April 1913, Furniture Manufacturers' Association Papers, B 5, F 3; Grand Rapids Guild Merger Papers, 1911, B 1, F 5; Kleiman, "Great Strike," chap. 5.

## CHAPTER 8
## FASHIONING THE MACHINE TOOL HUB: CINCINNATI

1. John A. LeBlond, "Reminiscences," *Fleur de Lis* 3 (1919–20): 48–50; William Lodge, "A Short History of the Machine Tool Business," *Cincinnati Magazine* 1 (November 1909): 8–9. The "Big Six" firms were Lodge and Shipley, Cincinnati Shaper, American Tool Works, Cincinnati Milling Machine, G. A. Gray, and LeBlond (George Wing, "History of the Cincinnati Machine Tool Industry" [D.B.A. diss., Indiana University, 1964], 139). *Fleur de Lis* was the LeBlond company magazine, bound volumes of which are held at the Cincinnati Historical Society (CHS). The inch sizing of lathes refers to the diameter of the turning head into which metal pieces were inserted for rotary cutting and the maximum size that the tool could hold. Heavy lathes for naval cannon or locomotive wheels were rated at 36″ or larger.

2. On bicycle manufacturing and technological change, see David Hounshell, *From the American System to Mass Production* (Baltimore, 1984), chap. 5.

3. LeBlond, "Reminiscences," 50–53.

4. No Cincinnati firm moved into the growing market for grinders until 1910, except for tool sharpeners, and until the 1920s none stepped into the related field of metal-forming tools (die presses, hammers, forging machines) like those produced in the Pittsburgh, Cleveland, or Chicago districts, or at Oberlin Smith's Ferracute Co. near Philadelphia. On grinding, see Charles Cheape, *Family Firm to Modern Multinational* (Cambridge, MA, 1985), a study of the Norton abrasives company.

5. James Schwartz, *Cincinnati Milacron, 1884–1984* (Cincinnati, 1984), 10–11, 214; George Barnwell, *The New Encyclopedia of Machine Shop Practice* (New York, 1941). Though Cincinnati never produced more than about 15 percent of all machine tools made in the United States, its share of the metal-cutting division's output was a quarter to one-third.

6. Cincinnati was a center of fine lithography and printing, as artifacts in the Cincinnati Historical Society's collections testify. By 1888, Gray had sold stone planers, as substitutes for use of secondhand metal planers, to lithographers and bank note companies in New York, Detroit, Buffalo, Chicago, Indianapolis, St. Louis, Baltimore, and, of course, Cincinnati. See G. A. Gray Papers, Collection BC-017, Box 25, Files 4, 5, 7, and 8, Cincinnati Historical Society.

7. Gray Papers, Box 25, Scrapbook, 17, 21. These inch measurements indicate the width of the table moving through the "throat" of the planer, where the cutting tools were fixed. Base model tools had a square throat, 22″ × 22″, for example, and the table length varied from 4′ to 20′ (later up to 48′). The combination of throat and table length options multiplied the number of models, though the range was scaled to the sorts of work planing involved. Thus there were no 22″ planers made with 20′ tables, or 60″ planers with 6′ tables, as users' production needs moved throat and table length upward in tandem. On the Sellers screw drive and much else relating to planer design, see J. Richards, "Evolution of Planing Machines," *American Machinist* 35 (1911): 199–201 (hereafter cited as *AM*), and Oberlin Smith, "Modern American Machine Tools," *Engineering Magazine* 8 (1894–95): 54–61 (hereafter, *EM*).

8. *Western Merchant and Manufacturer* 23 (October 13, 1888): 1–2. One of Gray's selling slogans was "Pig iron is cheap, and

we use plenty of it!" (Gray Papers, Box 25, Scrapbook, 54).

9. *AM* 26 (1903): 325–28.

10. *AM* 14 (May 14, 1891): 4. Castings order sheets may be found in Gray Papers, Box 25, Scrapbook, 68–69. Planers with common throat sizes, say 24", and different table lengths (6', 8', and 10'), would share parts for the throat-related components but not for the floor bed and table sections.

11. Gray Papers, Box 21, Order Books, passim. The order books list planer specifications, dates of receipt and shipment, client, agent (if any), price, and discount.

12. *AM* 14 (May 14, 1891): 4; Wing, "Cincinnati Machine Tool Industry," 140; Gray Papers, Box 25, Scrapbook, 53, 58, 61; Box 21, Order Book no. 3, 21–58 (for 1897). Cincinnati Milling Machine, which engaged the same European agents as did Gray, also won a gold medal at Brussels, and two other Cincinnati tool builders garnered silver awards.

13. Gray Papers, Box 21, Order Book no. 3, 21–58.

14. Ibid.

15. Gray Papers, Box 21, Order Books, nos. 3–5 (1897, 1900, 1903, 1906, 1909, and 1912). On standard versus special tools, see Joseph Horner, "Specialized Machine Tools," *Cassier's Magazine* 26 (1904): 404–19.

16. Gray Papers, Box 21, Order Books, vol. 3, inside front cover. Pasted inside the front cover are a set of sheets tracing changes in Gray's discounts to dealers from 1894 through 1929. Copies of new price lists, which from 1892 until 1907 reflected addition or deletion of models and options, not price changes, may be found in the Box 25 Scrapbook, on 21, 85, 155, and 239.

17. Gray Papers, Box 25, Scrapbook, 228–29. Gray's estate was assessed at $387,000, roughly a quarter of it being stocks of regional railways and banks and nationally prominent industrials (American Tobacco, Procter and Gamble).

18. Gray Papers, Box 21, Order Books, 4:260–300 and 5:57–83. The absence of surviving correspondence limits the reconstruction of this sequence, but records of planer specifications document the dual pricing of identical models. Dealers appear to have quoted the higher prices to new clients.

19. 1906–7 were the peak years for Cincinnati machine tool sales between 1890 and World War I (*Iron Age* 85 [1910]: 843). *Iron Age*'s weekly machinery market reports from Cincinnati proved to be exceptionally valuable sources.

20. Gray Papers, Box 21, Order Books, vol. 3, inside front cover. Sales dropped sharply in 1908 in reaction to the late 1907 panic, but the discounts remained intact until 1913, and list prices were not revised until June 1917.

21. Wing, "Cincinnati Machine Tool Industry," 140; Gray Papers, Box 21, Order Books, vol. 6.

22. This excludes grinding, of course, to focus on cutting edges. Later, tungsten carbide and ceramic tools were devised that could cut steel. Abrasive grinding shaped the alloy cutters themselves. For details, see Zay Jeffries, "The Present Status of Cemented Tungsten Carbide Tools and Dies," *Metals and Alloys* 1 (1929–30): 222–25.

23. Harless Wagoner, *The U.S. Machine Tool Industry from 1900 to 1950* (Cambridge, MA, 1966), 8–10 (hereafter, *USMT*). For technical detail, see O. M. Becker and Walter Brown, "High-Speed Steel in the Factory," a seven-part series in *EM* 29–31 (1905–6). For a worker's critique, see *EM* 31 (1906): 93–96.

24. G. A. Gray Co., "High-Speed Planing," 1906, in Box 25, Scrapbook, Gray Papers. Into the post–World War I years, most machine tools were belt driven, though shafting was increasingly powered by electricity. Individual motors for each tool, which Gray had introduced in the early 1890s, became general only during the 1920s, after issues of standard voltages, direct versus alternating current, and fixed versus variable speed motors were largely resolved. See Wagoner, *USMT*, 11–18.

25. "Cincinnati High Power Milling Machines," *AM* 31–2 (1908): 401–9. For a time after 1900, *AM* used one volume number per year, with two separately paginated parts. For other design innovations at Cincinnati, see *AM* 31–1 (1908): 935–37; 33 (1910): 483–87, 630–32; 34 (1911): 287–90, 527–30; *Iron Age* 80 (1907): 1210–13; 82 (1908): 1867–68; 84 (1909): 319–23; 85 (1910): 1516–17. CMM's redesigned millers drew major orders from the auto industry, pushing employment to 600 and

sales beyond $1 million in 1910 (Schwartz, *Milacron*, 26, 214). CMM's sales engineering was "widely copied" in the trade (Schwartz, *Milacron*, 31; Wing, "Cincinnati Machine Tool Industry," 253). For a technical overview of milling machine development, in which Brown and Sharpe and CMM were leaders, see Robert Woodbury, "History of the Milling Machine," in idem, *Studies in the History of Machine Tools* (Cambridge, MA, 1972), 58–100.

26. Henry Wood, "Shop System for Greater Output," *Iron Age* 88 (1911): 268–71; E. Stubbs, "The Lodge and Shipley Stores System," *AM* 37 (1912): 141–43; Hugo Diemer, "System in the Lodge and Shipley Machine Shop," *Iron Age* 89 (1912): 24–28. Wood and Stubbs were L&S managers, Diemer an engineering consultant. Bickford Drill's H. M. Norris detailed a "unit system" for tracking work-in-progress in *AM* 34 (1912): 251–57, 297–301.

27. Morrell Gaines, "Tabulating Machine Cost-Accounting for Factories of Diversified Product," *EM* 30 (1905–6): 364–73, quotation from 366.

28. This is "product costing," promoted vigorously by Church as a crucial procedure "in situations where a diverse line of products used factory resources at widely varying rates" (H. Thomas Johnson and Robert Kaplan, *Relevance Lost: The Rise and Fall of Managerial Accounting* [Boston, 1987], 53). Church published extensively in *Engineering Magazine*, a key journal in metalworking and heavy industry, and *EM* reprinted his arguments in two volumes: *The Proper Distribution of Expense Burden* (New York, 1908) and *Production Factors in Cost Accounting and Works Management* (New York, 1910).

29. Diemer, "System," 24–25; A. R. Erskine, "Going Cost Systems and Monthly Closings for Industrial Corporations," *Journal of Accountancy* 8 (1909): 168–72. When Diemer visited the lathe builder's plant, "260 separate jobs were on the tracing board" (26).

30. *AM* 34 (1911): 490–93; Gaines, "Tabulating Machine," 365; JoAnne Yates, "Coevolution of Information-Processing Technology and Use: Interaction between the Life Insurance and Tabulating Industries," *Business History Review* 67 (1993): 1–51; Arthur Norberg, "High-Technology Calculation in the Early Twentieth Century: Punched Card Machinery in Business and Government," *Technology and Culture* 31 (1990): 753–79. *AM* reported that clerks punched cost information on 600 cards per hour, but this is dubious for each card coded 10–15 bits of data, a task unlikely to be steadily accomplished every six seconds. For a preview of Lodge's costing approach, see his comments on a paper concerning high-speed tools, in *Transactions of the American Society of Mechanical Engineers* 32 (1910): 774–76 (hereafter, *Transactions ASME*).

31. The expense of gathering useful statistical information was a constant complaint by users of product cost accounting systems, a point at the heart of Erskine's critique ("Going Cost Systems," n. 107). Norberg ("High-Technology Calculation," 767) cites the 1913 case of an industrial executive who asked for five-year comparative data on costs. A three-year analysis based on tabulated cards was ready in one day, but adding the two earlier years would have taken his clerks a month of handwork to prepare and thus was dropped.

32. H. M. Norris, "Actual Experience with the Premium Plan," *EM* 18 (1899–1900): 572–84, 689–96; *Iron Age* 83 (1909): 470–71, 1204; *AM* 37 (1910): 1017–19; 34 (1912): 251–57, 297–301. For confirmation that the stability of piece-rate or premium scales was collectively agreed upon in Cincinnati metalworking, see E. F. DuBrul's remarks in *Transactions ASME* 25 (1904): 88.

33. Diemer, "System," 25, 27–28.

34. Lodge initially had imagined sharing five cents per hour saved with foremen (*Iron Age* 83 (1909): 470–71).

35. *Iron Age* 75 (1905): 1030, 1115; 85 (1910): 608.

36. Lodge, "Short History"; Schwartz, *Milacron*; LeBlond, "Reminiscences"; Herbert Merritt, "Formative Years of Cincinnati Milacron" (typescript, 1989), 25, Rare Book Room, Cincinnati Public Library; *Iron Age* 77 (1906): 706.

37. *AM* 26 (1903): 325; *Iron Age* 75 (1905): 1332. LeBlond charged Champion a fee for using its plans, but the new firm was also framing its own designs for larger lathes.

38. Thomas Egan, "Manufacturing Advantage" (address to the Commercial Club, 1899), Commercial Club Papers, Speeches, Box 1, on deposit, CHS.

39. These included woodworking machines, boilers, coal and ice machinery, beer and distillery outfitting, electric motors, tobacco and food processing machines, and much else that drew to varying degrees on castings for components.

40. *Iron Age* 76 (1905): 1328, 1559; 77 (1906): 454, 1785, 2012; 80 (1907): 1184; 81 (1908): 89; 84 (1909): 1348, 1433.

41. *Iron Age* 75 (1905): 874, 1703, 1806, 2018; 80 (1907): 586. When overtime could not be avoided, workers' rates were increased 25 percent (75 [1905]: 1030), and in some boom periods attempts were made to establish night shift operations (77 [1906]: 1292).

42. These two-page ads promoted thirty-two Cincinnati builders whose products were plumped as being "celebrated for their quality, speed, and economy." For an example, see Scrapbook, 328–29, Box 25, Gray Papers, CHS.

43. *Machinery* 8 (1902): 26–28; *Iron Age* 77 (1906): 2092; 78 (1906): 72–73; 79 (1907): 300; 85 (1910): 94; Schwartz, *Milacron*, 26–27. The colony group included Bickford (owned by a Geier cousin), Cincinnati Planer, the new Cincinnati Lathe and Tool (started by a CMM manager), Cincinnati Balcrank, and Triumph Electric, which among other things built motors for machine tools.

44. Naomi Lamoreaux, *The Great Merger Movement in American Business* (New York, 1985). The only significant machine tool merger was Niles-Bement-Pond, an alliance of the nation's heaviest tool builders oriented largely toward railway supply. Arguably the array of collective institutions erected in Cincinnati provided valuable services that reinforced proprietary attitudes of indifference or hostility to mergers.

45. C. J. Hobart, "The Employment Bureau," *Bulletin of the National Metal Trades Association* 2 (1903): 321–30. The CMTA's antiunion fervor and its members' solidarity remained constants on the local scene, as the association figured largely in defeating the next major machinists' union initiative in 1919 (see Wing, "Cincinnati Machine Tool Industry," chap. 8). By 1906, William Lodge served as treasurer for the NMTA (*Open Shop* 5 [1906]:342). Machine tool firms constituted just under half of the CMTA's sixty-eight members.

46. *Iron Age* 76 (1905): 179–80; 77 (1906): 454; 79 (1907): 1680–81; 80 (1907): 1033, 1103; 86 (1910): 88; 87 (1911): 64. Creating the Associated Foundries of Cincinnati seems to have been a step toward confronting unionized molders, the region's only strongly organized metalworkers, who had won a 15 percent increase in day wages earlier in 1910 (85 [1910]: 1497).

47. Albert Steigerwalt, *The National Association of Manufacturers, 1885–1914* (Grand Rapids, MI, 1964), 17–33; Wagoner, *USMT*, 73–74; Charles Goss, *Cincinnati: The Queen City, 1788–1912* (Chicago, 1912), 2:13. Lodge's partner Charles Davis was a member of the invitation committee that set up the initial 1895 National Association of Manufacturers convention. Egan, the pivotal organizer, headed the committee and served as the NAM's Ohio coordinator for many years.

48. Wagoner, *USMT*, 73–75. Early on, many firms outside Cincinnati failed to appreciate the value of collective actions on pricing, but the NMTBA conducted aggressive "education" campaigns to convince laggards of the importance of price maintenance and the demoralizing effects of cuts, essentially teaching firms that "demand for machine tools was price inelastic," as Wing explained, insensitive to price increases and inexpansive with price decreases ("Cincinnati Machine Tool Industry," 257).

49. *Iron Age* 76 (1905): 970; 77 (1906): 228; 78 (1906): 633. The NMTBA issued a carefully worded 1906 statement claiming that there had been "no concerted action . . . *in convention* to make any advance or agreement on prices" (emphasis added), evidently to deflect criticism that might lead to restraint of trade investigations. For membership, see *Machinery* 16 (1909–10): 233.

50. Wagoner, *USMT*, 75; *Iron Age* 85 (1910): 312; Scrapbook, Box 30, Gray Papers, CHS (for price lists). In 1921, responses to a Federal Trade Commission inquiry indicated that there were 150 open price associations active. See Joseph Foth, *Trade Associations* (New York, 1929), 230, and Arthur Eddy, *The New Competition* (New York, 1915).

51. Wagoner, *USMT*, 76–92, 97–101. The NMTBA featured Alexander Church at its 1911 convention as an extension of Geier's Uniform Cost Accounting Committee's effort to construct a tool builder's

cost system that would "eliminate the possibility of anyone selling machinery unknowingly at a price below the cost of production" (76).

52. Cincinnati Enquirer, *Leading Industries of Cincinnati* (Cincinnati, 1900), 26; Wing, "Cincinnati Machine Tool Industry," 176–82. In a 1902 letter to Cincinnati Shaper's Perrin March, Lodge noted that "good machinists" were also being drawn away to "Pittsburgh, Chicago, [and] St. Louis" (182). The city's antiunion tenor was duplicated in virtually all metal trades centers so likely had little effect on labor supply.

53. *AM* 19 (1896): 1191; Wing, "Cincinnati Machine Tool Industry," 186; "How Cincinnati Trains Its Workers," *Cincinnati Magazine* 1 (June 1910): 16–19; *Iron Age* 87 (1911): 771; *System* 9 (1906): 255–57. The apprentices in the Woodward classes were predominantly machinists but also included draftsmen and pattern makers. Geier had first organized the "apprentice school" at CMM in May 1907, and its instructor headed the public school program in 1909 (*Machinery* 15 [1908–9]: 299–300; Wagoner, *USMT*, 90–91).

54. *Machinery* 16 (1909–10): 223–24; Wing, "Cincinnati Machine Tool Industry," 188; *Iron Age* 87 (1911): 534; *AM* 34 (1911): 591–93. Echoing the university cooperative engineering course, the technical high schools in 1911 added an alternating-weeks class-and-shop option for students in the final two years of the four-year course.

55. *Iron Age* 82 (1908): 1478; 87 (1911): 224, 340; *Cincinnati Industrial Magazine* 3 (July 1909): 17–18.

56. David Noble, *America by Design* (New York, 1977), 184.

57. *Iron Age* 77 (1906): 1503–4. Lodge and Shipley, LeBlond, and Bickford were prominent among the initial supporters, but Geier (CMM) was absent, though he later became an enthusiast. Worker training at UC dated to 1901, when a technical school founded in 1886 was transferred to its care (*Bulletin: NMTA* 3 [1904]: 552–53).

58. *AM* 30–32 (1907): 364–65; *Cincinnati Magazine* 1 (April 1910): 4–5.

59. Few working-class youths had the resources or support to complete high school before the 1920s. See John Modell, *Into One's Own* (Berkeley, CA, 1989), chap. 3.

60. *Iron Age* 87 (1911): 709; Wing, "Cincinnati Machine Tool Industry," 188–92; Noble, *America by Design*, 187.

61. For a critical review, authored by a director of a rival independent trade school, see *Machinery* 18 (1911–12): 524–25.

62. Wing, "Cincinnati Machine Tool Industry," 192–203; Schwartz, *Milacron*, 214. The Cincinnati Shaper records Wing used have not been preserved, nor are there any detailed accounts of profits in collections at the CHS. Until 1919, machine tools were not listed separately in Census of Manufactures categories, so no estimates of profitability can be drawn from tabulations for the Cincinnati district in this period.

63. *Iron Age* 76 (1905): 430; 77 (1906): 455; 85 (1910): 312; Wagoner, *USMT*, 79; Johnson and Kaplan, *Relevance Lost*, 129–35. See also William Hooper, *Railroad Accounting* (New York, 1915), 86–89, for the ICC's 1907 mandatory accounting practices, and J. K. Butters, *Effects of Taxation: Inventory Accounting and Policies* (Cambridge, MA, 1949), for later developments. Johnson and Kaplan chiefly blame public accounting requirements for audited financial reports for the collapse of product costing, but they also mention federal taxation. As few specialty producers were involved in financial markets, here tax requirements seem most relevant.

## CHAPTER 9
## BACK EAST: THE ELECTRICAL EQUIPMENT INDUSTRY

1. For a valuable firm-level study, see Matthew Roth, *Platt Brothers and Company: Small Business in American Manufacturing* (Hanover, NH, 1994).

2. *Machinery* 10 (1903–4): 579–84; Jeremy Brecher, *Brass Valley* (Philadelphia, 1982); Charles Venable, *Silver in America: A Century of Splendor* (New York, 1994); Textile Machinery Works papers (Reading), Accession no. 904, Hagley Museum and Library; Oliver Carsten, "Work and the Lodge" (Ph.D. diss., University of Michigan, 1981); New York Metropolitan Commission, *Reports* (on apparel, jewelry, and printing) (New York, 1923–26); *Metal Industry*, monthly industrial reports on Connecticut cities, 1905–13; Floyd Parsons, ed., *New Jersey: Life, Industries and*

*Resources* (Newark, 1928); John S. Gordon, "Man of Steel," *Audacity* 3 (Fall 1994): 28–41; Philip Scranton and Walter Licht, *Work Sights* (Philadelphia, 1986).

3. Elihu Thomson, "Personal Recollections of the Development of the Electrical Industry," *Engineering Magazine* 29 (1905): 563–64 (hereafter cited as *EM*).

4. Harold Passer, *The Electrical Manufacturers: 1875–1900* (Cambridge, MA, 1953), 281; David Nye, *Electrifying America* (Cambridge, MA, 1990), 37–39; W. Bernard Carlson, *Innovation as a Social Process* (New York, 1991), 282. Nye reports the Electricity Building as being "as long as two football fields and as wide as one" (39).

5. To oversimplify, dc could not be transmitted long distances effectively but ac could. Dc had been the current of choice for individual, usually industrial, installations, whereas ac was, despite some drawbacks, far more effective for central stations delivering power over a wide network area. By showing that ac could be converted into dc at the point of use, Westinghouse laid the groundwork for building dc industrial applications into ac-based area power systems. At the time, General Electric had less ac experience or production capacity. See Carlson, *Innovation*, chaps. 5, 6. On the fair as utopian, see James Gilbert, *Perfect Cities* (Chicago, 1991).

6. Passer, *Electrical*, 283–87. GE and Westinghouse shared the later contracts about equally.

7. Carlson, *Innovation*, 282–95; Passer, *Electrical*, 270; John Hammond, *Men and Volts* (Philadelphia, 1941), 257. EG's profit on sales in 1891 was 13 percent, whereas T-H's profit on sales soared that year to 27 percent.

8. Federal Trade Commission, *Electric Power Industry: Supply of Electrical Equipment and Competitive Conditions*, 70th Cong., S. Doc. 46 (Washington, DC, 1928), 14–24; Proctor Reid, "Private and Public Regimes: International Cartelization of the Electrical Equipment Industry in an Era of Hegemonic Change" (Ph.D. diss., Johns Hopkins University, 1989), 22–69; Carlson, *Innovation*, 269. The "big stuff" is a term here borrowed from Stephen Meyer, "Technology and the Workplace: Skilled and Production Workers at Allis-Chalmers, 1900–1941," *Technology and Culture* 29 (1988): 839–64. For the A-C buyout of Bullock, see *Electrical World and Engineer* 43 (1904): 510 (hereafter, *EWE*).

9. *Electric World* 23 (1894): 4.

10. Discussion of H. L. Gantt, "A Bonus System of Rewarding Labor," *Transactions of the American Society of Mechanical Engineers* 23 (1902): 364–65 (hereafter, *Transactions ASME*).

11. As Carlson noted, citing Passer, an electrical system "was a capital good whose purchasers had to make complex calculations concerning both original investment and operating costs; consequently, they frequently chose equipment on the basis of its efficiency and reliability, rather than simply its price" (*Innovation*, 233).

12. See works by Passer, Carlson, Hammond, and Reid, cited above; Thomas Hughes' essential and prize-winning *Networks of Power* (Baltimore, 1983), which focuses on technical innovation and the construction of electric power systems; Forrest McDonald, *Insull* (Chicago, 1962); Leonard Reich, *The Making of American Industrial Research* (New York, 1985); and Andre Millard, *Edison and the Business of Innovation* (Baltimore, 1990). There has been no full-scale study of either company in the last half-century, in some measure owing to problematic access to corporate archives (but see David Nye, *Image Worlds: Corporate Identities at General Electric* [Cambridge, MA, 1985], ix–xiii, and Reich, *Making*, xv). Electrical trade journals and published transactions of electrical engineering societies also too rarely offer discussions of production practice, costing, and pricing. Both GE and Westinghouse published house organs, *General Electric Review* (hereafter, *GE Review*) and *Electric Journal* (hereafter, *EJ*), respectively. The latter, designed in part for use by the firm's apprentices and engineers, is more valuable for present purposes.

13. Carlson, *Innovation*, 304–6, 336–39; Reich, *Making*, 78.

14. *EWE* 34 (1904): 872; *The Schenectady Works of the General Electric Company* (Schenectady, NY, 1901), 7; John Broderick, *Forty Years with General Electric* (Albany, NY, 1929), 31, 62–4; Passer, *Electrical*, 271–74, 306, 315, 334.

15. Reich, *Making*, 78, 92; Broderick, *Forty Years*, 34, 95; FTC, *Supply*, 42–43, 46, 56; Hammond, *Men and Volts*, 340–44, 349.

16. In 1923, GE sold $218 million worth of electrical goods, Westinghouse an estimated $110 million, whereas in 1960, GE sales stood at $4.2 billion, Westinghouse at $2.0 billion (FTC, *Supply*, 74; Jules Backman, *The Economics of the Electrical Machinery Industry* [New York, 1962], 61).

17. Carlson, *Innovation*, 293, 297–99; *EJ* 1 (1904): 179; *Machinery* 8 (1901): 33; FTC, *Supply*, 22–23. On proprietary capitalists' creation of new enterprises, see Philip Scranton, "Build a Firm, Start Another," *Business History* 35 (1993) 115–51. For more on Westinghouse's managerial style, see Francis Leupp, *George Westinghouse* (London, 1919), chaps. 17–18, and Charles Scott, ed., *Anecdotes and Reminiscences of George Westinghouse* (Westinghouse Corp. and ASME, 1939; bound typescript), Hagley Library, especially contributions by L. M. Aspinwall, August Buchholtz, A. G. Christie, J. L. Crouse, G. B. Griffin, and E. H. Heinrichs (GW's staff aide, 1889–1912), each essay separately paginated.

18. *Machinery* 8 (1901): 33; Passer, *Electrical*, 334; *EWE* 43 (1904): 53–54. On Disston, see Harry Silcox, *A Place to Live and Work* (University Park, PA, 1994), and for Roebling, Clifford Zink and Dorothy Hartman, *Spanning the Industrial Age* (Trenton, NJ, 1992). Westinghouse's approach at Trafford is consistent with the "fraternal paternalism" discussed in Philip Scranton, "Varieties of Paternalism," *American Quarterly* 36 (1984): 235–57.

19. It is plausible that George Westinghouse channeled profits from his railway-related enterprises into building his electrical capabilities, but I have not found direct evidence for this.

20. Passer, *Electrical*, 315, 330–34; Carlson, *Innovation*, 282–83; George Westinghouse, "The Electrification of Railways," *Transactions ASME* 32 (1910): 945–79; *GE Review* 1 (July 1903): 5–8; W. S. Murray, "The Log of the New Haven Electrification," and "Discussion," *Transactions of the American Institute of Electrical Engineers* 27 (1908): 1613–1720. For an earlier joint venture in electric rail, see David Barnes, *Electric Locomotives: Westinghouse Electric and Baldwin Locomotive Works* (Philadelphia, 1896).

21. Thomas Hughes has termed this latter difficulty a "reverse salient," drawing on a military analogy in which the advance of the front is constrained by spirited defense at one point along its span. See Hughes, *Networks of Power*, chap. 4.

22. "Progress in Motor Development and Application," *Electrical Manufacturing* 43 (May 1949): 75–89. Baldwin equipped its tools with individual motors (82), a more complex shift than having existing shaft and belting setups powered by a central electric driver. Crowded conditions at Baldwin's plant may have encouraged this choice (see John Brown, *Baldwin Locomotive Works* [Baltimore, 1995]). Gray Planer Co. records indicate that Baldwin ordered new tools with ac drive by the late 1890s (Order Books, Gray Papers, Cincinnati Historical Society).

23. For induction motors, see *Electrical Engineering*, International Library of Technology, vol. 134B (Scranton, PA, 1922), secs. 25, 26, 32, 37, and for a concise discussion, Hughes, *Networks of Power*, 111.

24. Passer, *Electrical*, 296–300.

25. Ibid., 299; *Electrical Manufacturing* 43 (May 1949): 89.

26. Passer, *Electrical*, 301–5; 310–13; Hammond, *Men and Volts*, 275–84; *Machinery* 10 (1903–4): 1–5; G. R. Parker, "The Relation of the Steam Turbine to Modern Central Station Practice," *GE Review* 13 (1910): 62–66. *Machinery*'s correspondent regarded turbine development as "the most important development in steam engine work since the time of George H. Corliss" (10 [1903–4]: 1). Turbines with 200,000 kW capacity were built by the 1930s, and the first cost of early ones was roughly half that of steam engines with identical power output.

27. *EJ* 2 (1905): 496; Passer, *Electrical*, 312; *Machinery* 6 (1899–1900): 123; 8 (1901–2): 134–39; 13 (1907–8): 295–99; *Factory* 1 (1907–8): 66–67. For heroic work in meeting performance contracts, see Hammond, *Men and Volts*, 278–84.

28. G. I. Stadeker, "Winding of Dynamo-Electric Machines," *EJ* 7 (1910): six parts, June–November; *American Machinist* 34 (1911): 976–81 (hereafter, *AM*). On the transition, see David Hounshell, *From the American System to Mass Production* (Baltimore, 1984).

29. Alfred Chandler, *The Visible Hand* (Cambridge, MA, 1977), 433.

30. George Stratton, "The Management of Production in a Great Factory," *EM* 34

(1907–8): 569–76; *Machinery* 17 (1910–11): 841–42.

31. Chandler, *Visible Hand*, 430; Broderick, *Forty Years*, 79–81.

32. George Stratton, "Labor-Cost Distribution at the General Electric Shops," *EM* 34 (1907–8): 956–64; *Machinery* 17 (1910–11): 842; *AM* 31-1 (1908): 411–13. One premium plan that has come to light was in part of women's coil-winding at Westinghouse; see Elizabeth Butler, *Women and the Trades* (New York, 1909; reprint, Pittsburgh, 1984), 217–18.

33. Yet none of these centers became "company towns" in the textile or coal industry sense. Workers in both Schenectady and East Pittsburgh, for example, elected Socialist local governments in the prewar years (Ronald Schatz, *The Electrical Workers* [Urbana, IL, 1983], 36).

34. Both, however, seem to have used stable price lists for basic models, with variable discounts. For comparison with Gray Planer, see Passer, *Electrical*, 300.

35. Stratton, "Management," 576. The addition of turbogenerator capacity at Lynn in the early 1900s moved it beyond the realm of motor production and development, noted by Chandler (*Visible Hand*, 429) and Carlson (*Innovation*, 303) as its initial focus.

36. John Broderick, "The Standardization of Electrical Apparatus," *EM* 22 (1901–2): 24–30.

37. *EJ* 9 (1912): 409–15, 573–76. Westinghouse's basic Standards Book ran 350 pages, including dimensions for 137 sizes of "Taper Pins and Reamers" made in the company tool room.

38. John Van Deventer, "Extreme Variety versus Standardization," *Industrial Management* 66–67 (1923–24), seven parts, extended quotation from 66 (1923): 254, emphasis in original. *IM* was the renamed *Engineering Magazine*.

39. *Machinery* 12 (1905–6): 399–401; 13 (1906–7): 1–5; 18 (1911–12): 619–23; *AM* 33 (1910): 1–6.

40. *AM* 33 (1910): 723–24, 36 (1912): 85–87; Butler, *Women*, 215. Butler reported 500 apprentices among E&M's 10,000 workers in 1907, which likely included both boys and apprentice engineers in their early twenties.

41. *AM* 36 (1912): 87.

42. David Noble, *America by Design* (New York, 1977), 171–74.

43. Henry Prout, *A Life of George Westinghouse* (New York, 1921), 279–81; *Westinghouse: Past, Present, and Future* (Pittsburgh, 1936), 12–15, Pamphlet Collection, Hagley Library; E. H. Heinrichs, "George Westinghouse: Anecdotes and Reminiscences," 30–43, in Scott, *Anecdotes*; FTC, *Supply*.

## CHAPTER 10
## THE PERILS OF PROVIDENCE: JEWELRY'S ERRATIC COURSE

1. See Eileen Boris, *Home to Work* (New York, 1994), chap. 4.

2. *Manufacturing Jeweler* 12 (1893): 648–49, 809; 13 (1893): 39 (Hereafter *MJ*). In this era, *MJ* published two volume numbers each year. Each year as well, the journal released a separately paginated Anniversary Issue in October or November. These will be cited, for example, as *MJ* 12 (AI—October 21, 1893): 45–48. For a useful profile of work at Gorham, see *American Jeweler* 17 (1897): 191–94.

3. William Cobb and Co. Papers, Cost Book, 1893–1914, Rhode Island Historical Society, Providence, RI; see esp. 10–21. As penciled annotations suggest, Cobb used his trade experiences to quote initial prices, then tracked costs to discover whether or not he had made a decent profit. This helped him defend prices on repeat orders from downward pressures, as did his frequent notations of the tools costs for initial orders that had not been figured in earlier. For an overview of Bonnett's and Quarters' careers and those of other platers, see *MJ* 25 (AI—October 19, 1899): 52–57, and 37 (AI—November 16, 1905): 50–54, and for a lucid introduction to the trade's technical aspects, see John Urquhart, *Electro-plating: A Practical Handbook* 5th ed. (London, 1920).

4. For a profile of the Mossberg tool firm, see *Machinery* 4 (1897): 9–10.

5. *The Jobber's Handbook for 1895: A List of Manufacturing Jewelers and Kindred Trades* (Providence, 1895), 52–60, 154–76; *MJ* 12 (1893): 395–96.

6. *MJ* 24 (1899): 110, 124, 171; 26 (1900): 56; 28 (1901): 40; 29 (1901): 522. In earlier articles, I accepted Nina Shapiro-Perl's argument that the 1890s depression created widespread homework and deskilling within the jewelry trades. Further re-

search in industry sources suggests dating this decay to the years after 1905. See Nina Shapiro-Perl, "Labor Process and Class Relations in the Costume Jewelry Industry: A Study in Women's Work" (Ph.D. diss., University of Connecticut, 1983).

7. *MJ* 50 (1912): 1154. For the "dishonest" making of silver novelties, see *MJ* 28 (1901): 40.

8. A journal report later estimated that 90 percent of all makers worked through jobbers (*MJ* 43 [1908]: 6), though smaller firms surely predominated in this group.

9. For an account of a small firm's rapid expansion due to jobbers' beauty pin orders, see *MJ* 28 (1901): 76.

10. *MJ* 26 (1900): 231–32; 38 (1906): 795–96.

11. In the thirty years before World War I, the trade press reported only a few notable patent lawsuits, each of which concerned a findings innovation (in clasps or separable studs, for example). Advertisements on occasion contained the word "patented" in reference to designs, perhaps to warn off imitators, but to my knowledge no court tests of jewelry design infringements took place. On the insignificance of patents, see *MJ* 24 (1899): 316.

12. *MJ* 24 (1899): 41–42, 59; 25 (1899): 244; 26 (1900): 56, 124; 35 (AI—November 10, 1904): 58. Returns (other than for defects) were uncommon in other fashion trades but enervated book publishers. Dating ahead involved shipping an order and dating the invoice sixty or ninety days later, at which time the six-month calendar for payments commenced. Findings houses seemed most involved with scheme goods, especially chains and studs, which provided them an outlet beyond supplying manufacturers.

13. *MJ* 25 (1899): 30; 27 (1900): 32; 30 (1902): 29; 35 (AI—November 10, 1904): 26–34. Of ten buildings profiled in the 1904 report, only one had space unleased. Three others were also constructed at Attleboro in 1899 (*MJ* 26 [1900]: 56).

14. *MJ* 24 (1899): 162, 241, 261–62, 266.

15. Ibid., 284–86, 317–18, 360. International Silver had a rocky start, passing several preferred stock dividends in 1900 and being forced to reverse its price advances in 1901 (*MJ* 26 [1900]: 528; 28 [1901]: 98).

16. *MJ* 24 (1899): 286; 25 (1899): 183–84; 26 (1900): 191–92; 29 (1901): 527–28; 43 (1908): 710–14. The Rhode Island School of Design was founded with a small fund left over from the Centennial Exposition and supported in the late nineteenth century by the Metcalf family. NEMJA sponsored an annual prize competition for jewelry designs but had few other relations with the school in this period (*MJ* 32 [1902]: 447).

17. *MJ* 32 (1903): 336; 33 (1903): 66; 43 (1908): 1082D.

18. Overtime increments to hourly rates were not paid, manufacturers asserting that they got no more for the jewelry just because it was made after dark.

19. *MJ* 43 (1908): 1082D. This residual in Providence was $4 million in 1899 and $5 million in 1906.

20. *MJ* 33 (1903): 66; 37 (1905): 416; 40 (1907): 602. These figures omit an unknown number of firms that started after 1893 and expired before 1903.

21. *MJ* 31 (1902): 447, 512, 612–12, 634, 694, 712, 768; 31 (AI—November 6, 1902): 4, 6; 32 (1903): 24.

22. The Leominster, Massachusetts, cluster of horn and celluloid specialists, who produced combs and hair ornaments, represent another set of "outliers." *MJ* regularly provided updates on their situation in its fall Anniversary Issue but usually ignored them otherwise.

23. *MJ* 32 (1903): 44, 52, 70, 84, 267, 288, 293, 378, 462, 511–12, 516, 567.

24. *MJ* 32 (1903): 567–68, 651–52; 33 (1903): 128, 162, 172B–172D, 192–98, 228, 239–40, 242, 291, 300B. *MJ* provided surprisingly evenhanded coverage of events, encouraging owners to compromise and printing extended interviews with organizers and union officials. Remnants of Local 9 presented an eight-hours demand in 1905, which was ignored (*MJ* 36 [1905]: 397–98).

25. *MJ* 30 (1902): 447, 506, 516, 574; 31 (1902): 629, 729; 32 (1903): 651–52; 34 (1904): 32, 420; 35 (1904): 219–20, 241, 292; 36 (1905): 456–64, 526, 682–84; 37 (1905): 650–52; 38 (1906): 14–18, 370–72, 795–96; 39 (1906): 180; 41 (1907): 664. Regarding the RISD endowment, a Metcalf heir pledged the $50,000 on the condition that it be matched by other donors. After months of pleas in *MJ*, area jewelry manufacturers contributed less than $4,000, whereas Brown and Sharpe alone donated

$5,000 and Gorham $2,000. Enforcement provisions for the Stamping Act were inadequate, all parties agreed.

26. *MJ* 38 (1906): 693–94. Taunton was to be disappointed; by 1911 daily "shop trains" seven and eight cars long left the town to carry hundreds of workers to jewelry factories in Attleboro (*Metal Industry* 9 [1911]: 404 [hereafter, *Metal*]).

27. Scattered reports and the Cobb records suggest that bench workers earned $0.25/hour (and up to $0.40), or $15/week in these years, averaging ten months employment on a sixty-hours basis, including three long-hours months and three of short time, plus a two-week (unpaid) summer shutdown and about six weeks of layoff in the spring. The most able sample makers worked year-round, earned $1,000 or more, and were tempted by entrepreneurial possibilities.

28. *MJ* 39 (1906): 319–20; 40 (1907): 55–56. As usual, proposals for a specialty trade school, this time at Attleboro, failed in 1907 (*MJ* 40 [1907]: 297–98, 316, 338, 814).

29. *MJ* 37 (1905): 99–100, 416; 38 (1906): 617–18; 39 (1906): 830. There is no mention in the contemporary sources consulted of the possibility that jewelry work might be "unmanly" when contrasted with other metal trades occupations.

30. *MJ* 39 (1906): 415–16, 760; 40 (1907): 28, 369–70.

31. *MJ* 41 (1907): 516, 657, 672, 950, 1040, 1080.

32. *MJ* 41 (1907): 672, 950, 1080; 42 (1908): 228, 656, 734. "Garret" firms were tiny new enterprises that often occupied attic rooms in manufacturing buildings.

33. *Metal* 7 (1909): 85, 192, 425, 463; 8 (1910): 96; 9 (1911): 489; *MJ* 43 (1908): 6; 44 (1909): 407–8; 49 (1911): 575; 50 (1912): 480, 792, 1064, 1154. It is not clear why jobbers commenced refusing to purchase initial seasonal inventories, for they already had long credits from the makers and return privileges.

34. Margaret Abels, "Jewelry and Silverware," in Commonwealth of Massachusetts, *Forty-Fifth Annual Report on the Statistics of Labor: 1914* (Boston, 1914), V-102–3; *Metal* 8 (1910): 359. Though annual totals were feeble, Abels added that the "most usual . . . 10 cents an hour [earnings] is a high one for home work and would permit an individual working nine hours a day to make a living" (107). Boris, in her study of motherhood and industrial homework (*Home to Work*), did not discuss these anomalous women contractors in the jewelry sector. They are well worth pursuing, if this is at all possible.

35. Abels, "Jewelry," 93–102, 107; *MJ* 44 (1909): 451–53, 476–82, 535–36.

36. Findings companies by 1910 could provide frames as well as rings; some, like Providence's Metal Products Corporation, had their own designers (*Machinery* 17 [1910–11]: 181–84). Such capabilities reduced entry costs to below a hundred dollars.

37. *MJ* 50 (1912): 702; *Metal* 9 (1911): 363.

38. *Metal* 9 (1911): 528; 10 (1912): 51. This tactic was also termed "hand-to-mouth" buying.

39. *Metal* 10 (1912): 181, 224, 438; 11 (1913): 54, 145, 233, 319–20, 447, 539; 12 (1914): 47, 92, 179, 317, 530.

40. This pattern, including a lessened proportion of outwork, continued into the 1920s. See Children's Bureau, U.S. Department of Labor, *Industrial Home Work of Children: A Study Made in Providence, Pawtucket, and Central Falls, R.I.* (Washington, DC, 1922). By the 1920s, the F. W. Woolworth's chain alone purchased 10 percent of the nation's cheap jewelry (*MJ* 68 [1921]: 84). For an approving examination of current-day intensified competition, see Richard D'Aveni, *Hypercompetition: Managing the Dynamics of Strategic Maneuvering* (New York, 1994).

## CHAPTER 11
## WORKSHOP OF THE WORLD: PHILADELPHIA

1. Uptown manufacturing in Manhattan could not easily expand facilities on adjacent parcels. Thus space-demanding firms frequently built new plants in the outer boroughs, Yonkers (Otis Elevator), or across the Hudson in northern New Jersey (Tiffany moving to Newark in 1896). The island's staggering land prices also limited development and fostered severe congestion. See Edward Pratt, *Industrial Causes of Congestion of Population in New*

*York City*, Columbia Studies in History, Economics, and Population 43:1 (New York, 1911).

2. For further information, see Philip Scranton, *Proprietary Capitalism* (New York, 1983); *The Philadelphia Textile Manufacture* (Philadelphia, 1984); *Work Sights* (Philadelphia, 1986) (with Walter Licht); *Figured Tapestry* (New York, 1989); "Large Firms and Industrial Restructuring," *Pennsylvania Magazine of History and Biography* 116 (1992): 419–65; "Build a Firm, Start Another," *Business History* 35 (1993): 115–51; and "Pennsylvania's Multiple Industrializations," *Pennsylvania History* 61 (1994): 6–17.

3. For the shoe trade, see *Report of the Industrial Commission*, 57th Cong., 1st sess., H. Doc. 183 (Washington, DC, 1901), 14:290–310, 321–49 (hereafter cited as *RIC*) and Judy Goldberg, "Strikes, Organizing, and Change: The Knights of Labor in Philadelphia, 1869–1890" (Ph.D. diss., New York University, 1984).

4. "Testimony of Frank Leake," *RIC* 14:273–78.

5. "Leake," *RIC* 14:279; "Testimony of Robert Dornan," ibid., 310–15; Scranton, *Figured Tapestry*, 112–15, 131–32 (hereafter cited as *FT*). Interestingly, Dornan credited the ingrain substitution in large part to silver matters. When, after the repeal of the Sherman silver purchase statute in 1893, the federal government ceased buying silver at ninety cents or more an ounce, its market price dropped to fifty-eight cents, giving Providence jewelers their opportunity. However, import valuations of goods made in silver-currency nations like China, whence came the bulk of straw mats, also dropped sharply. Even under the Dingley tariff of 1897, mats could be landed at a U.S. gold standard price of seven cents/square yard. With a three-cent flat rate and wholesalers' profits figured, they could be sold to stores for under fifteen cents/yard, half the price of cheap ingrains. Thus even by 1900, ingrain output rebounded only to two-thirds of its 1893 level.

6. Scranton, *FT*, 140–48.

7. "Leake," *RIC* 14:274, 279; Scranton, *FT*, 136–40, 153–60, and "Build a Firm."

8. Scranton, *FT*, 197–225. In knitting, where women were two-thirds of all workers, the strike folded rapidly. Forfeit money represented sums, proportional to firm size, deposited with the Textile Manufacturers' Association to guarantee that companies would resist workers' demands, decline to negotiate separately, and shun the press.

9. Ibid., 230–77; "Large Firms," 430. The Bromley's Quaker Lace Co. was most successful in brand naming, in part through mass distribution of annual fashion catalogs featuring lace used both in women's clothing and for home decoration.

10. Scranton, *FT*, 237–41. Similar fiddling was noted in brass foundries at this time, as rising copper prices ran into buyers' resistance against increased casting charges. See *Metal* 8 (1910): 258–59.

11. Howell Harris, "The Rise and Fall of the Open Shop: Philadelphia's Metal Trades" (draft MS, 1991–94), chaps. 5, 6; and "Employers' Collective Action" (typescript, 1990), extended quotation from p. 9. Harris's study is under contract to Cambridge University Press. My thanks to the author for permission to reference his early findings.

12. Thomas Savery Journals, 1894–1902, Accession no. 291, Hagley Museum and Library, Wilmington, DE (esp. entries for March 9–April 1, 1895; June 1–2, 1897; August 3, 1900; August 10–13, 1901; and June 23–25, 1902). See also Rice, Barton, and Fales, *A Line of Men One Hundred Years Long* (Worcester, MA, 1937).

13. Harris, "Rise and Fall," chaps. 5, 6; *RIC* 14:350.

14. Their "rule" was one apprentice for every eight journeymen, and "one for the shop." Thus an enterprise employing eighty molders would have at any time no more than eleven apprentices. In any year, only two or three of these would complete their terms and join the regular labor market.

15. This preceded and outlasted the well-known Murray Hill agreement between organized machinists (IAM) and the national Metal Trades Association.

16. Harris, "Rise and Fall," chap. 3.

17. Harris, "Employers," and "Getting It Together: The Metal Manufacturers' Association of Philadelphia, c. 1900–1930," in *Masters to Managers*, ed. Sanford Jacoby (New York, 1991), 111–131.

18. William Vogel Jr., *Precision, People and Progress* (Philadelphia, 1949); *Machin-*

*ery* 13 (1906–7): 58–59; Scranton and Licht, *Work Sights*, 222–25, 248.

19. John Macfarlane, *Manufacturing in Philadelphia, 1683–1912* (Philadelphia, 1912).

20. *The Manufacturers' Club of Philadelphia: Business Classification of Members* (Philadelphia, 1895); *The Manufacturer* 6–9 (1893–96). *The Manufacturer* was the club's weekly journal, which reported sectional activities, notable guests, and debates among members on political topics; it also listed the evenings that members would be "in house" to receive friends and business colleagues.

21. Scranton and Licht, *Work Sights*, 133–36; Philip Scranton, "Between Firm and Market" (paper presented at the International Working Group on Regional Industrial Restructuring, Osaka, Japan, August 1993).

22. *Public Ledger Almanac for 1897* (Philadelphia, 1896); "The Bourse in History" (panel exhibition, The Bourse, 1992); *The Bulletin 1939 Almanac and Year Book* (Philadelphia, 1939); *RIC* 7:129–32; *Eighth Annual Report of the Commissioner of Labor, 1892: Industrial Education* (Washington, DC, 1893), 105–7; Edward France, "The Philadelphia Textile School," in *Fifth Annual Report of the Alumni Association of the Philadelphia Textile School* (Philadelphia, 1906), 31–39; and "A Quarter Century of Technical Education in Textiles," *Ninth Annual Report* (Philadelphia, 1910), 42–49. For the School of Industrial Art and women's trade education, see Nina deAngeli Walls, "Art, Industry, and Education" (Ph.D. diss., University of Delaware, 1994).

23. *RIC* 14:351–52; Edgar Marburg, "A Historical Sketch of the Engineers' Club of Philadelphia," *Proceedings of the Engineers' Club of Philadelphia* 18 (1901): 61–67; Scranton, "Between Firm and Market"; *The Manufacturer* 13 (February 1, 1900): 53; *Public Ledger Almanac for 1898* (Philadelphia, 1898), 63; *Bulletin Almanac: 1939*, 287; *Commercial America* 1–9 (1905–12), passim. A complete set of *Commercial America* is held at the Library of Congress.

24. Scranton, *FT*, 46, 311; Harris, "Getting It Together," quotation from 129. MMA members paid ten cents per month per skilled worker as dues, a rate that rested unchanged for over twenty years. The largest firms (Baldwin, Cramp, Midvale) had their own means for labor recruitment and did not join, thus avoiding fees that would have run to several hundred dollars monthly.

25. J. Roffe Wike, *The Pennsylvania Manufacturers' Association* (Philadelphia, 1960), 18–36, 79–90, 117–19; Harris, "Getting It Together," 127. The workmen's compensation debate ran from 1911 through 1915. PMA added a fire insurance service for manufacturers in 1919.

26. *Tenth Annual Report of the Alumni Association of the Philadelphia Textile School* (Philadelphia, 1911), 13–22.

27. In its early decades as solely an undergraduate school, Wharton drew heavily on the Philadelphia area for students. Sixty percent of graduates, 1894–1914, went immediately into business positions; "half [of them] joined family firms, many starting out as proprietors or corporate officers." See Steven Sass, *The Pragmatic Imagination* (Philadelphia, 1982), 137. A Penn professor started the Engineers' Club, and it filled with alumni working in metal trades firms (Marburg, "Historical Sketch"). Financier Anthony Drexel founded the institute in 1891 to provide "education in the practical arts" in day and evening classes. It later adopted the Cincinnati cooperative engineering curriculum, students alternating three-month periods of study and work in "the highly developed industries of Philadelphia" (*Bulletin Almanac: 1939*, 275).

28. Henry Spangler, "Training in the Engineering Trades in Philadelphia," *Proceedings of the Engineers' Club of Philadelphia* 26 (1909): 113–34. Spangler also passed over the capture and free-rider problems encountered when apprentices bolted to opportunities at other firms before indentures ended, easily able to double third-year earnings that averaged under six dollars weekly. Link-Belt's James Dodge argued strenuously for the value to workers of deferring employment to enter trade school training, but to little effect (*Machinery* 10 [1903–4]: 203–4).

29. Spangler, "Training," 113–34; Harris, "Getting It Together," 128; Walter Licht, *Getting Work: Philadelphia, 1850–1950* (Cambridge, MA, 1992). Half the boys starting in Philadelphia schools at age six or seven dropped out before their eleventh birthdays, on average completing four years (figures for 1906–8 at Spangler,

"Training," 116–17). On Williamson, see *Machinery* 17 (1910–11): 877–80; and Department of Commerce and Labor, *Twenty-Fifth Annual Report of the Commissioner of Labor, 1910: Industrial Education* (Washington, DC, 1911), 41–45, 757–61.

30. Given the remarkably thin research on manufacturers' collective efforts, this is a perilously tentative judgment.

31. For details and analysis that underlie these paragraphs, see John Brown, *The Baldwin Locomotive Works, 1831–1915* (Baltimore, 1995). Brown's study represents the most thoroughly developed research effort into heavy capital equipment production in two generations.

32. In the decade after 1898, Cramp constructed thirty-three commercial ships, five battleships, six cruisers, plus many smaller vessels. The family enterprise had incorporated in 1892, selling stock to fund an earlier round of technical updating, but retaining a controlling interest.

33. This and preceding paragraphs on shipbuilding are based on Thomas Heinrich, *Ships for the Seven Seas* (Baltimore, 1997), esp. chaps. 6, 7.

34. Albert Churella, "Corporate Response to Technological Change: Dieselization and the American Railway Locomotive Industry during the Twentieth Century," *Business and Economic History* 25, no. 1 (1996): 27–31.

35. Philip Scranton, "Market Structure and Firm Size in the Apparel Trades: Philadelphia, 1890–1930," *Textile History* 25 (1994): 243–73.

36. Figures drawn from Department of the Interior, *Compendium of the Eleventh Census* (Washington, DC, 1895), 932–33. For Snellenberg, see *The Clothing Gazette* 11 (June 1891): 48. His outlet became one of the city's six major department stores in later years. Gross profits do not include salaries to principals and office staff, which obviously were smaller at custom shops, but do reflect expenses for labor, insurance, materials, power, and other basic manufacturing costs.

37. *The Clothing Gazette* 12 (May 1892): 65–66; *The Clothing Designer and Manufacturer* 9 (August–September 1916): 11.

38. *Fifth Annual Report of the Factory Inspector of the Commonwealth of Pennsylvania* (Harrisburg, PA, 1895), 360–417; *RIC* 7:49.

39. U.S. Census Office, *Census Reports: Twelfth Census of the United States*, vol. 8, pt. 2, *Manufactures, States and Territories* (Washington, DC, 1902), 784–91; *Thirteenth Annual Report of the Factory Inspector of the Commonwealth of Pennsylvania* (Harrisburg, 1903), 52–406; *The Clothing Gazette* 29 (May 1901); Philip Scranton, "Apparel Arts and the U.S. Men's Clothing Industry," in *Apparel Arts: Fashion Is the News*, ed. Gianinni Malossi (Milan, 1989).

40. One manufacturer commented in 1901: "The advantages of operating one's own plant are almost numberless.... In former years we gave out large quantities of work to contractors, and we were always in hot water for one reason or another." He abhorred the "continued theft of trimmings" and "spool-silk" but was most exercised that "the tailors slighted the work in every possible manner that would not lead to quick detection"; some "would alter the lines of the garments and thereby cause us great annoyance because the sizes were not as intended" (*Garment Manufacturer* 6 [November 1901]: 35). See ibid., 24, for quotation in text; figures drawn from *Thirteenth Annual Report—Factory Inspector* (n. 39) and *Second Industrial Directory of Pennsylvania* (Harrisburg, PA, 1916), 1179–1373.

41. *Garment Manufacturer and Buyer* 16 (March 1906): 89–98; *Manufacturing Clothier* 2 (June 1915): 38, for extended quotation; Rosara Passero, "Ethnicity in the Men's Ready-Made Clothing Industry, 1880–1950: The Italian Experience in Philadelphia" (Ph.D. diss., University of Pennsylvania, 1978); Louis Levine, *The Women's Garment Workers* (New York, 1924), 234–36, 277–79.

42. For a national perspective on these shifts, see Bernard Smith, "A Study of Industrial Development: The American Clothing Industry in the Late Nineteenth and Early Twentieth Centuries" (Ph.D. diss., Yale University, 1989).

43. For a fuller treatment, see Philip Scranton, "Webs of Productive Association in American Industrialization" (paper presented at the European Social Science History Conference, Leeuwenhoek, The Netherlands, May 1996).

44. Jewelry workers' unions in the New York–Newark district eventually had some regional effect in this regard but failed to penetrate Providence and the Attleboros. The reluctance of jewelry firms to relocate outside the metropolitan

region helped stabilize the high-end jewelry trades. Comparable organizing in the New York and Philadelphia apparel industry, ca. 1915–37, along with municipal and state regulation of workplaces, helped end the worst sweatshop conditions but proved incapable of preventing the longer-term flight of clothing manufacturing toward outlying regional towns, southern locales, and ultimately non-U.S. production sites.

45. Even the Oakley Colony created its own separate, collective foundry company.

46. Donald Tulloch, *Worcester: City of Prosperity* (Worcester, MA, 1917); Joshua Chasan, "Civilizing Worcester" (Ph.D. diss., University of Pittsburgh, 1974); Roy Rosenzweig, *Eight Hours for What We Will* (New York, 1983); Charles Washburn, *Industrial Worcester* (Worcester, MA, 1917).

47. Thin research to date on post-1890 industrial Newark or New York makes this judgment tentative. Newark did sponsor a technical high school that became the city College of Engineering in 1919 (and eventually the New Jersey Institute of Technology) and completed its ship channel to Port Newark in 1919, projects that manufacturers from multiple sectors promoted through the Board of Trade. John Hyatt's technical novelties, celluloid (1869) and a lathe for turning perfect spheres (1885), had weighty implications for industrial specializations, but not in Newark (John Cunningham, *Newark* [Newark, NJ, 1966], 180–81, 212, 247–48).

48. Linda Ewing, "Industrial Dualism and Sector Structure: A Historical Study of Michigan's Tooling Industry" (paper presented at the Social Science History Association, New Orleans, 1991).

49. Scranton, *FT*; Jeremy Brecher, *Brass Valley* (Philadelphia, 1982); *Engineering Magazine* 19 (1900): 493–510; 661–82; Roy Kelly and Frederick Allen, *The Shipbuilding Industry* (Boston, 1918).

## CHAPTER 12
## WAR, DEPRESSION, AND SPECIALTY PRODUCTION INTO THE 1920s

1. John Milton Cooper, *The Warrior and the Priest* (Cambridge, MA, 1983); Commission on Industrial Relations, *Final Report and Testimony*, 64th Cong., 1st sess., S. Doc. 415, vols. 3, 4 (Washington, DC, 1916).

2. *American Machinist* 46 (1917): 62 (hereafter *AM*); Ronald Schaffer, *America in the Great War* (New York, 1991).

3. George Soule, *Prosperity Decade* (New York, 1947), chaps. 4, 5; Moses Abramovitz, *Inventories and Business Cycles* (New York, 1950); National Industrial Conference Board, *The Cost of Living in the United States, 1914–1930* (New York, 1931).

4. *AM* 37 (1912): 787; 55 (1921): 721.

5. Harless Wagoner, *The U.S. Machine Tool Industry from 1900 to 1950* (Cambridge, MA, 1968), 25, 100 (hereafter cited as *USMT*); *AM* 37 (1912): 956; 38 (1913): 46–50, 119.

6. *AM* 37 (1912): 283–84; 38 (1913): 665; 39 (1913): 670, 1031–32; 40 (1914): 565. Congress lowered ad valorem tariff rates on machine tools to 15 percent in the Underwood schedules, but builders anticipated little foreign penetration. Quotation is from William Lodge's valedictory series, "Managerial and Manufacturing Experience," *AM* 38 (1913): 1067. Lodge's other articles, initially written as instructions for his successor as president of Lodge and Shipley, appeared at 38 (1913): 1015–19 and 39 (1913): 139–42, 177–79, before being published as William Lodge, *Rules of Management* (New York, 1913). Lodge died four years later, aged sixty-nine (*AM* 46 [1917]: 835). Oberlin Smith, his older, insightful colleague in metal-forming tool design, outlived him, passing away in 1926 at age eighty-six (*AM* 65 [1926]: 228a). For a complete exposition on the workings of a machine-hour-rate costing system at Westinghouse Electrical and Manufacturing Co., see *AM* 44 (1916): 367–71.

7. *AM* 41 (1914): 166.

8. George Wing, "The History of the Cincinnati Machine Tool Industry" (D.B.A. diss., Indiana University, 1964), 140, 199, 208, 241; *AM* 50 (1919): 1167–69, plus unpaged inserted map. In 1923, by contrast, the NMTBA estimated that one-quarter of all machine tool companies had left the field or gone bankrupt as a result of the postwar depression (*Machinery* 30 [1923–24]: 767).

9. *AM* 49 (1918): 906. The writer "Chordal" was a machinist, design engineer, and, later, patent attorney, who wrote memorable columns on metalwork-

ing shop life for *American Machinist* during the 1880s. After a thirty-year break, he recommenced his witty contributions in July 1918 and continued them until shortly before his death, at sixty-nine, in 1920. For much of the late nineteenth century, he worked for or in association with the Niles Tool Works at Hamilton, Ohio, near Cincinnati, where George Gray was superintendent until 1886, before Gray relocated to the city and opened his own firm (*AM* 51 [1920]: 380).

10. G. A. Gray Papers, Collection BC-017, Box 21, Order Books, vol. 6, 199–248, Cincinnati Historical Society. Among the others were Cincinnati Planer, American Tool Works, Betts Machine, Sellers, Niles-Bement-Pond, Chandler Planer, and the Hamilton Machine Tool Company (Box 30, Scrapbook: Other Firms).

11. Gray Papers, Box 30, Scrapbook: Other Firms, 16.

12. Gray Papers, Box 21, Order Books, vols. 5 and 6 (for 1912 and 1917).

13. Gray Papers, Box 21, Order Books, vol. 6, 199–248. Cincinnati Planer similarly advanced its prices on standard 22"/5' models from $775 in August 1915 to $2,160 three years later, appreciably more aggressive hikes than at Gray. See Cincinnati Planer price lists, Gray Papers, Box 30, Scrapbook: Other Firms, 12i–12m.

14. The Machine Tool Section of the War Industries Board complained early in 1918 about "scalpers," dealers who added 40–50 percent to the "fair price" of tools, but it focused only on used tool marketers in the report (*AM* 48 [1918]: 425–27).

15. Gray Papers, Box 21, Order Books, vol. 6.

16. Ibid.

17. *AM* 43 (1915): 573.

18. Gray Papers, Box 21, Order Books, vols. 6 and 7; Wing, "Cincinnati Machine Tool Industry," 208; *AM* 56 (1922): 711. Inquiries made by the National Recovery Administration showed that, in terms of output and value added, machine tools were the hardest hit among thirty-four sectors investigated concerning the impact of the 1920–21 depression (see "Consequences of Depression," NRA Records, RG 9, Box 3498, File 103-8, National Archives, Washington, DC). Price lists suggest that firms dropped tool prices about 20 percent in 1920–21, whereas Ford effected a 30 percent cut in its car and truck prices (*AM* 53 [1920]: 699). Once demand recovered somewhat, builders hiked prices sharply, on average 45 percent between 1922 and 1923, "according to NMTBA estimates" (Wagoner, *USMT*, 144).

19. *AM* 55 (1921): 396.

20. *AM* 55 (1921): 894–95; 59 (1923): 492, 822; 63 (1925): 523–25, 551–52.

21. *AM* 52 (1920): 584b, 1120; 53 (1920): 68.

22. Wing, "Cincinnati Machine Tool Industry," 230–33; *AM* 53 (1920): 246, 386, 686.

23. Such is Wing's interpretation ("Cincinnati Machine Tool Industry ," 233–35).

24. *AM* 50 (1919): 762–63; Wing, "Cincinnati Machine Tool Industry," 241.

25. *AM* 54 (1921): 1030–31, 1114–18; Wing, "Cincinnati Machine Tool Industry," 212–15. CMM seems to have pioneered locally in workplace lunchrooms before the war, the firm fitting up the space and purchasing equipment before turning it over to the workers, "who appointed a committee to purchase their own food and set their own prices," an arrangement "with which the management was not connected in any way" (*AM* 46 [1917]: 851). In 1925, Lodge and Shipley built its base model lathes in "lots or runs, as they are called. . . . In busy times practically every run brought through is sold completely months before the parts are finished. . . . Any orders coming in after the run is completely sold out must wait to be scheduled in the next." This strategy combined batch production with "banking" orders to avoid accumulating inventory (*AM* 62 [1925]: 31).

26. *AM* 59 (1923): 339–40; 62 (1925): 151–54; 63 (1925): 407–9, Wing, "Cincinnati Machine Tool Industry," 210–11; Frederick Geier, *The Coming of the Machine Tool Age* (New York, 1949), 22.

27. *AM* 34 (1911): 434; 59 (1923): 75.

28. *AM* 57 (1922): 359; 63 (1925): 492a–492b; *Machinery* 29 (1922–23): 22. The Yale expositions continued into the late 1920s; for a report on the 1927 show, see *Machinery* 33 (1927): 865.

29. *AM* 61 (1924): 560a–560c; 63 (1925): 524a–524e.

30. *AM* 65 (1926): 432a, 510a–510c, 580a–580e; *Machinery* 33 (1926–27): 330, 332, 865; 34 (1927–28): 42.

31. *AM* 63 (1925): 427; 67 (1927): 374a,

375–77, S2–S9, 526a–526i. Seven metal trades journals also secured exhibit space. After a year's hiatus, the NMTBA exhibit was again presented at Cleveland in 1929, drawing 30 percent more exhibitors and 80 percent more "mechanical executives and engineers" than in its first incarnation (*Machinery* 35 [1928–29]: 772; 36 [1929–30]: 186). Though slimmed down during the Great Depression, the exhibition survived to become a stable element of the trade's annual calendar and its industrial culture.

32. But see *AM* 63 (1925): 532 for continuing concern on this point.

33. *AM* 55 (1921): 249–51; 59 (1923): 263; *Machinery* 31 (1924–25) 367; 32 (1925–26): 238, 359; Wagoner, *USMT*, 202–5. See also *Machinery* 30 (1923–24): 852 for commentary on the folly of aping mass producers in the machine tool trade. On the continuing need for "all-round machinists," see *AM* 62 (1925): 330.

34. Quoted matter from *AM* 51 (1919): 286; 62 (1925): 805–7. For other comments on the limits of standardization, see *AM* 50 (1919): 1143–44; 61 (1924): 630, and, for advocacy, Taylorite Carl Barth's unheeded "Standardization of Machine Tools," *Journal of the American Society for Mechanical Engineering* 38 (1916): 968–71.

35. *Machinery* 33 (1926–27): 510; *AM* 58 (1923): 746; 62 (1925): 944a; Gray Papers, Box 30, Scrapbook: Other Firms. The excess profits tax repeal came in 1921 (W. Elliot Brownlee, *Federal Taxation in America* [New York, 1996], 59). Once the Great Depression arrived, shared price lists fade away, though only further research will suggest whether this was a result of crisis-induced secrecy, firm failures, or other sources. On the promised end to the excess profits tax, see *AM* 55 (1921): 987–88. The New Deal's NRA officials believed that the "dominant part of the industry" had for some time undertaken "to 'administer' or 'control' prices," citing a standard "cost formula, 'firm' price policy, and uniform distribution trade practices" as evidence (see Alexander Sachs, "Material Bearing on the Machine Tool Industry," November 1933, 21 in NRA Records, RG9, Box 3498, File 103-22, National Archives).

36. These three paragraphs are based on *AM* 63 (1925): 525–32. In providing an occasion for machine tool builders' critique of automobile firms' complaints, *American Machinist* clearly was reacting to a fall 1924 series in *Machinery*, in which builders replied in December to a set of automobile makers' criticisms presented in the October and November issues. For what seem rather bland responses by tool firms, see *Machinery* 21 (1924–25): 263–67.

37. *AM* 63 (1925): 525. The degree to which the Great Depression transformed auto–machine tool trade relations cannot be treated here. Wagoner's *USMT* is silent on this point.

38. *AM* 62 (1925): 929–31. Ironically, though repeatedly pointing to the small batch scale of tool production, Schlesinger (or his translator, as the piece was first published in German) could not help but adopt the prevailing industrial rhetoric when referring to tool builders' gauge and assembly practices as a "standardized" system for "interchangeable mass production." A historian of technology with some feel for textual criticism could have a field day examining the prescriptive uses and misapplications of such terms in various fields of American production since the turn of the century.

39. A. L. Faulkner to William Wilson, secretary of labor, October 26, 1915, 2, File 33/106; Raymond Doyle, Interview Transcript, November 26, 1917, File 33/698, Federal Mediation and Conciliation Service Records (hereafter, FMCS), RG 280, National Archives; W. A. Viall, "Strike Report," 1–4, 33, Brown and Sharpe Papers, Series C.2, Rhode Island Historical Society. Doyle, an employment clerk, explained that upon hiring a worker, "[w]e would send down an entrance card to the Metal Trades Association. If the man had a bad record somewhere else he will be discharged. . . . We wouldn't be allowed to hire a striker as a member of the Metal Trades Association. If we did we would be put out of the organization."

40. "Daily Shop Tally," September 20–November 6, 1915, Transcript of "Proceedings at Strikers' Meeting," October 30, 1915; Brown and Sharpe Papers, Series C.1, Labor Relations, 1915–18; *Providence Journal*, September 24, 1915, 1–2; Viall, "Strike Report," 30; Faulkner to Wilson, 4, File 33/106, FMCS. Evidently, Brown and Sharpe sent a stenographer to the strike meeting to take down a verbatim account. Consistent with trade customs, Viall met several times with Finnell, an

employee, to urge that workers return, but refused all contacts with outsiders. The MTA's control of employment access meant that strikers could get work only outside the state.

41. *AM* 48 (1918): 565–68, 855–58; 54 (1920): 230–32, 649–52, 1231; 55 (1920): 671; 58 (1923): 602–4; 59 (1923): 635, 965; 62 (1925): 993–96; *Machinery* 32 (1925–26): 692–93. Viall, surveying 350 former apprentices in 1925 for the *Machinery* article, found this distribution of their current occupations: 44 proprietors, superintendents, and managers, 32 in sales, 18 engineers, 89 foremen, 42 draftsmen, 104 skilled workers, and 21 others.

42. *Metal Industry* 21 (1923): 423, 463; 22 (1924): 380, 423 (hereafter cited as *Metal*); Anna Weinstock to H. L. Kerwin, January 16, 1930, File 165/826, FMCS; Brown and Sharpe Manufacturing Company, *Small Tools: Catalog No. 31* (Providence, RI, 1929), x, 8–58, 222, 236–37; *AM* 52 (1920): 1231–32.

43. Faulkner to Wilson, 1, File 33/106, FMCS; *AM* 46 (1917): 847–50, 931–32; 52 (1920): 1231–36; 62 (1925): 994. The new routing system, which replaced work tags and piece chasers in 1913, achieved 76 percent on-time job completion for 25,000 orders that year. By 1918, with 67,000 orders, the on-time rate reached 85 percent (*AM* 46 [1920]: 1236). Statistical analysis done by the employment office, which hired 6,800 workers during 1917, nominally replacing the entire workforce, showed that actually just under 2,000 positions (the "jobs made vacant") were filled over and again by a transient workforce, two-thirds of which stayed fewer than three months. About 4,000 positions never came open, being held by long-term employees. In addition, of those newly hired, 34–38 percent had worked previously at Brown and Sharpe, contrasted with Magnus Alexander's estimate of 25 percent in industry generally (*AM* 49 [1918]: 855–58).

44. Charles Venable, *Silver in America, 1840–194): A Century of Splendor* (New York, 1994), 226–29, 238–39.

45. *Manufacturing Jeweler* 53 (1913): 322; 83 (October 25, 1928): 22 (emphasis in original, hereafter cited as *MJ*). In July 1924, *MJ* shifted from volume to issue pagination. Hence issue dates will be cited after vol. 74. The high-end trades in Newark and New York charted a more positive path, retaining much of their earlier high-skill, mid- to high-price goods character, focused on gold, diamonds, and increasingly platinum and its alloys, which appealed to custom and high-fashion markets. Their trade organizations functioned far more effectively, and, on the same account, labor unions had better success in organizing, though not in sustaining, strikes.

46. *MJ* 52 (1913): 174; 54 (1914): 1043; 55 (1914): 60, 1340; 60 (1917): 763; 62 (1918): 1035; 78 (January 7, 1926): 22.

47. *MJ* 53 (1913): 1390–92; 54 (1914): 1044; 55 (1914): 1006, 1142; 56 (1915): 838, 924; 57 (1915): 430–33; 58 (1916): 647; 60 (1917): 360; 63 (1918): 233, 982, 1000–1001, (AI—November 7): 5; 67 (1920): 1272B; 70 (1921): 934; *AM* 47 (1917): 229–32. Brass, crucial in cheap jewelry, advanced from $0.13 to $0.36 per pound, 1915–16, and "the little mirrors that jewelers use in vanity cases" from "$3.50 to $30.00 per thousand" (*MJ* 58 [1916]: 181). *MJ* separately paginated its anniversary issues.

48. *MJ* 52 (1913): 1338, 1376; 53 (1913): 556–58; 54 (1914): 270, 280–81. The 90 eastern manufacturers who joined 168 New York and Chicago wholesalers in this scheme were implicitly acknowledging that it was better to be "in the hands" of jobbers than to face open market competition. Any maker not joining could only sell direct to such retailers, department stores, and mail-order houses if they could manage without any goods from the larger producers. Presumably, the collaborators judged that this would be difficult, as the complaints from those left outside, triggering the suit, demonstrated. Those who regularized their marketing in the old channels would be gradually admitted to the associations, narrowing retail buyers' options to play off wholesalers against manufacturers, thus building a base for uniform terms and price firmness. Of course, since history matters, the organizers had a huge trust deficit to overcome and their aggressive tactics only confirmed outsiders' judgment that particular rather than general interests were being advanced, hence the latter's appeal to the state.

49. *MJ* 57 (1915): 88, 138–39, 510, 613, 688, 858–64; 60 (1917): 298, 300, 946; 62 (1918): 367, 437; 63 (1918): 364, 439, 550,

644. A revival of the publicity idea in 1917 went nowhere. The jewelry tax remained in place until 1926 (*MJ* 78 [March 11, 1926]: 34).

50. *MJ* 56 (1915): 61; 57 (1915): 1205–6; 58 (1916): 199; 60 (1917): 1198.

51. William Blackman to Wilson, June 30, 1917, File 33/509; Walter Lochner to Wilson, August 14, 1917, File 33/643; John Schwartz to National War Labor Board, September 10, 1918, File 33/2310, FMCS; *MJ* 63 (1918): 84, 143, 156, 196, 232.

52. Commonwealth of Massachusetts, *Forty-Fifth Annual Report of the Statistics of Labor* (Boston, 1914), pt. 5, p. 100; *MJ* 60 (1917): 274–75, 324–26; 74 (1924): 553; 75 (December 18, 1924): 22; U.S. Department of Labor, Children's Bureau, *Industrial Home Work of Children* (Washington, DC, 1922), 28–59. This last is a study of Rhode Island conducted in 1918, by which time the mesh bag problem had diminished. About one-third of the 2,300 children working at home did tasks associated with jewelry production (e.g., stringing beads, setting stones, carding jewelry). The quoted description of the Attleboro machines suggests that they were modeled on "automatic" textile looms that stopped when a warp thread broke.

53. *Metal* 21 (1923): 462–63; *MJ* 52 (1913): 1398; 53 (1913): 651, 1142, (AI—October 30, 1913): 8, 114; 62 (1918): 461–62.

54. J. Ellery Hudson, *Twenty-Eighth Annual Report of Factory Inspection* (Providence, 1922); *MJ* 64 (1919): 849; 65 (1919): 1070; 76 (May 14, 1925): 20; 77 (December 3, 1925): 8; 79 (November 11, 1926): 10; 82 (May 10, 1928): 29.

55. Hudson, *Twenty-Eighth Report*, 28; *MJ* 68 (1921): 296; 79 (November 11, 1926): 10.

56. *MJ* 66 (1920): 1068; 69 (1921): 170, 1066; 71 (1922): 1073; 72 (1923): 361, 618, 684; 73 (1923): 536–40; 75 (July 24, 1924): 34, (October 2): 31; 76 (April 2, 1925): 24; 77 (July 30, 1925): 5; 80 (June 9, 1927): 19; *Metal* 22 (1924): 41.

57. *MJ* 74 (1924): 500, 972, 975; 75 (November 6, 1924): 28; 77 (September 17, 1925): 18, (November 19): 16–21; 78 (Februrary 18, 1926): 24; 79 (July 1, 1926): 6, (October 7): 7.

58. *MJ* 72 (1923): 188; 73 (1923): 1429; 76 (March 19, 1925): 18–22.

59. *Metal* 22 (1924): 130; 23 (1925): 41; *MJ* 66 (1920): 386; 67 (1920): 1342; 69 (1921): 962; 73 (1923): 925; 74 (1924): 191; 77 (October 1, 1925): 30.

60. *MJ* 80 (May, 12, 1927): 28, (June 9): 18–19; 82 (March 29, 1928): 5; 83 (June 7, 1928): 14, (August 2): 18.

61. Furniture Manufacturers Association, Minutes of Board and Association Meetings, 1912–18, Box 5, Folders 3–4, FMA Papers, Collection no. 84, Grand Rapids Public Library; *Furniture Manufacturer and Artisan* 71 (1915): 298–99 (hereafter cited as *FMA*).

62. Frank Ransom, *A City Built on Wood* (Grand Rapids, MI, 1955), 55; *FMA* 77 (1918): 251–52; 80 (1920): 100–101. The bomber's wingspread was 100′, the fuselage 63′ long, and it carried a crew of six. Seven trucks were dedicated to hauling lumber to factories and parts to the final assembly site. A 1918 estimate indicated that only 100 furniture firms were active on war production work, 7 percent of all enterprises (77 [1918]: 148, 159).

63. *FMA* 71 (1915): 129, 298–99; 72 (1916): 245; 73 (1916): 62, 76, 284; 76 (1918): 35, 212; 78 (1919): 22; Ransom, *City*, 57. Scattered evidence suggests that women earned about twelve dollars weekly, about a third less than male workers.

64. Ransom, *City*, 55, 58; *FMA* 78 (1919): 22, 186–87, 224–25, 266; 79 (1919): 4, 203, 230–32, 308.

65. *FMA* 83 (1921) 208, 260; Minutes, 1919–21, FMA Papers, Box 5, Folders 4–5, GRPL. Jobbers reportedly confined their business to staple furniture (Chicago's specialization) and to odd lots and close-outs (77 [1918]:150). The FMA also introduced a workers' health and life insurance program and a reduction to forty-eight hours weekly in 1919, authorized reintroducing piecework bonus plans to speed production, and sharply reprimanded Imperial's Foote for negotiating with cabinetmakers during a brief walkout late in September 1919. That year homeownership in Grand Rapids stood at 50 percent, second among all American cities.

66. *FMA* 82 (1921): 60, 202, 247; 83 (1921): 262; 88, pt. 1 (1924): 234; 89, pt. 1 (1925): 10–11; Ransom, *City*, 61–62. *FMA* switched in 1924 to two-part, annual volume numbers, separately paginated, e.g., 88, pt. 1, and 88, pt. 2. In 1926, it adopted issue pagination at volume 90, skipped to volume 92 at midyear, then continued two annual volume numbers thereafter.

Go figure. The regional proportions of furniture shipments nationally were as follows: East—46 percent, West—31 percent, and South—23 percent (88, pt. 1 [1924]: 68). The costing volume mentioned is H. D. Potter, *Cost Finding Principles for Furniture Factories* (Grand Rapids, MI, 1924). Though there are a few trade journal intimations that FMA members shared a common costing system, no mention of this is contained in the association's papers.

67. *FMA* 82 (1921): 246; 83 (1921): 207; 84 (1922): 85, 97; 85 (1922): 211; 86 (1923): 115; 88, pt. 2 (1924): 204; 89, pt. 1 (1925): 68; 95 (March 1928): 51; Ransom, *City*, 57; Minutes, December 24, 1924, FMA Papers, Box 5, Folder 6; Grand Rapids Market Association Records, FMA Papers, Box 2, Folder 22, GRPL.

68. *FMA* 71 (1915): 86–87; 79 (1919): 5, 90; 81 (1920): 164; 83 (1921): 58.

69. *FMA* 83 (1921): 110–11; 164–67.

70. *FMA* 87 (1923): 16, 247; 88, pt. 1 (1924): 219, 268; Francis Campau, "Draft Narrative of the Controversy," 1926, FMA Papers, Box 14, Folder 20; "Brief for Petitioners, Berkey and Gay Furniture Company, et al. v. FTC," U.S. Circuit Court of Appeals, Sixth Circuit, undated, FMA Papers, Box 14, Folder 19.

71. Campau, "Draft," 4; "Official Report of Proceedings before the FTC: Trade Practice Submittal—Furniture Industry," September 8, 1924, 63, FMA Papers, Box 14, Folder 17.

72. Campau, "Draft," 6–15; Minutes, April 27, 1927, FMA Papers, Box 5, Folder 7; Campau, "Disposition of the Case," 1930, Box 14, Folder 20; *FMA* 92 (August 1926): 50; 94 (July 1927): 51.

73. "Brief for Petitioners," 13–19, 22–23, 34 (emphasis in brief); Campau, "Disposition," 4–7; "Berkey and Gay v. FTC, Petitions to Review Orders of FTC, Decision," Sixth Circuit, U.S. Court of Appeals, June 28, 1930, 1–3, 5, FMA Papers, Box 14, Folder 19; *Grand Rapids Spectator*, July 26, 1930, 3–5, 15.

74. *FMA* 83 (1921): 110–11, 164–67; 85 (1922): 16, 60–61; 88, pt. 1 (1924): 18–19, 130–36; 89, pt. 1 (1925): 164–66, 269.

75. Campau to Robert Irwin, January 11, 1926, FMA Papers, Box 15, Folder 22; *Chicago Journal*, April 25, 1925; *New York World*, April 26, 1925; *St. Louis Post Dispatch*, April 26, 1925; *Grand Rapids Press*, May 26, 1925; *Grand Rapids Herald*, May 30, 1925; *New York Times* July 11, 1925; *Cincinnati Enquirer*, July 19, 1925 (Argus Press Agency clippings in Box 15, Folder 25).

76. *St. Louis Star*, July 18, 1925; *Atlanta Constitution*, July 26, 1925; *Rocky Mountain News*, July 26, 1925 (Argus Clippings) FMA Papers, Box 15, Folder 25; Julius Amberg to Francis Campau, May 13, 1926, Box 15, Folder 22. Brown's sentence remained unserved in 1928, when the cases ended, but Amberg's reference to the "printed Digest of Evidence" in the letter cited indicates that alliance materials did not long elude the prosecution. Seventy years later, archived FMA files still held clusters of telegrams and letters inquiring after Grand Rapids' price movements in this period. Typical is Hubbard, Eldredge, and Miller's December 1920 note to Campau from Rochester, New York, affirming that "we want to act along the same lines as the Grand Rapids manufacturers and whatever they decide to do we will follow" (Box 15, Folder 22).

77. Mark Foote, "Draft Report on FTC Furniture Findings," January 17, 1924, Box 15, Folder 22; *FMA* 93 (February 1927): 41, 84. See also Richard May, "The Trade Association and Its Place in the Business Fabric," *Harvard Business Review* 2 (1923–24): 84–97. Grand Rapids' Charles Sligh chaired the Manufacturers' Defense Committee.

78. "Joint and Several Demurrers of Certain Defendants, U.S. v. Aulsbrook and Jones Furniture Company, et al., No. 13833," U.S. District Court, Northern District of Illinois, Eastern Division, 1926, FMA Papers, Box 15, Folder 23; Campau to Robert Irwin, April 16, 1926; Campau to Buffalo Retail Furniture Association, March 31, 1927, Box 15, Folder 22; *FMA* 93 (February 1927): 41, 50; (April 1927): 35–38, 50.

79. *FMA* 95 (April 1928): 39; Campau to A. C. Brown, June 1, 1925, FMA Papers, Box 15, Folder 22; *Grand Rapids Press*, March 6, 8, and 9, 1928, Box 15, Folder 25.

80. For a similar perspective on a different industry, see Gerald Berk, "Communities of Competitors: Open Price Associations and the American State, 1911–1929" (unpublished paper, 1995; forthcoming in *Social Science History*).

81. See Linda Ewing, "Kent County's Woodworking Sector: Interview Find-

ings" (unpublished paper, 1990, in author's possession).

82. Edmund Day and Woodlief Thomas, *The Growth of Manufactures, 1899 to 1923,* Census Monograph no. 8 (Washington, DC, 1928).

83. A $40 million value-added threshold was established for these tables, based on this measure's capacity to cover 92 percent of all industrial value added in American manufacturing, a share comparable to the 1909 survey.

## CHAPTER 13
## LOOKING AHEAD

1. Robert Rydell, *World of Fairs* (Chicago, 1993), 5, 6, 36, 158–65; Russell Weigley, ed., *Philadelphia: A Three-Hundred-Year History* (New York, 1982), 571–75. Only the city's African-American newspapers printed the text of Randolph's fiery speech, and the official history of the fair entirely omitted reference to his presence.

2. Dexter Kimball, "The Engineer," in *Congress of American Industry* (Philadelphia, 1926), 59.

3. *Industrial Management* 65 (1923): 32; 66 (1923): 253–64; 67 (1924): 36–46, 343–47 (emphasis in original, hereafter cited as *IM*); Richard Tedlow, *New and Improved* (New York, 1990), 311–26; Stephen Meyer, "Technology and the Workplace: Skilled and Production Workers at Allis-Chalmers, 1900–41," *Technology and Culture* 29 (1988): 839–64; Jeffrey Meikle, *Twentieth Century Limited* (Philadelphia, 1979), 40, 42, 103–4; Walter Paterson, *An Industrial Heritage: Allis-Chalmers Corporation* (Milwaukee, 1976).

4. Samuel Vauclain, *Steaming Up!* (New York, 1930), 165–68, 276–77; John Brown, *The Baldwin Locomotive Works* (Baltimore, 1995), 223–33. According to an NRA study, locomotive builders nationally operated at 29 percent of capacity in 1929, versus an astonishing 1.2 percent of capacity in 1933. See NRA Papers, RG 9, Box 3189, "Supplemental Code of Fair Competition for the Locomotive Manufacturing Industry," 2:4, National Archives.

5. Thomas Heinrich, *Ships for the Seven Seas* (Baltimore, 1997); Philip Scranton and Walter Licht, *Work Sights: Industrial Philadelphia, 1900–1950* (Philadelphia, 1987).

6. G. R. Simonsin, ed., *The History of the American Aircraft Industry* (Cambridge, MA, 1968), 73–105. For the complexity of aircraft design, see Walter Vincenti, *What Engineers Know and How They Know It* (Baltimore, 1990).

7. Tom Lilley et al., "Conversion to Wartime Production Techniques," in Simonsin, *Aircraft Industry,* 121–37; Aircraft Industries Association of America, "Aircraft Manufacturing in the United States," in Simonsin, *Aircraft Industry,* 165.

8. Lilley, "Conversion," 137–39. Similar flexibility emerged in aircraft corporations' shift to missile hardware manufacturing, though their long-term dependence on military contracting fostered crises at the close of the Cold War comparable to those Cramp and other warship builders faced after World War I. See G. R. Simonsin, "Missiles and Creative Destruction in the American Aircraft Industry, 1956–61," in idem, *Aircraft Industry,* 228–41. For an effective comparison of wartime aircraft production in the United States, Great Britain, and Germany, see Jonathan Zeitlin, "Flexibility and Mass Production at War," *Technology and Culture* 36 (1995): 46–79.

9. *IM* 70 (1925): 239–41; Philip Scranton, *Figured Tapestry* (New York, 1989); Gavin Wright, *Old South, New South* (New York, 1986); Annette Wright, "Strategy and Structure in the Textile Industry: Spencer Love and Burlington Mills, 1923–62," *Business History Review* 69 (1995): 42–79.

10. For a typical discussion of this durable buyers' market, see *IM* 73 (1927): 344–47.

11. Frank Ransom, *A City Built on Wood* (Grand Rapids, MI, 1955); Hugh Depree, *Business as Unusual: People and Principles at Herman Miller* (Zeeland, MI, 1986); Philip Scranton, "Manufacturing Diversity," *Technology and Culture* 35 (1994): 476–505; Federal Writers Project, *They Built a City: 150 Years of Industrial Cincinnati* (Cincinnati, 1938), 179–91, 281–99; David Noble, *Forces of Production* (New York, 1984), 82, 202; James Schwartz, *Cincinnati Milacron, 1884–1984* (Cincinnati, 1984); Victor Strauss, *The Printing Industry* (New York, 1967); Albert Hinrichs, *The Printing Industry in New York and Its Environs,* Regional Plan Monograph no. 6 (New York, 1924); Emily Clark Brown, *Joint Industrial Control in the Book and Job*

*Printing Industry*, Bureau of Labor Statistics Bulletin no. 481 (Washington, DC, 1928), 80–81.

12. Michael Storper and Richard Walker, *The Capitalist Imperative* (Cambridge, MA, 1989), chaps. 4, 7; Annalee Saxenian, *Regional Advantage* (Cambridge, MA, 1994); David Lampe, ed., *The Massachusetts Miracle* (Cambridge, MA, 1988); Susan Rosengrant and David Lampe, *Route 128* (New York, 1992); Allen Scott, *Metropolis* (Berkeley, CA, 1988); Robert Cringeley, *Accidental Empires* (Reading, MA, 1992).

13. *IM* 65 (1923): 314–17; B. M. Selekman, Henriette Walter, and W. J. Cooper, *The Clothing and Textile Industries in New York and Its Environs*, Regional Plan Monographs nos. 7–9 (New York, 1925); Joel Seidman, *The Needle Trades* (New York, 1942), chaps. 10–13; Kenneth Dameron, *Men's Wear Merchandising* (New York, 1930); Charles Goodman, *The Location of Fashion Industries*, Michigan Business Studies, 10:2 (Ann Arbor, 1948); Leonard Drake and Carrie Glasser, *Trends in the New York Clothing Industry* (New York, 1942); U.S. Department of Labor, *Cost Savings in the Clothing Industry* (Washington, DC, 1955); René Konig, *A La Mode: On the Social Psychology of Fashion* (New York, 1973).

14. See Roger Waldinger, *Through the Eye of the Needle* (New York, 1986), chap. 3.

15. For a classic characterization of auxiliaries' situations in the 1920s, see *IM* 68 (1924): 133–34.

16. Michael Storper and Susan Christopherson, "Flexible Specialization and Regional Industrial Agglomerations: The Case of the U.S. Motion Picture Industry," *Annals of the Association of American Geographers* 77 (1987): 104–17; Storper and Walker, *Capitalist Imperative*, 150; author's interview with David Horowitz, former head of Time/Warner Communications, April 18, 1996. Mr. Horowitz commented that films are "blind" productions, made in advance of market demand, and television series' pilots much resemble product samples awaiting buyers' interest. As in styled furniture and fabrics, novelty involves variations of known features, and though copying is endemic, standardization is anathema. See also David Horowitz and Peter Davey, "Financing American Films at Home and Abroad," *Columbia-VLA Journal of Law and the Arts* 20 (1996): 461–93.

17. Extensive interaction with users was customary in the "office equipment industry" at least from the 1920s. See James Cortada, *Before the Computer* (Princeton, NJ, 1993), 266–73.

18. See Paul Carroll, *Big Blues: The Unmaking of IBM* (New York, 1993), and Cringely, *Accidental Empires*.

19. That, despite critiques, this view remains in force is suggested in a summary statement in a recent journal article: "The essence of the Second Industrial Revolution was the fusion of mass production with mass marketing." Though the editors also allow that "most industries do not tend to evolve into structures dominated by big businesses," the essentialism of the main claim rests unmodified (editors' introduction to Harris Corporation, "Founding Dates of the 1994 Fortune 500 U.S. Companies," *Business History Review* 70 [1996]: 70–71).

20. On Worcester, see George Alden, "The Washburn Shops of the Worcester Polytechnic Institute," *Journal of the American Society of Mechanical Engineers* 37 (1915): 391–94, and Herbert Taylor, *Seventy Years of the Worcester Polytechnic Institute* (Worcester, MA, 1937).

# INDEX

(Note: Certain terms, such as diversity, firms, demand, flexibility, capital, et al., are so frequently used as to limit their value as index headings and have been omitted.)

## ABOUT THE AUTHOR

PHILIP SCRANTON is the Kranzberg Professor of the History of Technology at the Georgia Institute of Technology. His most recent book is *Figured Tapestry: Markets, Production, and Power in Philadelphia Textiles, 1885–1941*.